Quinton

VIENNA CIRCLE COLLECTION

VOLUME 10

EDITOR: HENK L. MULDER

SELECTED PAPERS IN
LOGIC AND FOUNDATIONS, DIDACTICS, ECONOMICS

KARL MENGER

SELECTED PAPERS IN LOGIC AND FOUNDATIONS, DIDACTICS, ECONOMICS

D. REIDEL PUBLISHING COMPANY

DORDRECHT : HOLLAND / BOSTON : U.S.A.

LONDON : ENGLAND

Library of Congress Cataloging in Publication Data

Menger, Karl, 1902-
Selected papers in logic and foundations, didactics, economics.

(Vienna circle collection ; v. 10)
'Bibliography of works by Karl Menger': p.
Includes index.
1. Mathematics–Philosophy–Collected works. 2. Logic, Symbolic and mathematical–Collected works. 3. Mathematics–Study and teaching –Collected works. 4. Uncertainty–Collected works. 5. Diminishing returns–Collected works. I. Title. II. Series.
QA8.4.M46 510'.1 78-12374
ISBN 90-277-0320-5
ISBN 90-277-0321-3 pbk.

Published by D. Reidel Publishing Company,
P. O. Box 17, Dordrecht, Holland

Sold and distributed in the U.S.A., Canada, and Mexico
by D. Reidel Publishing Company, Inc.
Lincoln Building, 160 Old Derby Street, Hingham,
Mass. 02043, U.S.A.

Printed in the Netherlands

KARL MENGER

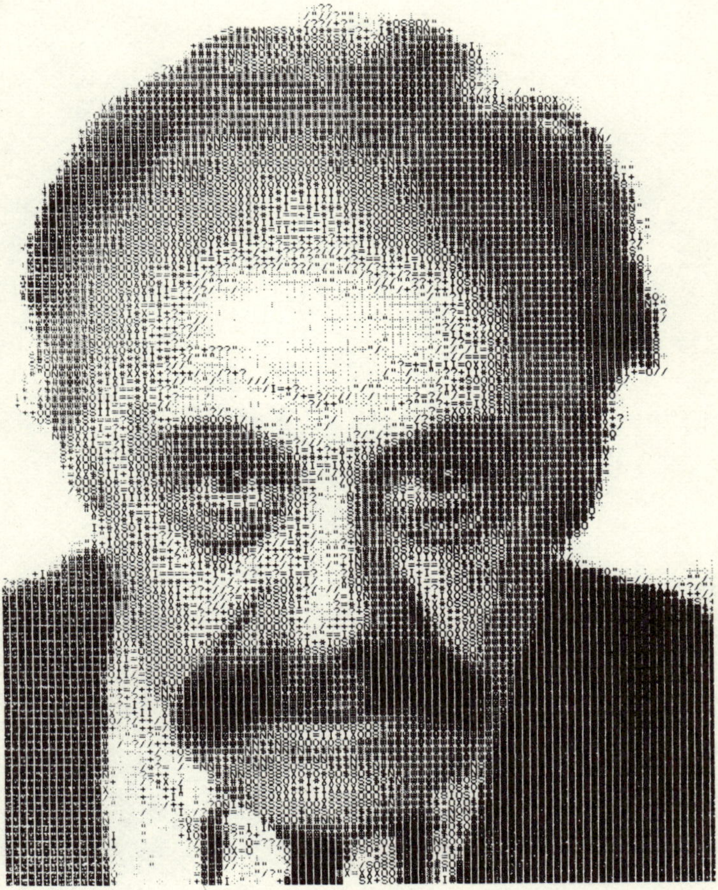

The computer-generated portrait was made in
The Computer Place, Atlanta, Georgia (November 1977)

TABLE OF CONTENTS

ACKNOWLEDGMENTS xi

INTRODUCTION 1

PART I. PAPERS INTRODUCING LOGICAL TOLERANCE

	Logical Tolerance in the Vienna Circle	11
Chapter 1	The New Logic (1933)	17
	Appendix (1937)	42
Chapter 2	On Intuitionism (1930)	46

PART II. OPUSCULA LOGICA

Chapter 3	Meaningfulness and Structure (1930)	61
	Appendix (1978)	63
Chapter 4	A New Point of View on the Logical Connectives (1978)	68
Chapter 5	An Intuitionistic-Formalistic Dictionary of Set Theory (1928)	79
Chapter 6	Ultrasets and the Paradoxes of Set Theory (1928)	88
Chapter 7	A Logic of the Doubtful. On Optative and Imperative Logic (1939)	91

PART III. FUNDAMENTAL CONCEPTS IN PURE AND APPLIED MATHEMATICS

Chapter 8	A Counterpart of Occam's Razor (1960, 1961)	105
Chapter 9	A Theory of the Application of the Function Concept to Science (1970)	136
Chapter 10	Variables, Constants, Fluents (1961)	144
Chapter 11	Wittgenstein on Formulae and Variables (1978)	153

PART IV. DIDACTICS OF MATHEMATICS

Introduction 161
Chapter 12 A New Approach to Teaching Intermediate Mathematics (1958) 163
Chapter 13 Why Johnny Hates Math (1956) 174
Chapter 14 On the Formulation of Certain Questions in Arithmetic (1956) 185
Chapter 15 On the Design of Grouping Problems and Related Intelligence Tests (1953) 189
Chapter 16 The Geometry Relevant to Modern Education (1971) 199

PART V. PHILOSOPHICAL RAMIFICATIONS OF SOME GEOMETRIC IDEAS

Chapter 17 On Definition, Especially of Dimension (1921–1923, 1928) 207
Chapter 18 Square Circles (The Taxicab Geometry) (1952, 1978) 217
Chapter 19 The Algebra of Geometry (1978) 220
Chapter 20 Geometry and Positivism. A Probabilistic Microgeometry (1970) 225

PART VI.

Chapter 21 My Memories of L. E. J. Brouwer (1978) 237

PART VII. ECONOMICS. META-ECONOMICS

Chapter 22 The Role of Uncertainty in Economics (1934) 259
Chapter 23 Remarks on the Law of Diminishing Returns. A Study in Meta-Economics (1936) 279

PART VIII. GULLIVER'S INTEREST IN MATHEMATICS

Chapter 24 Gulliver in the Land without One, Two, Three (1959) 305

Chapter 25 Gulliver's Return to the Land without One, Two, Three (1960) 315
Chapter 26 Gulliver in Applyland (1960) 320

Karl Menger: Principal Dates 324

FIELDS OF RESEARCH 325

BIBLIOGRAPHY OF WORKS BY KARL MENGER 327

INDEX OF NAMES 339

ACKNOWLEDGMENTS

The Author and Publishers gratefully acknowledge the permission given by the Editors, Publishers and Copyright Holders of the following publications to reproduce in this book, in the original or in amended form, the articles mentioned below:

'The New Logic', *Philosophy of Science*, vol. 4 (1937), pp. 299–336. Copyright © 1937 The Williams and Wilkins Co., Baltimore.
'New Approach to Teaching Intermediate Mathematics'. *Science*, vol. 127, pp. 1320–1323, 6 June 1958.

'Der Intuitionismus,' *Blätter für deutsche Philosophie*, vol. 4 (1930), pp. 311–325. Copyright © 1930 Deutsche Buchvertrieb, Schmidt & Co.

'On the Formulation of Certain Questions in Arithmetic' and 'Why Johnny Hates Math'. *The Mathematics Teacher*, vol. 49 (1956), pp. 528–530 and 578–584.

'A Logic of the Doubtful. On Optative and Imperative Logic'. *Reports of a Mathematics Colloquium*, *University of Notre Dame* (Indiana), vol. 1 (1939), pp. 53–64.

'What is Dimension?' *American Mathematics Montly*, vol. 50 (1943), pp. 2–7.

'Variables, Constants, Fluents', from Feigl (ed.), *Current Issues in the Philosophy of Science*. Copyright © 1961 C.B.S. International Publishing, N.Y.

'Bemerkung zu Grundlagenfragen. I und II'. *Jahresbericht der deutschen Mathematiker Vereinigung*, vol. 37 (1938) and
Dimensionstheorie (1928) pp. 74–77. Copyright 1928 B. G. Teubner GMBH, Stuttgart.

'The Logic of the Laws of Return'. *Economic Activity Analysis* (1954), pp. 419–481. Original version Copyright 1954 John Wiley and Sons, Inc. New York.

'On the Design of Grouping Problems and Related Intelligence Tests', appeared (in part) in *Journal of Educational Psychology*, vol. 44 (1953). Reprinted by permission of the American Psychological Association.

'The Role of Uncertainty in Economics' from *Essays in Mathematical Economics: In Honor of Oskar Morgenstern*, Edited by Martin Shubik, Copyright © 1967 by Princeton University Press. Reprinted by permission of Princeton University Press.

'Gulliver in the Land Without One, Two, Three'. *Mathematical Gazette*, vol. 43 (1959) pp. 241–250.

'Gulliver in Applyland'. *Eureka* (1960). Mathematical Society, Junior Branch of the Mathematical Association, Cambridge University, Cambridge, England.

'Gulliver's Return to the Land Without One, Two, Three'. *American Mathematics Monthly*, vol. 67 (1960), pp. 641–648.

INTRODUCTION

This volume brings together those papers of mine which may be of interest not only to various specialists but also to philosophers. Many of my writings in mathematics were motivated by epistemological considerations; some papers originated in the critique of certain views that at one time dominated the discussions of the Vienna Circle; others grew out of problems in teaching fundamental ideas of mathematics; still others were occasioned by personal relations with economists. Hence a wide range of subjects will be discussed: epistemology, logic, basic concepts of pure and applied mathematics, philosophical ideas resulting from geometric studies, mathematical didactics and, finally, economics. The papers also span a period of more than fifty years. What unifies the various parts of the book is the spirit of searching for the clarification of basic concepts and methods and of articulating hidden ideas and tacit procedures.

Part I includes papers published about 1930 which expound an idea that Carnap, after a short period of opposition in the Circle, fully adopted; and, under the name "*Principle of Tolerance*", he eloquently formulated it in great generality in his book, *Logical Syntax of Language* (1934), through which it was widely disseminated. "The New Logic" in Chapter 1 furthermore includes the first report (1932) to a larger public of Gödel's epochal discovery presented among the great logic results of all time. Chapter 2 is a translation of an often quoted 1930 paper presenting a detailed exposition and critique of intuitionism. The former paper may even today serve nonspecialists as an initiation into the history of logic up to the mid-1930's.

Part II contains minor logic papers of some epistemological significance. Chapter 3 is the translation of a little note of 1930 connecting meaningfulness and structure of logicomathematical expressions, while an appendix presents later reflections on parenthesis-free notation discovered by Łukasiewicz. (Wittgenstein had claimed parentheses to be indispensable). Chapter 4 contains a summary and an expansion (not heretofore published) of papers which developed between 1951 and 1965 a new (algebraic) point of view on the logical connectives, *not*, *or*, *implies*, etc. The famous theorem that all connectives can be expressed in terms of a single one

("Sheffer's stroke") is split into several propositions: With regard to some algebraic operations of expressing, actually at least three connectives are needed; and only a certain combination of several such operations makes it possible to uphold the assertion of the famous theorem from the point of view of what I called the *algebra of functions*. Chapter 5 is a translation of a 1927 paper through which some of the outstanding set-theoreticians of that period, as they told me, for the first time understood Brouwer's (to quote Fraenkel, all but incomprehensible) definition of a set. The clarification culminates in a dictionary-like translation of Brouwer's ideas into well-known concepts of traditional set theory. Chapter 6 presents the most elementary motivation imaginable for the distinction between sets and classes, now maintained in most axiomatics of set theory. Chapter 7 is a reprint of a 1938 note on a topic about which a vast literature has since developed under the name of deontics: a logic of wishes, in this paper based on a logic of the doubtful.

Part III is devoted to some of the most fundamental ideas of pure and applied mathematics. They all are connected, in one way or another, with the notion of a *variable* – according to Russell "certainly one of the most difficult to understand." The clarification of the underlying problems attempted in Chapter 8 also shows *why* it has been so difficult to understand or to say what a variable is. The reason is the existence of an immense medley of totally diverse and in part unrelated things (in the widest possible meaning of this word), which have been indiscriminately designed as variables: things mathematical and nonmathematical, abstract and concrete, generic and specific, symbolized things and symbols. Just as in medieval methodology masses of superfluous notions connected with single ideas called for Occam's Razor to shave off what was unnecessary, so in mathematico-scientific methodology the confused conglomerate of ideas, all of them designated by the same term 'variable', calls for a *Counterpart of Occam's Razor* to widen a framework that is insufficient. The paper supplies this for a two-fold use: in ontology and in semantics. Chapter 9 contains a detailed analysis of those functional connections with which Mach wanted to replace causal explanations. Without being either theoretical science or applied mathematics these studies launch something that might be called a *theory of the application of mathematics to science*. Because of the fundamental character of these topics, I have included in Chapter 10, at the risk of some repetition, parts of a third paper on variables, constants and fluents. (The reader will notice that where a topic recurs it is usually treated in a somewhat different perspective.) During the years when I attended the meetings of the Vienna Circle (from 1927 to Schlick's assassination in 1936), the discussions never touched on the problems of variables here considered. Nor did Wittgenstein deal with them in the *Tractatus*; and in his writings, posthumously published before I wrote on the Counterpart of Occam's Razor, I discovered only one passage bearing a direct relation to variables as discussed in Part III of this book. It is a rather odd remark, but contained in a part of his work to which Wittgenstein seems to have paid particular attention. In Chapter 11, that remark is the object of a discussion which I announced long ago but have not published heretofore.

Part IV begins with three short papers applying the results of Part III to didactics, written in the late 1950's at the time of the great debate about the teaching of mathematics. Chapter 12, "A New Approach to the Teaching of Intermediate Mathematics" brings out conceptual and notational difficulties of calculus in its traditional form and solves them by what has been called "a sweeping new method." Chapter 13, "Why Johnny Hates Math", describes, from the point of view of a student in a secondary school, a few of the dozen conceptually different meanings of the term variable and of the operationally different rules according to which the letter x is used. Chapter 14, written jointly with H. J. Curtis, studies questions such as "$9 \times 4 = ?$", which arise on the most elementary level of arithmetic and yet have analogues even in advanced mathematical instruction. Occasionally, an intelligent youngster, though of course knowing that the expected response is "36", may yet say "4×9" in order to find out why this obviously valid answer is unacceptable. But it is unlikely that the teacher will explain. Arithmetic questions rather set the pattern for studying some parts of mathematics by guessing what is meant or expected but not said or asked. Chapter 15 deals with intelligence tests. Analysis shows that many of these tests actually probe the ability to guess what someone (namely the author of the test) happens to have in mind and, in the presence of time limits, to guess fast – an often very valuable faculty indeed which, however, certainly has nothing to do with logical thinking. Yet it is well known – and after what is said in Chapters 12, 13, 14 will not be surprising – that the achievements of students in elementary and secondary mathematics as presently taught are highly correlated with their performance in such intelligence tests. Chapter 16 is an excerpt of a paper, "The Geometry Relevant to Modern Education", dealing with a dilemma that has arisen because not enough time is available for teaching on a secondary level all of Euclidean geometry rigorously: Shall the material be presented without modern rigor or ought geometry not to be taught at all? The paper shows a way out of this dilemma.

Part V is devoted to the philosophical ramifications of certain geometric ideas. Chapter 17 contains remarks about scientific definitions that I developed in connection with the idea of curves and the concept of dimension. They were presented to the early Circle and then often quoted by Carnap, Waismann and others in their talks and publications. Carnap considered them to be paradigms of what he called the *explication* of concepts as well as of the treatment of explications as such. Chapter 18 deals with my taxicab geometry, which recently has found its way into several textbooks. Round squares or square circles have been considered as the pinnacle of absurdity and the quintessence of impossibility by practically all philosophers and philosophical schools in antiquity, in the middle ages and in modern times. In fact, their inconceivability has been one of the few points of universal consensus, even after Minkowski in the early 1900's proved their existence in a new non-Euclidean geometry, then considered as ivory tower mathematics. But, years ago, I brought square circles literally to the man in the street. I showed that in a modern city the end points of all taxi rides starting at the same point and having the same length (as proved by the fares) – in other words, the points of a circle in the taxicab geometry – lie on a square. Chapter 19 is an excerpt of a paper that in 1928 introduced a theory synthesizing numerous branches of mathematics, in particular various geometries and Boole's algebra of classes. It is a theory of joining and intersecting that I later called *algebra of geometry*. (Boolean algebra is the special case for which distributive laws hold.) The paper also contains the definition of what G. Birkhoff later called *normed lattices*. From 1928 on, Viennese students of mine, especially G. Bergmann (subsequently turned philosopher), greatly elaborated the theory. In the excerpt, without entering into technicalities, I emphasize one detail. It concerns the initial words of Euclid's *Elements*, "A point is that which has no part." They are nowhere else used in that great book, nor *could* Euclid use this definition in his deductions since he had no theory of the concept of part. The algebra of geometry, however, does include a definition and a deductive theory of parts. I therefore could use Euclid's definition thereby awakening, in the field of deductive geometry, the famous initial sentence of the *Elements* from its Sleeping Beauty rest of 2000 years. Chapter 20 is based on a critique of Euclid's *Elements* that is more radical than that of his assumptions about space in the large (about extensibility and parallelism of lines). The latter critique culminated in the

19th century in the development of non-Euclidean geometries as alternatives to the geometry of the *Elements*. The former challenges the assumptions made in Euclidean as well as non-Euclidean geometries about space in the small, and in fact the applicability to very small regions of the entire conceptual framework of points, lines, congruence, etc. For example, instead of points we may have to study lumps; and instead of a number describing the rigid distance between two elements all I assume is the probability that this distance belongs to a given interval. Such a probabilistic geometry reconciles the relations

$$A = B, \qquad B = C, \qquad A \neq C,$$

a conjunction which Poincaré mentioned again and again as characteristic of the physical in contrast to the mathematical continuum. The resulting probabilistic geometry is furthermore applicable to psychophysics – to the continuum of sensations superimposed on a physical continuum of stimuli. But we may well have to be satisfied with studying the probability that two lumps are indistinguishable. The paper concludes with a related idea outlined in my note on *ensembles flous* (1951). (Recently, others have extensively treated such entities under the name of *fuzzy sets*.) All that I assume to be known about an element and a subset of some universe is the probability that the element belongs to the subset. If for each element of the universe this probability is either 1 or 0, then and only then is the subset a set in the classical sense.

Part VI describes my memories of L. E. J. Brouwer, mentioned in Parts I, II and V in connection with logic and geometry, whom I knew well from 1925 until an unfortunate controversy ended our relations in 1929. The inclusion of my reminiscences in these Selected Papers is motivated by the inclusion of some references to me in Brouwer's recently published Collected Works. Moreover, Chapter 21 sets forth many of the methodological ideas that motivated my early work and describes its genesis.

Part VII is devoted to questions connected with economics. Chapter 22, on "The Role of Uncertainty in Economics", begins with a discussion of a game of chance that has fascinated many intellects since it was formulated in 1713 in St. Petersburg. In order to reconcile the concept of expectation with normal human behavior in this game, Cramer and Daniel Bernoulli introduced ideas concerning a person's evaluation of monetary gains, which anticipate the concept of marginal utility; they are marred only by the use of precise and uniform evaluation formulae where in reality there is no precision and in fact great diversity. But, as demonstrated in my essay, their treatment of the Petersburg game, though universally regarded as a solution (even by Poincaré), is purely *ad hoc* and does not achieve a reconciliation of theoretical expectation and human behavior in games of chance in general. My paper proposes, however, a more radical modification of the concept of expectation involving not only the amounts but also the probability of gains; and this idea in combination with the introduction of several parameters allowing for various traits of individuals leads indeed to a satisfactory description of actual human behavior (normal as well as abnormal). But what I regard as the essential point of that early paper is the introduction of probabilistic ideas into this branch of economics, whereas Bernoulli had introduced an economic idea into the theory of probability. Chapter 23, on the laws of return, deals with the logical relations between some famous propositions of economics rather than with the validity of these propositions, thus with what, applying modern methodological terminology, I call *meta-economics* rather than with economics. The intent of the paper was to familiarize economists (especially of nonmathematical schools) who claimed to *prove* certain assertions about returns, with what logicians demand of proofs. Only very elementary mathematics is involved in the demonstrations; but a modicum of mathematics is indispensable in order to present a method that economists (and indeed all social scientists) interested in proofs of quantitative laws may find useful in checking conjectures: the method of demonstrating that a proposition β is *not* a consequence of a proposition α by exhibiting a function that satisfies α but not β.

Part VIII once more returns to variables, functions and fluents – this time in a satirical vein. Chapter 24 describes a trip of Gulliver (not recorded by Swift) to an island where the numerals are somewhat comparable to our designations of functions. The people lack symbols for one, two, three, though they possess higher numerals, just as we lack designations for the identity and constant functions though we have symbols such as *log* and *cos*. Chapter 25 tells the story of Gulliver's return after the islanders had introduced a "modern arithmetic". Chapter 26 is a sort of diary of Gulliver in a land, called Applyland, where the traveller is amazed by people's odd methods of applying arithmetic to objects. Yet they show a strong resemblance to our methods of applying calculus to fluents.

I thus hope that philosophers and methodologists, geometers and economists, scientists who apply mathematics and epistemologists who study its foundations, as well as historians of philosophy and geometry and teachers of mathematics on all levels will find in this volume something worthy of their attention.

Large parts of the book can be read without any technical knowledge of mathematics. These sections include: the history of logic (Chapter 1); parts of the discussion of intuitionism (Chapter 2); the Introduction to the Counterpart of Occam's Razor (Chapter 8); what is said about definitions in general and about dimension and the curve concept in particular (Chapters 17 and 21); the taxicab geometry (Chapter 18), the critique of traditional (Euclidean and non-Euclidean) geometries in the small and ideas concerning a positivistic frame of geometry (Chapter 20); the discussions of intelligence tests (Chapter 15) and of the Petersburg game of chance (Chapter 22); the problem of meta-economics and the critique of the proofs of the law of diminishing returns in the economic literature; in Part II, the remarks about the antinomies of set theory (Chapter 6) and about frontal notation (Chapter 3, Appendix). Only the most elementary mathematics is needed for the reading of Chapters 10, 13 and 14; and not much more for the understanding of Chapters 8 and 10 and of the Gulliver stories in Part VIII. The scientist who has used even a modicum of mathematics will find explicit rules for the application of the function concept and of analysis – rules that for centuries have been used but, it seems, never were articulated before the publication of the papers collected in Part III (Chapters 8, 9, 10). Even accomplished mathematicians may want to familiarize themselves with the treatment of the logical connectives as an example of the algebra of functions (Chapter 3), the idea of a probabilistic microgeometry (Chapter 20) and the theory of the application of mathematics to science (Chapters 9, 10). They may also enjoy the satirical stories at the end of the volume. Finally, historians of philosophy and mathematics will find material concerning the development of basic ideas of logic and geometry as well as numerous remarks about the Vienna Circle.

Some material included is new, in particular the note on Wittgenstein (Chapter 11) and my reminiscences of Brouwer (Chapter 21). The algebras of the logical connectives and of geometry as well as the taxicab geometry, while excerpted from previous writings of mine, have not heretofore been presented quite in the form in which they appear in this volume (Chapters 4, 18, 19). In republishing the other material I have made no significant changes in content except for occasional additions indicated by bold-face brackets. I have tried, however, to improve throughout the manner of expression, especially of the older papers. In this task I have had the invaluable suggestions and untiring assistance of Dr. Phyllis Kittel.

For the author's views on ethics the reader is referred to Morality, Decision and Social Organization, Volume 6 in the present series.

I also wish to thank my friends, Prof. B. Schweizer for compiling the bibliography and Prof. A. Sklar for many important suggestions concerning my work during the past two decades. Mrs. P. Townsend-Johnson and Mr. L. W. Johnson have compiled the Index, Mr. R. Kowalski helped with reading of the proofs. I am particularly grateful to Dr. E. Mulder for bringing up to date and editing the bibliography and for many suggestions.

PART I

PAPERS INTRODUCING LOGICAL TOLERANCE

LOGICAL TOLERANCE IN THE VIENNA CIRCLE

"It makes no sense to ask for justification, or to question the legitimacy, of linguistic forms," V. Kraft writes in his book *The Vienna Circle* (p. 60). "One should instead pay homage, in the logical analysis of language, to a principle of tolerance." And in Note 65 of his book, the historian of the Circle continues: "This was emphasized for the first time by K. Menger in the controversy about the foundations of mathematics ('Der Intuitionismus', 1930)" – herein Chapter 2 – "and thereafter K. Popper endorsed this point of view as a regular tenet of the logic of science (*The Logic of Scientific Discovery*, 1935)."

Actually, the germ of the idea that Carnap adopted in 1934 and so aptly called "Principle of Tolerance" can be traced back even farther. In the spring of 1927, at the end of a paper introducing an intuitionistic-formalistic dictionary of set theory (herein Chapter 5) I briefly emphasized that the word "constructivity" probably is capable of definitions in various ways and in various degrees. When elaborating this idea in [1930, 9] I further pointed out

(1) that "for each of the various versions of constructivity one could develop a corresponding deductive mathematics", especially systems more restricted than intuitionistic mathematics – a remark envisaging possibilities which in the past few decades have fully materialized in a proliferation of (especially hyperintuitionistic) systems of mathematics;

(2) that "the insistence on a particular idea of constructivity, the designation of the corresponding developments as meaningful and the rejection of transcending results as meaningless *have not the least cognitive content*," and are to be relegated "from logic and mathematics to the biography of the proponent";

(3) that what matters in mathematics and logic is exclusively the question as to the propositions into which certain other propositions can be transformed according to certain rules – I spoke of an *implicationist* point of view – while the justification (*Begründung*) of propositions or rules of transformation by appeals to intuition "*are nothing but empty words*".

In further developing these ideas in a lecture "The New Logic" (1932) – herein Chapter 1 – I emphasized

(4) that the entire activity of the mathematician consists in transforming propositions (which can be chosen in various ways but must be clearly listed) into other propositions by means of rules of transformation (which can be chosen in various ways but must be clearly listed);

(5) that this simple statement of fact is all that mathematics and logic can say about that activity, which is neither in need nor capable of receiving a foundation (*eine "Begründung"*).[2]

In the fall of 1927, when joining the Schlick Kreis – the group (named in Vienna after its founder) that became known abroad as the Vienna Circle – I soon realized that the multiplicity of constructivity definitions was intimately connected with another matter. Under the influence of Wittgenstein's *Tractatus*, all members of the Circle, in particular Schlick, Carnap and Waismann, but even Hahn made constant references to *the* language and *the* logic. In a report about the intuitionistic-formalistic dictionary mentioned above during one of the first meetings of the Circle that I attended, I seriously questioned the uniqueness of language and logic. And, while rarely participating in the general discussions of the Circle, I pursued this topic in numerous conversations with most of its members thereafter. But Schlick and Waismann at that time refused to take my skepticism seriously; Hahn was not favorably disposed; Carnap, too, at first shook his head; Neurath was not very interested in the question, while Kraft remained silent; and Husserl's disciple, my personal friend F. Kaufmann, who regularly attended the meetings of the Circle, was dead set against the idea.

With regard to "the language," however, Carnap in the course of the year not only abandoned his belief in uniqueness but began to put strong emphasis on the multiplicity of languages between which one might choose. He described his conversion in *The Logical Syntax of Language* (1934) as follows: 'The earlier positivism of the Vienna Circle was essentially that of Wittgenstein. . . . On that view it was a question of '*the* language' in an absolute sense; it was possible to reject both concepts and sentences if they did not fit into *the* language." Even more explicitly, Carnap wrote in *Testability and Meaning* (1936): "I was wrong in thinking that the language I dealt with was *the* language, i.e. the only legitimate language – as Wittgenstein, Schlick, and Lewis likewise seem to think concerning the language-forms accepted by them."

But with regard to "the logic" – a much profounder topic than "the language" in the loose sense in which the latter term was used in 1930 – Carnap's belief in uniqueness remained unshakeable up to the time of his first visits in Vienna after he had left for Prague in 1931. So I kept insisting on this point whenever I saw him[4] and defended logical tolerance in discussions with other members of the Circle. But only Gödel when present supported me with a nod of assent.

It thus came as a pleasant surprise to me when in 1934 Carnap completely adopted this point of view in his book *Logische Syntax der Sprache* (*The Logical Syntax of Language*). In it, he coined the term *Principle of Tolerance* in the field of logic, giving a broad and very eloquent presentation of the underlying idea, which was to have a great impact on philosophers. In the Foreword to the book, Carnap described the effect of this principle in overcoming historical barriers and opening the view on "the boundless ocean of unlimited possibilities"; and he wrote: "The elimination of the standpoint that language form must be proved to be 'correct', together with . . . wearysome controversies which arise as a result of it is one of the chief tasks of this book," In conformity with the above-quoted point (1), Carnap developed two theories based on logics corresponding to two degrees of constructivity[5] stressing that "everyone is at liberty to build his own logic. All that is required of him is that he must state his methods clearly. Both the forms of construction for sentences and the rules of transformation may be chosen quite arbitrarily. Only the question of the consequences to which one or the other of the choices leads [arises]. No question of justification arises at all." (See (3) and (5) above.)

As to the genesis of these ideas, Carnap mentioned in *Logical Syntax* that the attitude of tolerance had been emphatically represented in my above quoted paper "Intuitionism"[6], but not without first remarking that this attitude was tacitly shared by the majority of mathematicians. My comment on this remark was[7] that "I should be glad if Carnap were right; but since the prominent mathematicians (Poincaré, the Paris School at the beginning of this century, Hilbert, Weyl, Brouwer [and I might of course have added Russell's name]) who have dealt with the foundations of mathematics have explicitly expressed opinions which, however they otherwise diverge, are all diametrically opposed to the above mentioned principle, I am afraid that I must bear the responsibility alone."[8]

Thirty years later, in his 'Intellectual Autobiography' (1963)[9], Carnap finally gave an account quite at variance with the passages quoted from *Testability and Meaning* and *Logical Syntax* (written when his Viennese period was still fresh in his mind) and altogether oblivious of the genesis of the idea of logical tolerance. He mentions the principle in the section "The Beginning of my Work in Philosophy" adding that the attitude it expressed "has remained the same throughout my life." Clearly Carnap's memory was faulty on this point.

Applying logical tolerance I turned against the rather loose employment of the word *meaningless* (*sinnlos*), which was especially rampant in the Circle in the years 1927–32. As far as I remember, Neurath never indulged in its use, at least not literally – he rather used "metaphysical" – and Hahn was the first to abandon it; Carnap and Feigl followed; but Schlick continued the usage occasionally; and Waismann had at that time the habit of designating as meaningless everything that went beyond whatever it was that he was asserting, the latter usually being based on the latest of the kaleidoscopically changing suggestions of Wittgenstein. Despite my high regard for Waismann, I must confess that I found this habit of his somewhat irritating; and I often said to him, "If you disagree with such and such sentences, then don't say them, or say that you won't say them. But don't state that they are meaningless – especially in absence of precise rules concerning what is meaningful!" And even after the Circle identified the meaningful with what can be verified or somehow be tested I remained reserved. For what about set theory? What even about metaphysics as such? While in full agreement with the well-founded criticism of the known metaphysical theories I could not join in the dogmatic rejection of metaphysics in the lump.[10] Only years later, when studying Popper's writings, did I realize that at the time of the Circle I had been struggling with what Popper called the *Problem of Demarcation*.

At first sight, the polemic against the use of the word 'meaningless' may itself appear as intolerance of a kind. Actually, however, the loose employment of that term as it occurred in the early Circle and the dogmatic rejection of anything stigmatized by it are themselves acts of intolerance based, in the absence of precise rules concerning the term meaningful, on personal taste or on vague psychology. What has been mentioned above as point (2) of the paper "Intuitionism" represents a tolerance of second order – as it were, a tolerance by double negation.

In Chapter 2 of [1934, 8], this attitude toward meaninglessness is developed with regard to propositions about the "essence" of good and evil[11]. In keeping with my suggestions to Waismann, all that is said about these propositions is the emphatic statement that the author will neither affirm nor negate them. Furthermore, it is pointed out there that this view can be extended to value judgments in general, including some mathematicians' arbitrary and dogmatic rejection of certain well-defined procedures as meaningless. "For woe to him who declares statements to be meaningless!

He resembles a man who notices his mortal enemy inescapably mired in a swamp and instead of going his way, rushes toward him with drawn dagger."

Still another aspect of logical tolerance is the rejection of the claim that all mathematical propositions are tautologies. Between 1927 and 1931, I found that this thesis was asserted by all the other members of the Circle. It was usually attributed to Wittgenstein – not quite correctly: the *Tractatus* expounds the somewhat different, though equally untenable doctrine that the propositions of mathematics are equations. In the matter of tautologies, I found it difficult to communicate with Schlick, Waismann, Carnap and even Hahn. For the members of the Circle tried to reassure me that their thesis was purely technical and not an offense to mathematics[12]. I, on the other hand, was not concerned about the survival of mathematics, no matter what might be said about it, and was merely trying to find out exactly what they meant by tautology. But beyond Wittgenstein's concept (See below p. 21) which, while clear, was restricted to the very elements of logic and utterly inadequate to comprise all mathematical propositions and procedures, I never got a clear answer. In the 1937 version of 'The New Logic' I inserted the following footnote:

"The frequently heard statement that mathematics is a great tautology or a system of tautologies does not seem to me an adequate description of the situation. The concept of tautology has so far been defined only within the calculus of propositions. This concept might, it is true, be defined beyond that calculus, but it might be defined in different ways, whereas the sentence 'Mathematics is a system of tautologies' suggests a reference to an absolute logic."

NOTES

[1] In all quotations, italicization is the same as in the original.

[2] Subsequently, I stressed this point of view on several occasions, especially in [1934,9], pp. 27 sq., [1974, 1], pp. 20 sq. and [1936, 8].

[3] I had planned to present to the Königsberg meeting on the *Epistemology of the Exact Sciences* in September 1930 a short paper anticipating points (4) and (5) of my 1932 lecture (see above) and designating point (5) as a *behavioristic attitude towards the questions of foundations of mathematics and logic*. When it became clear that because of my departure for the United States I would not be able to go to Königsberg I decided to ask my student and assistant G. Nöbeling to read the paper before the meeting. But Kaufmann, with whom I discussed the plan, almost passionately entreated me not to carry it out because he was convinced that I would change my opinion. Unfortunately, I gave in and found only two years later the time and an opportunity to resume the topic in public (that is, also outside conversations in the Circle).

[4] From K. Popper's later writings (especially "Replies to my Critics" in *The Philosophy of Karl Popper*, 1974, p. 968) it appears that Popper objected to Carnap's dogmatism

before 1932 with arguments some of which were identical with the criticism (then unknown to Popper) that I expressed in the Circle ever since 1927.

[5] Carnap's "Language I" is based on a version of intuitionistic logic; "Language II" uses the traditional logic, that is, the limiting case without any restrictions of constructivity.

[6] On p. 52 of *Logical Syntax* Carnap writes: "In the conflict over the logical foundations of mathematics, [the tolerant] attitude was represented with especial emphasis (and apparently before anyone else) by Menger."

[7] Cf. [1937, 6], p. 335.

[8] Carnap (*loc. cit.* p. 52) qualified his statement about the majority of mathematicians embracing the attitude of logical tolerance by adding the remark "as far as special mathematical calculi are concerned," without which that statement in the early 1930's would have been completely untenable. Of course, what was being discussed in my writings on the subject were *philosophical* problems concerning logic and mathematics – not special mathematical calculi – and questions connected with value judgments and ethics (See the next to the last paragraph of this Introduction). On the other hand, in "Intuitionism" I myself had pointed out how valuable were, for example, the writings of Sierpiński about what does and what does not follow from the various cases of the axiom of choice in contrast to the endless discussions as to whether this postulate should or should not be assumed, whether it was or was not evident, and the like.

[9] *The Philosophy of Rudolf Carnap* (Library of Living Philosophers, 1963).

[10] After my above-quoted lines commenting on a remark of Carnap's, I added the following paragraph in a footnote: "I take this opportunity to emphasize that I do not agree with the pronouncements on metaphysics which come from the group of which Carnap is a member. It is true that an expert logician can easily find logical errors, and very elementary errors at that, in many metaphysical theories, just as he can in many economic and sociological ones. Such errors are, however, objections only to special existent systems. If all metaphysics is rejected because its statements cannot be tested, then, I am afraid, very extensive parts of mathematics must likewise be discarded. Thus one arrives at arbitrary statements which resemble some of the critiques of classic mathematics discussed in this paper" (Cf. [1937, 6], p. 335).

[11] What is said there about essences is very close to Popper's criticism of what he calls essentialism (in *Poverty of Historicism*, *Open Society*, *Conjectures and Refutations*, and *loc. cit.*[4]); and I am glad to see a complete agreement of Popper's views on the futility of defining a science, the arbitrariness of precisely circumscribing its object, and the like, with those expressed in [1934, 8].

[12] The German word *Tautologie* has an even stronger connotation of triviality than the English *tautology*.

CHAPTER 1

THE NEW LOGIC

In the spring of 1932, the economic and political situation in Austria had reached a point where it was virtually impossible to find academic employment even for the most talented students. Mathematics was especially hard hit. I keenly felt the general frustration in my Mathematical Colloquium in which I enjoyed the collaboration of, among others, F. Alt, K. Gödel, G. Nöbeling, and A. Wald.

At this juncture, I thought of Vienna's extraacademic intelligentsia who, between the two world wars, had reached a remarkable intellectual level. Until Hitler's invasion brought it all to an abrupt end, Vienna was teeming with physicians and engineers, lawyers and public servants, businessmen and bankers, seriously interested in the ideas and the philosophy of science – I have never found the like anywhere else. It occurred to me that many of these people might be willing to pay a relatively high admission to a series of interesting lectures on basic ideas of science and mathematics; and the receipts might subsidize the research of young talents.

I first discussed this plan with H. Hahn, whom I found most favorably disposed. We approached our friend, the physicist H. Thirring, who was equally enthusiastic and proposed that we speak with the chemist H. Mark, just returned to the university of his native Vienna. In a joint meeting, we outlined the program of a series: Mark was to open it with a lecture 'Classical Physics, Shaken by Experiments'; Thirring should continue with a talk on 'The Changes of the Conceptual Frame of Physics'; then Hahn would discuss 'The Crisis of Intuition'; and I should conclude the series with a lecture on 'The New Logic' (*Die Neue Logik*). Since a fifth talk seemed desirable I suggested that after Hahn's talk G. Nöbeling deliver a lecture on 'The Fourth Dimension and the Curved Space'. As a catchy title for the whole series, I proposed 'Crisis and Reconstruction in the Exact Sciences'.

The speakers all served the project without compensation; students volunteered to send out programs and to sell tickets; the university let us use an auditorium. The admission fee was comparable to that of the *Staatsoper* (Vienna Opera). Within a few days all seats were sold, and many late-comers had to be turned away. The talks, delivered in five consecutive weeks during the spring of 1932, were received with enormous interest. The Viennese publisher F. Deuticke brought them out in a booklet, [1933, 1], titled *Krise und Neuaufbau in den Exakten Wissenschaften* (like the lecture series), which received favorable reviews all over the world. We were particularly gratified by P. W. Bridgman's at Harvard (*Review of Scientific Instruments*, vol. **4**, 1933): " . . . As may be inferred from the titles and the known interests of the philosophical group at Vienna, the emphasis is placed on the reactions which recent activities in physics and mathematics bring about in the general philosophical outlook. The discussion goes much more into detail and makes much greater demands on the concentrated attention of the audience than would be possible in popular lectures in this country, but the Vienna audience seems to have survived and, in fact, it is stated [in a brief Preface to the booklet, see [1933, 1a]] that every seat was filled and that the lectures will be made an annual event [at the request of the audience]. No popular exposition in English lays such emphasis on the significance of recent developments in mathematics; in view of the greatly increasing use of mathematics in physics I believe that many physicists will find in these lectures stimulating fresh points of view to ponder. The last lecture [*Die Neue Logik*] on recent changes in points of view toward logic, is particularly to be recommended."

In the fall of 1936, the chemist James K. Senior of the University of Chicago, a man with a keen interest in and deep understanding for the philosophy of science, began to translate *Die Neue Logik*. I rearranged some of the final parts of the paper and added a few paragraphs [1937, 6]. In the following version, the arrangement, the formulations and even the emphasis by italics of the 1932 original are carefully restored (except for the omission of references to the other lectures in the series "Crisis and Reconstruction"). Most of my additions to the 1937 translation are herein included – but all of them between bold face brackets or in an Appendix. Notes [1] to [12] to [1933, 1] are interspersed with Notes [1a] to [12a] from [1937, 6]. Of particular interest among the additions are probably the remarks comparing the prescriptions and interdictions propounded by intuitionists of various shades with the dietetic rules decreed by various philosophical schools in antiquity (p. 39).

THE NEW LOGIC

A 1932 LECTURE

That the exact sciences are subject to crises and reconstruction and that even geometry has undergone changes is widely known. Moreover, as far as experience and intuition are concerned, anyone, even though he be unfamiliar with the details, can well understand that new empirical discoveries may overthrow the most venerable of ancient theories, and that intuition, if it ventures too far afield, may be forced to retire from positions erroneously occupied. One subject, however, is generally supposed to be unchanging and unshakable. That subject is logic. Hence, to anyone who is not an adept in the field, a discussion of crisis and reconstruction in connection with logic may seem not only strange but incomprehensible.

As a matter of fact, for two thousand years, logic has been the most conservative of all the branches of knowledge. Aristotle, who is considered the father of the subject, assumes as a starting point that every proposition ascribes a predicate to some subject. The propositions are classified on the one hand as affirmative or negative, on the other hand as universal or particular. "All cats are mammals" is universal and affirmative; "Some mammals are cats" is particular and affirmative; "Some mammals are not cats" is particular and negative; "No cat is a fish" is universal and negative. According to Aristotle, all inference consists in deriving a third proposition from two propositions of the given form. For example, from the premises that all cats are mammals and all mammals are vertebrates, it follows that all cats are vertebrates. From the premises that all cats are mammals and no mammal is a fish, it follows that no cat is a fish. All the inferences which Aristotle considered possible he arranged in three figures divided into altogether fourteen varieties. In the Middle Ages, these modes of inference were expanded to four figures including altogether nineteen varieties, and were designated by the names Barbara, Celarent, etc. If, from the premises that to every (respectively no) M belongs the predicate P, and that to every S belongs the predicate M, it is inferred that to every (resp. no) S belongs the predicate P, then the inference is said to be according to the mode Barbara (resp. Celarent). The three principles of identity, contradiction and excluded middle, which were later called the fundamental principles of logic, were also formulated by Aristotle, curiously enough not in his logic but in his metaphysics. The kernel of his logic, the above mentioned theory of subject-predicate propositions, was essentially all that was regarded as pure logic for two thousand years thereafter.

In the Middle Ages, it is true, the scholastics undertook a number of important logical investigations. But during the Enlightenment, it became customary to regard the logical work of the medieval scholars as hair-splitting while their results fell into oblivion and were replaced by less fundamental considerations. Similarly, Leibniz's views on logic, which were far in advance of their time, also remained without direct effect. It was perfectly clear to Leibniz that a mere treatment of subject-predicate propositions was inadequate and must be supplemented by a logic of relations. Furthermore, he treated the logical principles and their mutual relations more systematically than his predecessors, and devised the project of a *lingua characteristica*, which should permit all scientific propositions to be stated in precise form, and of a *calculus ratiocinator*, which should contain and treat by computations all the methods of inference. But the clearest evidence that Leibniz's views, like the works of the scholastics, found no echo, is the famous dictum of Kant in the introduction to the second edition of the *Critique of Pure Reason* (1787) "That logic has trodden this sure path since the earliest times, can be seen from the fact that, since Aristotle, it has not been obliged to take a single backward step. . . . A further noteworthy fact is that until the present it has been able to take no step forward, and so seems apparently to be finished and complete." And in Kant's *Logic* occurs the statement "The logic of to-day is derived from Aristotle's Analytics. . . . Moreover, since Aristotle's time, logic has not gained much in content, and from its very nature it cannot. . . . For Aristotle omitted no item of reason."

A crisis in the old logic was brought about by mathematics, When confidence in geometrical intuition had been shaken, a purely logical reconstruction of geometry (and later of arithmetic) began with a complete enumeration of all *mathematical hypotheses* from which the whole system of theorems was deduced by strictly logical methods. But this trend in mathematics naturally also carried with it a search for clarity in regard to all *principles of inference* used in the deduction of mathematical propositions from axioms. And in the course of this search, it became evident that the old logic was inadequate for the demands of modern mathematics in both precision and completeness. It was therefore chiefly mathematicians who undertook and carried out the necessary reconstruction of logic. A short sketch of the chief stages of this reconstruction follows, but it should be understood that the result of the reconstruction (called *Logistik* in German and occasionally referred to as *logistics* in English), although much farther reaching than the Aristotelian logic, is today counted as a part of the classical content of logic. It will appear that the real problems of the new logic only begin where logistic stops.

The *first* step in the reconstruction of logic was the development of the so-called *calculus of classes* (also called algebra of logic) particularly by Boole, Peirce and Schröder in the second half of the nineteenth century[1]. The Aristotelian logic, as stated, deals mostly with the question: If the relations of two classes to a third class are known, what can be said about the relations of the two classes to each other? The calculus of classes systematically studies the relations of any number of classes. Aristotle only investigates inclusions – whether of two classes one is entirely or partly contained in the other; or whether one is entirely or partly outside the other. The calculus of classes investigates many other relations between classes besides inclusions, as well as operations with classes. For example, it treats systematically of the join and intersection of two classes A and B – the class of all elements in either A or B, and the class of those in both A and B, respectively. In contrast to Aristotle, the empty class containing no member is also considered. For instance, the intersection of the class of all cats and the class of all fish is the empty class. The calculus of classes, starting from a certain few propositions, gives a systematic treatment of all relations between classes. Among its theorems, there are nineteen which correspond to the Aristotelian-Scholastic modes of inference[2].

But in the calculus of classes, these ninteteen theorems are not the only ones; nor do they occupy in it a distinguished position. They are not among its initial propositions (in the first place because the whole of this calculus can be based on much fewer than nineteen propositions), and so they are not necessary for the founding of a systematic calculus of classes. But even if one were to assume so needlessly large a number of initial propositions, the particular nineteen Aristotelian ones would not be sufficient for the construction of the whole calculus of classes. This calculus is therefore a definite advance over the old logic of class inclusion.

A *second* step leads further. The calculus of classes is a theory deduced from a few initial propositions concerning the relations of classes, just as Euclidean geometry is deduced from a few initial propositions (axioms or postulates) concerning the relations of points, lines, and planes[3]. The calculus of classes is thus a special mathematical theory; it is, however, far from containing the whole of logic, for logic does not confine itself to the considerations of classes. If for example, propositions are derived from any set of propositions, this derivation is called logical inference. The subject matter of this inference is, however, not classes but propositions. And yet logic is expected to deal with the rules of inference (the rules of transforming and combining propositions so as to yield new propositions). The second step in the expansion of logistics, which goes back principally to Peirce and Schröder, was therefore the development of a *calculus of propositions*. This calculus studies how propositions are combined by words like 'and,' 'or,' 'not' and similar particles[4]. If p and q are two propositions, then a compound of particular importance is 'q or not p', which is briefly expressed in logic by the words 'p implies q' or 'q is implied by p'. If q is true, then the proposition 'q or not p' is surely true, whether p is true or false. A true proposition q is therefore implied by every proposition. If p is false, then *not-p* is true, and the proposition 'q or *not-p*' is surely true whether q is true or false. Thus a false proposition p implies every proposition q. It follows, for example, that '*not-p* implies p' is true if p is true, and false if p is false. Hence '*not-p* implies p' implies p, and vice versa.

Of special importance among the propositions composed of several statements connected by logical particles are those which are true in all cases, whether the component statements are true or false. For example, the proposition 'It is raining or it is not raining' is true, whether the component statement 'It is raining' is true or false. At Wittgenstein's suggestion, such complex propositions which are always true are called *tautologies*. The calculus of propositions is concerned with the listing of tautologies. As Frege proved, this calculus of propositions can be derived from some simple assumptions; that is, all tautologies can be obtained from a few simple tautologies.

It may be thought that the calculus treats only those transformations of propositions which result from the applications of the three Aristotelian principles (identity, contradiction and excluded middle). Such, however, is not the case. More correctly, the three Aristotelian principles play a role in the calculus of propositions similar to that played by the nineteen Aristotelian modes of inference in the calculus of classes. That is, the principles of identity, contradiction and excluded middle occur among the theorems of the calculus of propositions but they are not the only theorems of this calculus; nor do they occupy in it a distinguished position. In particular, they do not appear among those propositions actually chosen as initial propositions for this calculus; thus they are not a necessary part of its foundation. Neither are they sufficient for the derivation of the whole of it. The calculus of propositions is thus a decided advance over the old logic.

Since the whole calculus of propositions can be derived from certain simple initial propositions, and since, on the other hand, the principles of this calculus are supposed to be the principles of logical derivation it might perhaps be suspected that the calculus in question is founded on some sort of reasoning in a circle. Frege, however, carefully avoided this danger. His starting point are certain propositional formulae – in fact tautologies. According to Łukasiewicz, the three following may be chosen for this purpose.

(1) *p* implies (*not-p* implies *q*).
(2) (*not-p* implies *p*) implies *p*.
(3) (*p* implies *q*) implies [(*q* implies *r*) implies (*p* implies *r*)].

Of these, the first corresponds to the fact that a false proposition implies every proposition. This has been previously mentioned as has the second. The third formula is closely related to the syllogism; if *q* is implied by *p*, and *r* is implied by *q*, then *r* is implied by *p*.

The calculus of propositions is the totality of those propositional formulae which can be obtained from the three initial formulae by the use of two simple formative rules. *First*, in the initial formulae or in formulae already obtained from them, the symbols *p*, *q* and *r* may be replaced by other and possibly complex propositional symbols. For example, if, in the initial formula (1) one substitutes '*r* implies *s*' for *p*, then one obtains the formula

(*r* implies *s*) implies [*not*-(*r* implies *s*) implies *q*] which, according to the first formative rule, is to be taken into the calculus of propositions. *Second*, if the proposition *P* and the proposition '*P* implies *Q*' are two formulae in the calculus of propositions, then *Q* is also admitted as a formula of this calculus. It is thus clear that the calculus in question and the inferences used for its systematic development are kept carefully separate. The calculus consists of formulae; the development of the calculus is accomplished by the use of two rules for the construction of formulae. Both the initial formulae and the two formative rules are precisely stated. Obviously the latter are considerably simpler than the usual rules of logical inference.

But most common propositions, particularly those of mathematics, besides employing such words as 'and,' 'or,' 'not' and 'implies,' contain other logical particles, especially 'all' and 'some' or 'there are.' The exact rules for dealing with propositions containing these so-called logical quantifiers form a *third* chapter of the newer logic which, since the work of Peirce and Frege, has taken its place beside the calculi of classes and of propositions; it is called the *calculus of functions*. The historical origin of this name is as follows. Besides propositions, some of which are true like 'This charcoal is black,' and some of which are false like 'This charcoal is red,' there are also word combinations like '*x* is black' which are not propositions, but which become propositions when the name of a definite individual within a certain field is used to replace the symbol *x*, or when the combination is preceded by a quantifier. Such word combinations are called propositional functions. For example, the propositional function '*x* is black' becomes a true proposition if 'this charcoal' is substituted for *x*. It becomes a false proposition if 'this lime' is substituted for *x*. When mankind is chosen as the range of *x*, the proposition becomes false if the quantifier 'all' is placed before *x*, for then it becomes the false (universal) proposition 'All men are black.' Finally, if the quantifier 'some,' is placed before *x*, the proposition becomes the true (existential) proposition 'some men are black' or 'there exist black men.' Since the rules for logical operations on universal and existential propositions are derived from the theory of propositional functions, the study of operations with such propositions is called the calculus of functions.

The calculi of classes and of propositions together with the hitherto considered portions of the calculus of functions may be interpreted as mere refinements of the old logic – if the word 'refinement' is used in a very broad sense. But the *fourth* step in the development of the new logic is undoubtedly an extension of the content of the subject. The impetus to this step also came from mathematics; for the propositions which are the subject of mathematics, and which were first expressed by Peano in a general and rigorous symbolism, are only rarely statements about the membership of individuals in classes or about inclusions between classes. Neither are they often compounds of statements about classes connected by the words 'and,' 'or,' 'not,' 'implies,' 'all' or 'some'. Most mathematical propositions deal rather with *relations*, as Leibniz already recognized. The proposition '3 is less than 5' states a relation between two numbers; the proposition 'If, on a straight line, the point *q* lies between the points *p* and *r*, then *r* does not lie between *p* and *q*' is a general statement about a relation between the members of triples of points on straight lines. A logic useful to mathematicians must above all treat of relations.

What corresponds to a predicate is a class, namely the class of all those things which have that predicate. For instance, the class of all black things corresponds to the predicate 'black'. What corresponds to a relation between two things (called a dyadic relation) is a class of pairs of things, namely the class of all those pairs of things in which the first member of the pair has the given relation to the second. For example, the class of all pairs in which the first member is less than the second corresponds to the dyadic relation 'less than'. The extension of logic which treats propositions on dyadic, triadic, . . . relationships along with subject-predicate propositions can thus be characterized by the fact that besides classes of individuals it investigates classes of pairs of individuals, classes of triples of individuals, . . . The calculus of relations also starts with certain initial formulae, and with the help of a few exactly defined rules, derives from them a system of formulae.

Although logic is thus greatly increased in content, it is still inadequate to account for all the conclusions drawn in modern mathematics. In order to express the newer mathematics, particularly the theories of real numbers and of sets, a *fifth* step had to be taken. It was necessary to create an *expanded calculus of functions* dealing with classes of all sorts of classes of individuals, with classes of classes of classes and so on. This extremely important extension of logic is not only necessary to account for modern mathematics; but it is, according to Russell, also sufficient (when taken in connection with the portions of logic previously discussed) to serve as a basis for all of mathematics. That such is the fact will next be shown in a short sketch.[5]

Once in possession of this expanded calculus of functions, one can define, following Georg Cantor, two classes *A* and *B* to be equinumerate or of the same power provided they can be placed in one-to-one correspondence with one another – that is, if to every member of the class *A*, a member of the class *B* can be assigned in such fashion that each member of *B* is assigned to just one member of *A*. For example, the classes of crosses and dashes here shown

```
+   +   +   +   +
–   –   –   –   –
*   *   *   *
```

are said to be of equal power because they can be placed in one-to-one correspondence by assigning to each cross the dash directly below it. On the other hand, the classes of dashes and stars are not equinumerate, for, no matter how a star is assigned to each dash, there will be a star which is assigned to more than one dash. Children and savages actually count low-numbered classes by bringing the members of the class to be counted into one-to-one correspondence with the members of a class of the fingers of their hands – a fact that accounts for the importance of the number ten in the common number system.

It should not be imagined that the concept of equality of numbers presupposes a concept of number or an operation of counting. Without counting how many auditors or seats there are in a lecture room, the number of auditors and the number of seats may be shown to be equal by ascertaining that each seat is occupied and that each auditor has a seat; these facts establish the one-to-one correspondence necessary and sufficient for the class of all auditors and the class of all seats to be equinumerate. In the theory here outlined the concept of number rather depends on the concept of equality of number, for a number is defined as the class of all classes which are equinumerate with a given class.[5a] For instance, five is defined as the class of all those classes (such as the classes of crosses and dashes shown above) which are equinumerate with the class of the fingers of one hand. But Cantor went further. For each infinite set *S*, he combined into a class all those sets which are of the same power as *S*. One of his fundamental discoveries in the theory of sets is that different infinite numbers can be obtained in this way, for not every infinite set is of the same power as every other infinite set.

The natural numbers (1, 2, 3, . . .) are defined in purely logical terms – essentially by the fact that the principle of complete induction holds for these numbers. This important principle states that any proposition which holds for the number 1, and which, if it holds for any natural number n, also holds for the succeeding number $n + 1$, holds for every natural number.

In the original lecture, there followed at this point a sketch of the traditional introduction of the *integers* (positive, negative and 0) as pairs of natural numbers; the *fractions* as pairs of integers; the *real* numbers (rational numbers corresponding to fractions, and irrational numbers) as divisions ("cuts") of the ordered set of fractions; and the imaginary and *complex* numbers as pairs of real numbers. The *n-dimensional space* is defined as the set of all n-tuples of real numbers. The concepts of *limit* and *continuity* can then be introduced along with the other ideas of analysis.[5b]

In the words of Russell, after logic has been widened so as to include the expanded calculus of functions, *mathematics becomes part of logic*. This assertion is not by any means an arbitrary extension of the word logic to include all of mathematics. The historical development has above been described. All the conclusions drawn by modern mathematics were treated just as the most primitive conclusions were treated in the calculus of propositions. In other words, all these conclusions were derived from certain initial formulae with the help of certain formative rules. Thus there were obtained initial formulae which not only permit the derivation of the mathematical modes of inference but also suffice to derive all mathematics.

Such was the culmination of logistics, which at this point contained all that was needed. After two thousand years of petrification, logic had, in less than half a century, been entirely reconstructed by the mathematicians; and in the year 1900 a lecture on logic might have terminated with this happy result.

About the turn of the century, however, there came an entirely unexpected repercussion of the worst kind. The newly introduced unlimited operations with classes and classes of classes led to nothing less than an *antinomy*. Now an inner contradiction is unbearable even in the special theory of a particular domain of knowledge. For logic, however, the appearance of a contradiction in its structure is catastrophic.

Before explaining the newly discovered antinomy, it should be mentioned that paradoxes had appeared even in ancient logic. The one about the liar is well known. It was also called the paradox of the Cretan, because the Cretans were supposed to be particularly mendacious. In ancient times, the statement 'All Cretans are liars' in the mouth of a Cretan was regarded as a logical paradox. The best way to illustrate this sort of paradox with modern precision is to write on a board the following statement: 'The statement on this board is false'. Nothing further is written on the board, and the given statement is examined to see whether it is true or false. It will be proved to be neither. Assuming that the statement is true, it follows that the statement on the board is false, in contradiction with the assumption that it is true. Assuming that the statement is false, since this is precisely the statement on the board, it follows that the statement on the board is true, in contradiction with the assumption that it is false. Thus the assumption that the statement is false leads to the conclusion that it is true; and the assumption that the statement is true leads to the conclusion that it is false. In other words, the assumption of either truth or falsehood for the statement leads to a contradiction. Hence, the statement is neither true nor false, which is of course paradoxical, for, according to the principle of the excluded middle, every statement is either true or false, and any third possibility is excluded.

Precise analysis shows, however, that the words 'statement on this board' play an essential role in the paradox in which they appear; and these words are not of a logical nature. For this reason, the above and related paradoxes are (at Ramsay's suggestion) distinguished by a special name from purely logical antinomies and called *epistemological paradoxes*. [According to recent research by Tarski,[5c] what such paradoxes really show is that a self-consistent use or correct definition of the term 'true proposition' which is in harmony both with the laws of logic and with every-day language is impossible.]

What brought on the severe crisis in logistics was Russell's discovery in 1901 (following some related results obtained by Burali-Forti) of a *purely logical antinomy*, that is, an antinomy in which appear only the concepts of the logical calculus of functions, particularly the concept of class.

If *M* is the class of all men, then

(a) Every member of *M* is a man.
(b) Every man is a member of *M*.

The class *M* itself is not a man but a class of men, and therefore, according to (a) does not occur among the members of *M*. Similarly the class of all triangles in a plane is not a triangle and therefore does not appear among its own members. As these and many other examples show, there certainly exist classes which do not occur among their own members. On the other hand, if *N* is the class of all non-men, then

(a) Every member of *N* is not a man.
(b) Everything that is not a man is a member of *N*.

As *N* itself is not a man but a class of non-men, according to (b) *N* occurs among the members of *N*. Another example of a class which occurs among its own members is the class of all classes. If this class is called *K*, then

(a) Every member of *K* is a class.
(b) Every class is a member of *K*.

As *K* itself is a class (namely the class of all classes), according to (b), *K* occurs among the members of *K*.

Now let *L* be the class of all classes which do not occur among their own members. Then

(a) Every class which is a member of *L* does not occur among its own members. (For example, the classes *K* and *N* just mentioned are not members of *L*.)
(b) Every class which does not occur among its own members is a member of *L*. (For example, the class of all men and the class of all triangles in a plane are members of *L*.)

The question at issue is whether or not the class *L* occurs among its own members. First: It is impossible for *L* to be a member of *L*. For if *L* were a member of the class *L*, then *L* would contain as a member a class (namely *L*) which occurred among its own members, whereas according to (a) every class which is a member of *L* does not occur among its own members. Second: It is impossible for *L* not to be a member of *L*. For if *L* were not a member of *L*, then *L* would be a class which did not occur among its own members and still was not a member of *L*, whereas according to (b) every class which does not occur among its own members is a member of *L*. Hence it is impossible for *L* to be a member of *L*, and it is impossible for *L* not to be a member of *L*. This, however, is an antinomy, for, according to the principle of the excluded middle, every proposition is either true or false and, as has been shown, the proposition '*L* is a member of *L*' is neither true nor false.

[The effect of Russell's discovery was enormous. Most of what has since been done in logic has been more or less influenced by the fear of antinomies. It will be shown that the importance of the three methods pursued by modern logic – type theory, formalism, and intuitionism – far transcends that of the antinomies themselves, but historically the antinomies have been the most important stimulus for the recent developments.]

The *first* means of surmounting the crisis was found by Russell,[6] the discoverer of the purely logical antinomy. His solution is as follows: Every logical construction which occurs in mathematics must have as a starting point a certain field of individuals. Besides these individuals, there come under consideration classes of individuals, which must not be confused with the individuals themselves. Furthermore, there are classes of classes of such individuals (called classes of the second type) which are not to be confused with classes of individuals (called classes of the first type). More generally, for every natural number n, there are classes of the n-th type, and all these classes of different type must be carefully distinguished. Particularly, in speaking of all classes, it must always be indicated whether all classes of the first, second or n-th type are meant. A class which contains as members classes of different types must not be formed. The class of all classes mentioned above is a concept which cannot occur if this prohibition is respected. It does not belong in the hierarchy of all types. The class of all classes which do not occur among their own members is similarly forbidden and so this concept which opens the door to antinomies is excluded by type theory.

As a matter of fact, up to now, no antinomies have been discovered which are not excluded if the rules of type theory are observed. But [although this theory forbids certain concepts which have not led and probably will not lead to contradictions,] it gives no assurance that all possible but still undiscovered antinomies are excluded by it. For this reason, Poincaré, under similar circumstances, spoke of a shepherd who, to protect his flock from wolves, built a fence around it without, however, being sure that he had not enclosed a wolf within the fence.

A *second* program that has been followed since the discovery of the antinomies consists in the development of the formalistic or metamathematical method of Hilbert.[7] The thought underlying this method may be condensed as follows: For every mathematical or logical theory, it must first be stated how the undefined fundamental concepts of the theory are symbolized, and how the propositions of the theory are constructed as strings of these fundamental symbols. For instance, in an axiomatic theory of the Euclidean space, the fundamental concepts are called points, lines and planes. One of the fundamental relations is that of *lying on*. One of the rules by which geometric propositions are built up from these fundamental symbols is that, wherever the symbol for a point precedes the words 'lies on', the symbol for a line or a plane must follow. Secondly, the axioms of the theory must be formulated; that is, the strings of symbols corresponding to certain propositions are set down as initial formulae. And lastly, there must be given the formative rules whereby any strings of symbols corresponding to propositions of the theory can be combined and transformed into new strings such that the corresponding propositions are included in the theory. In the axiomatics of the calculus of propositions there are three such formulae and two such rules (cf. p. 22).

Thus the theory becomes a calculus, and the theory of this calculus is called the *metatheory* belonging to the original theory. This metatheory deals with the way in which the propositions of the original theory are connected and how they may be derived from one another, and it considers what propositions can be proved or refuted from the axioms. In the axiomatic Euclidean geometry, it is proved, for example, that the sum of the angles in any triangle is equal to 180°; that is, this theorem is deduced from the axioms with the help of the formative rules. Metageometry investigates the question which of the Euclidean axioms are necessary for the deduction of this theorem. The proof that the axiom of parallels is independent of the other axioms is metageometrical.

The metatheory belonging to a given theory is above all expected to prove that the theory contains no contradiction – in other words that, in the theory in question, it is never possible to prove an assertion and its negative. By metageometrical considerations it is proved that, if Euclidean geometry is free from contradiction, then non-Euclidean geometry is also free; and if the theory of real numbers is free from contradiction, then Euclidean geometry is likewise. Now Hilbert intended to prove by metalogic or metamathematics that a logic or mathematics founded on suitable axioms is free from contradiction. In this way he hoped not only to get logic and mathematics over the crisis caused by the actual discovery of the antinomies, but also to make these subjects secure for all time by the valid proof that within them any sort of contradiction is impossible.

Before going into what has been brought to light on this point by the use of metamathematics, it is well to mention a few general results that are of interest in this connection. Reminiscent of the various non-Euclidean geometries which differ from each other and the Euclidean geometry – each a system closed within itself – there have been constructed numerous logics which differ from one another and the traditional logic – each of them a system closed within itself. Some examples are the so-called many-valued logics which originated with Lukasiewisz and Post.[8] In the ordinary logic, all propositions are divided into two classes, the class of so-called true and the class of so-called false propositions, so that every proposition belongs to one and only one of the two classes, as expressed by the principles of the excluded middle (or third) and of contradiction. In the three-valued logic, propositions are divided into three classes, and a principle of excluded fourth holds. The assumption that through one point there are many parallels to a given line not only leads to an abstract system of ideas but can even be illustrated by models; similarly, it is possible to get an illustration of the three-valued logic by dividing propositions into surely true, uncertain and surely false ones. More generally, there is, for every natural number n, an n-valued logic which divides propositions into n classes and contains a principle of excluded $(n + 1)$th. The ordinary logic with its dichotomy of propositions comes under this classification as a two-valued logic. To each one of these n-valued logics, there belongs a mathematics; but, of these, only the one associated with the two-valued logic has been closely studied. However, the mere existence of different logics is of interest, for until recently any such state of affairs was regarded as out of the question. It was, in fact, explicitly designated by Poincaré as impossible. Apart from this, these many-valued logics may well be related to the theory of probability and may clarify some obscure points in its foundations[9].

[Another variant from the classical system of logistics is Lewis's[9a] theory of what he calls "strict implication". This term is applied to a relation between two propositions which comes nearer to the ordinary use of the word 'implication' than does the calculus of propositions. It avoids certain paradoxical sounding consequences of the latter, such as "a false proposition implies every proposition" or "a true proposition is implied by every proposition".]

If the problem at hand is to prove by a metatheory that a certain theory contains no contradiction, it must above all be clear what means of proof are permitted for the metatheoretical considerations. If any geometrical or other special theory is to be proved free from contradiction, all of logic can, if necessary, be used as a metatheoretical tool, so that the upshot of the whole argument is a proof by logical inference that, in the system of propositions of the theory in question, there is no contradiction – no assertion appears with its negative. What is meant, however, by a proof that logic itself and the mathematics involved in general logic are self-consistent, i.e., free from contradiction? In a proof of self-consistency, not all of logical inference can be used in the metatheory, otherwise the very system whose self-consistency was to be proved would be used as a means of proof. The object is naturally to prove self-consistency for as large a part as possible of logic and mathematics by the use of a minimum of metalogical methods of proof. For example, Herbrand succeeded in proving the self-consistency of the calculus of propositions and a part of the theory of natural numbers (containing the principle of the excluded middle) by the use of complete induction, but without the use of the principle of the excluded middle.

The program was to push forward to a proof of self-consistency for all of mathematics. Considering that a portion of mathematics is to be used in this proof, the fundamental problem may be formulated as follows: To prove the self-consistency of all logic and mathematics by the use of a part of logic and mathematics.

Such was the state of science until 1930, when a young Viennese mathematician, Kurt Gödel, made a completely unexpected and most significant discovery. He solved the fundamental problem, but in a negative sense, for he proved metalogically by the use of only the theory of natural numbers that *the self-consistency of mathematics and logic cannot be proved by a part of mathematics and logic.*[10]

Naturally, the question immediately arises whether this result is due to some flaw in the system of logical axioms, after the correction of which logic might be proved self-consistent. But this is not the case, since Gödel proved the following general theorem: *Any formal theory which contains the theory of the natural numbers cannot be proved self-consistent by means of principles which can possibly be expressed within the theory in question.* No matter how the system of logic is modified, provided it remains inclusive enough to serve as a foundation for the theory of the natural numbers, it still cannot be proved self-consistent by methods which can be expressed within the system. If there are added certain other metalogical methods of proof, which cannot be formulated within the logical-mathematical theory to be proved self-consistent, then the self-consistency of the theory under consideration can be proved; but then the methods of proof are [at least in certain respects] more inclusive than the theory whose self-consistency is to be demonstrated. For example, the theory of the natural numbers can be proved self-consistent, if operations with any class of natural numbers are taken for granted – that is, essentially, if operations with real numbers are admitted. What has turned out to be impossible is to prove with a portion of mathematics the self-consistency of a more inclusive portion of mathematics, provided the latter contains the theory of the natural numbers. What can be proved self-consistent by a portion of mathematics, is in general only a narrower or overlapping portion of mathematics. In other words, *for the proof of the self-consistency of a portion of mathematics, in general a more inclusive portion of mathematics is necessary.*

This result is so fundamental that I should not be surprised if there were shortly to appear philosophically minded non-mathematicians who will say that they never had expected anything else. For it should be clear to a philosopher that a theory, the structure of which contains no superfluous constituents cannot be founded on one of its parts, etc., etc. But when such general principles are applied to metalogical problems, they turn out to be not only not self-evident but false. For example (as already stated) it can be proved from the axioms of the theory of real numbers, that, from the self-consistency of this theory, there follows the self-consistency of *n*-dimensional Euclidean and non-Euclidean geometry, although *n*-dimensional geometry includes the whole theory of the real numbers as a part, namely as its one-dimensional special case. On the other hand, Gödel's investigation shows, for example, that it cannot be proved from the axioms of the theory of natural numbers (first systematized by Peano) that the self-consistency of the theory of real numbers follows from the self-consistency of the theory of natural numbers. Similarly, it cannot be proved by certain parts of the theory of natural numbers that the self-consistency of the whole theory follows from the self-consistency of these parts. This contrast shows clearly that Gödel's discovery is not a self-evident remark based on general principles, but a profound mathematical theorem which needs proof and which he did prove.

[In 1936, another important result concerning the self-consistency of mathematics has been obtained by Gentzen.[10a] He uses in his metamathematics a part of the theory of natural numbers together with certain transfinite methods. The part of arithmetic used includes complete induction but not what Gentzen calls questionable portions of arithmetic. By these means he proves that the whole of arithmetic including the questionable portions just referred to is free from contradiction. In this case a system of axioms leads to a proof of self-consistency for an overlapping system of axioms.]

Another illustration of the force of metamathematical methods is the second part of Gödel's discoveries. In order to explain this, it is necessary to digress. It was one of Euler's greatest discoveries that theorems concerning the natural numbers can also be proved by so-called transcendental methods – that is by the help of considerations which go beyond natural numbers and the principle of complete induction, in that they make use of the concepts of limit and contintuity as well as of operations with arbitrary real numbers and functions. For example, the theorem (discovered by Fermat and proved by elementary means) that every prime number of the form $4n + 1$ can be expressed, and in one way only, as the sum of the squares of two natural numbers has also been proved by transcendental means. No matter how much Euler's discovery was admired (and a branch of mathematics, the so-called analytical theory of numbers, has developed out of it), there still persisted the faith that all theorems about natural numbers could be proved by elementary means. For even when elementary theorems were formulated for which only proofs by transcendental means were found, this fact was ascribed to the circumstance that elementary proofs for these theorems had not yet been discovered. But Gödel has proved metamathematically that there are surely theorems and problems about the natural numbers which cannot be proved or decided by elemental methods.

[Among the undecidable problems there are questions concerning Diophantine equations – as to whether certain polynomial equations $P(x_1, x_2, \ldots, x_m) = 0$ with integer coefficients admit of integer solutions.[10b] For each integer n exceeding 2, Fermat's problem as to whether there exist three integers, x, y, z such that $x^n + y^n - z^n = 0$ is an example of a Diophantine equation. For many integers n, Fermat's question has been decided negatively; for all the other integers n it is as yet undecided. But none yields one of the Diophantine equations that Gödel proved to be undecidable. The theory of undecidable problems is in 1937 in the same state as the theory of transcendental numbers was in the middle of the last century. At that time Liouville had constructed extensive classes of transcendental numbers, but none of the numbers which had been dealt with in other parts of mathematics were known to be transcendental. Only some decades later was it proved that two of the best known numbers (e and π) are actually transcendental. For logic and the theory of knowledge it is of course unimportant that, so far, none of the most famous classical problems in mathematics has turned out to be undecidable. The important fact is that there certainly are questions of the same general type as Fermat's problems which can be formulated and proved to be undecidable.]

Again this result is not a consequence of an imperfection of the special assumptions about the natural numbers. For Gödel proved the general theorem[10c] that *in every formal theory which includes the whole theory of the natural numbers there occur problems which cannot be decided within the theory in question.* Just as there are propositions about natural numbers which can be proved only by methods taken from the theory of real numbers, so there are propositions about real numbers which can be proved only by methods taken from the theory of sets of real numbers.

And there are problems about sets of real numbers which can be solved only by assumptions about sets of higher power. In fact, in every formal theory including the whole theory of the natural numbers, there are statements *about natural numbers* which cannot be decided within the theory in question. In other words, a universal logic (such as Leibniz dreamed of) which, proceeding from certain principles, makes possible the decision of all conceivable questions, cannot exist.

Metamathematics is thus incapable of either proving the consistency of mathematics or developing decision procedures for all problems. But it would be a grave error to conclude that Hilbert's method has been unsuccessful. For it was metamathematics which besides yielding results of great intrinsic interest supplied Gödel with tools in his fundamental discoveries. In fact, metamathematics has so far proved to be the only way to deep insights into the foundations of logic and mathematics even though its application, particularly in Gödel's hands, has led to the destruction of some illusions. [It has replaced vague philosophic talk on fundamental subjects by precise demonstrations.]

Besides type theory and formalism (or metamathematics), there is a *third* program which has been followed since the discovery of antinomies precipitated a logical crisis. It consists in the development of a method which had already been vigorously pursued by the mathematician Kronecker in the 1880's, even before the antinomies had been found. It has been called by his followers *intuitionism.*[11]

Whereas Russell calls mathematics a part of logic, Kronecker takes his stand on the *primacy of mathematics over logic*. According to him, mathematical construction is fundamental. "The whole numbers were created by God; everything else is human handiwork." In particular, logical inference, if unaccompanied by mathematical constructions, can lead to mathematically false conclusions. Above all, Kronecker opposes indirect proofs in which from the fact that the non-existence of an entity with a certian property involves a contradiction it is inferred that an entity with that property exists, even though no method for the construction of such an entity need be given. As an example, suppose it were possible to deduce a contradiction from Fermat's conjecture that for every integer $n > 2$ and for every triple of integers x, y and z, the inequality $x^n + y^n - z^n \neq 0$ subsists. The classical mathematician would infer that there exist four integers $n(>2)$, x, y and z such that $x^n + y^n - z^n = 0$. Kronecker, on the other hand, would enunciate this existence theorem only if four such integers had been found, or if at least there had been given a method which would permit the finding of such integers in a finite number of steps. For example, if it were proved that integers of the desired kind existed among those with 1000 digits, then such integers could certainly be found if all the quadruples with 1000 digits were examined successively to see whether or not they satisfied the equality $x^n + y^n = z^n$. This process would, it is true, be so tedious that no man could carry it out, but nevertheless it would lead to the goal in a finite number of steps. The mere proof of the impossibility of the non-existence of integers of a desired kind, however, generally furnishes no procedure whatever for the discovery of such integers. In this latter case, Kronecker denies that the existence of an integer of the desired kind has been proved.

Poincaré directed his attacks against what he called *impredicative* definitions – definitions of an entity which refer to classes of which the entity to be defined is a member. The class of all classes is a typical impredicative concept. As a matter of fact, the concepts of the largest number of, and of an upper limit to, a set of numbers (concepts which are important for all higher mathematics) are also impredicative. Hence Weyl in his book *The Continuum* (1918) concluded that these concepts are to be discarded save when the entities in question can be determined operationally – that is, when they can be fixed by a definition which is not impredicative. [Early in this century, the question whether the so-called axiom of choice formulated by Zermelo was to be accepted or rejected played a great role in the discussions of the Paris school. In fact this question caused more argument than the axiom of comprehension which permits arbitrary subsets of any set to be formed, although this latter axiom leads, I believe, to greater difficulties than does the former, and in certain cases even condenses all the difficulties into one.] Complete induction, on the other hand, is regarded as valid by the Paris school, especially by Poincaré, who considered it a synthetic, a priori judgment in the sense of Kant.

Since about 1910, L. E. J. Brouwer has followed out the consequences of Kronecker's theses, and has shown how much of modern mathematics rests on indirect proofs of existence, and falls with these proofs. Complete induction, according to Brouwer is based on a primordial intuition (*Urintuition*) which, he says, also yields a concept of set. Investigations in the theory of sets which go beyond this concept he discards as meaningless. [For instance, he adopts Bergson's view that a continuum cannot logically be treated in the classic fashion as a completed aggregate of points.] Logic, according to Brouwer, gives only the rules of the language for mathematical inference about finite systems. To apply it to infinite totalities is meaningless.

In particular, the principle of the excluded middle, the source of the indirect proofs of existence, is not to be applied to infinite systems. Brouwer takes over Kronecker's restriction on the use of the words 'there exists'; they are to mean that there can be exhibited in a finite number of steps. And so (as long as Fermat's conjecture has been neither proved nor disproved) he denies the validity of the alternative: Either for every quadruple of integers $n(>2)$, x, y and z the inequality $x^n + y^n \neq z^n$ holds or there exist four integers $n(>2)$, x, y and z such that $x^n + y^n = z^n$. Only when the Fermat problem shall have been solved (that is, when either it shall have been proved that if $n > 2$, $x^n + y^n \neq z^n$ for all integers, or a quadruple shall have been found such that $x^n + y^n = z^n$) – only then will it be seen a posteriori that one of the two possibilities maintained by the alternative actually occurs. As long as Fermat's conjecture is neither proved nor disproved, not only is the solution unknown, but the very solubility of the problem is uncertain, so that a priori even the alternative cannot be asserted. On the other hand, it cannot [according to Brouwer] be maintained a priori that the Fermat problem or any given problem is undecidable.

How can classical mathematics be defended against these intuitionist attacks? This becomes apparent through a sober analysis of intuitionism. Here two things are to be distinguished: the mental activity of the intuitionist mathematicians, and their verbal and written reports of this activity. The intuitionists, indeed, take the position that mathematics consists only of their own constructive mental activity, whereas all external communications are merely more or less imperfect directions for the repetition of this mental activity by other human beings. Naturally, a sober critic can do nothing but stick to their external communications. These are, in part, of a mathematical nature, and contain not only constructions but also proofs which employ certain methods of inference in a perfectly regular fashion; in part they belong rather to a theory of knowledge, and maintain that intuitionist ideas have meaning, whereas farther-reaching methods of inference are meaningless.

First, as to assertions like "Certain methods of inference are based on intuition; others are meaningless." Such assertions are, in my opinion, nothing more than descriptions of subjective psychological processes or expressions of subjective tastes. They are therefore of interest only to biographers and historians; they do not belong to logic or mathematics. They are *value judgments* and thus at root expressions of feeling. According to one's own feelings they may be accepted or opposed by voluntary decisions concerning the methods of inference and construction actually to be used or not to be used. [To my mind they recall the dietetic rules enunciated by certain philosophical schools.]

Secondly, as to the residue of the intuitionist constructions when they have been purified and freed from their non-mathematical elements. This residue consists of the derivation of propositions by the help of certain methods and modes of inference; it is a system of inferences according to certain rules. But classical mathematics is nothing else. Intuitionists, to be sure, say that the whole of their mental constructions cannot possibly be formalized – cannot possibly be precisely stated in any system of axioms. Yet the inferences which the intuitionists, according to their verbal communications, actually draw, they draw with perfect regularity. These inferences can therefore be collected and traced back to a certain few principles. In fact, Heyting[11a] has actually set up a system of axioms for this intuitionist calculus of propositions and of functions. Consequently, the claim of the intuitionists can only mean that they reserve the right to use other inferences besides those contained in any system of axioms. At any given moment, however, the inferences actually drawn by them can be formulated. For example, up to now, in the calculus of propositions and functions, they have drawn no inference which is not included in Heyting's system. But should they, because of new discoveries or changes in conviction (*Evidenzerlebnisse*) use additional methods of inference, then it would merely be necessary to expand the systems of axioms, and again a formal system representing what had actually occurred would be obtained. In intuitionism, what is formulated is, so to speak, limited at the bottom – that is, a certain minimum is assured; it is left open at the top – that is, no maximum is fixed. But similarly, classical mathematics and logic do not by any means insist on the complete immutability of their underlying axioms, and they too reserve the right to introduce new ones. [In the sense of the simile used above, the dietetic prescription of the intuitionists does not enumerate all proper foods. It consists of examples of things to be eaten, together with the prohibition of certain forms of nutriment.]

The situations may be elucidated by considering the concepts of definability, constructivity, and provability which are discussed in intuitionism. All cases of applicability of these terms and all the rules regarding them have not yet been precisely fixed, not even in Heyting's system. The insistence of the intuitionists on their own constructions and their repudiation of farther-reaching "unconstructive" arguments would seem to me of mathematical importance, particularly in the theory of sets, on one condition –

namely, if there were only one way of defining constructivity; in that case the mathematics satisfying the postulate of constructivity would be distinguished from all other possible mathematics. Such however, is certainly not the fact. *The postulates for constructivity may be varied by degrees*. Hence all that intuitionism achieves is the realization of just one of these many possibilities, and certainly not the most stringent one; and it does so in a system of postulates which in some points are not clearly circumscribed. [All the imaginable postulates may be combined in such varied ways that it is not even possible to arrange the different combinations in a linear order. There are systems which partly overlap one another. Incidentally the same is true of diets. One code forbids flesh; a more restrictive one interdicts eggs also; a third, overlapping the second, permits eggs but neither flesh nor fish, and so on. Similarly, intuitionists differ as to what they allow and what they forbid, and even the same intuitionist may change his mind from time to time. Compare the theory of Lusin (based upon the views of Borel, Lebesgue, et al.) with that of Brouwer, or contrast the various stages of Brouwer with one another. There is no question that these various theories are of interest, but their interpretation by their respective authors is, I believe, untenable.]

Yet even some non-intuitionistically inclined mathematicians are of the opinion that intuitionistic mathematics being a part of classical mathematics (free of the law of the excluded middle) is more authentic or safer than the latter in its totality. In 1932 Gödel, however, found[12] that also conversely the whole classic calculus of propositions and theory of numbers (including the principle of the excluded middle) may be regarded as a part of the intuitionist theories, for every classical theorem within this field can be translated into an intuitionist theorem by the use of a simple dictionary. [Were there a contradiction between any two classical theorems, there would also be one between the two corresponding intuitionist theorems.] One of the rules for translation is that wherever, in the classical theorems, the words '*p* or *q*' appear, they must be replaced by the words 'it is impossible that *p* is impossible and *q* is impossible'. Hence, abandoning the principle of the excluded middle, but at the same time admitting (as intuitionists do) the impossibility of universal propositions, does not really limit but only renames the classical theorems in the calculus of propositions and the theory of numbers.

Actual restrictions can result only from Poincaré's renunciation of impredicative definitions. For as to the special concepts of the theory of sets, in 1927 I devised a dictionary [herein Chapter 5] by the use of which all the intuitionist concepts within this field may be translated into known special concepts of the classical theory. [What Brouwer calls "sets". "individualized sets" and "species" correspond respectively to analytic sets, Borelian sets and general subsets of analytic sets in theclassical theory.]. Here the difference between the intuitionistic and the classical developments lies only in the fact that the intuitionist dogmatically *limits* himself to the above mentioned special concepts. These he calls meaningful and constructive; farther-reaching ideas he designates as meaningless. This, however, is not part of mathematics but belongs to the value judgments mentioned above. [Moreover, there is no proof of the impossibility of contradictions within the restricted theory. The only certainty is that it excludes large parts of classical mathematics.]

It is possible though not very likely that one might develop an altogether narrow logic and a really *finitistic* mathematics which of course would be even much more restricted than the intuitionistic mathematics. The way to it would lead through critique of language and semantic considerations concerning the connection of logic and mathematics with empirical propositions. It has often been emphasized that no measurement can decide whether the length of a rod is rational or irrational; nor do purely empirical propositions include the quantifier, *all.* But while such a very narrow mathematics would be epistemologically interesting, the vast majority of mathematicians would take in it no greater interest than a geographer takes in the geography of his birthplace.

Having understood that logic and mathematics are not unchangeable or unshakeable; that the Aristotelian logic proved to be inadequate; that it was replaced by a mathematical logic, which in turn got entangled in paradoxes; that the theory of types, formalism and intuitionism are looking for a way out; that, in other words, logic and mathematics exist under the signs of crisis and reconstruction as do the sciences – having understood all this one is bound to ask what is the upshot of all these vicissitudes.

What interests the mathematician and all that he does is to derive propositions by methods which can be chosen in various ways but must be listed, from initial propositions which can be chosen in various ways but must be listed. And to my mind all that mathematics and logic can say about this activity of mathematicians (which neither needs justification (*Begründung*) nor can be justified) lies in this simple statement of fact. Which initial propositions and methods of inference the mathematician and logician choose, and what the relation of these propositions and methods is to so-called reality and to an inner feeling of conviction (*Evidenzerlebnis*) – these and similar questions belong to other and less exact sciences.

This activity of the mathematician is not in general contradiction-proof. He is not sure that he has not enclosed Poincaré's wolf within his fence. But that Poincaré lays this situation to the mathematicians' charge is due to the fact that he demands from mathematics a certainty surpassing that of all other human activities not only in degree but in essence. For if a shepherd builds a large fence around his flock he actually lacks an absolute guarantee that no wolf is anywhere inside the fence. Of course he looks out for wolves; but one may be hidden somewhere and only later suddenly appear.

It would transcend the scope of this lecture to go into all details and ramifications of the recent deep change in our conception of logic and mathematics. But one misunderstanding must be forestalled: that the mere transformation of certain propositions according to certain rules is nothing more than a game.

True, mathematics has aesthetic qualities, they are comparable to those of music – and no one reproaches music with the fact that its human importance is restricted to its aesthetic aspects. (That the aesthetic appreciation of mathematics is much rarer than that of music is essentially an unfortunate effect of poor teaching. For while the power of mathematical invention may be as rare as that of creative musical composition the gift of enjoying properly explained mathematics is hardly rarer than a musical ear.) But no matter how many-sided the aesthetic aspects of mathematics, its importance is not restricted to them, as a single example – and from a practical point of view not even the most important one – will demonstrate. In the early 1800's, what could have seemed to the superficial observer more like a mere game than deductions based on the assumption that in a plane there is more than one parallel to a given straight line through a given point? And yet it is known today that the paths of light rays and of mass points under the influence of gravitation behave like the straight lines in speculative systems which seem at first to be mere games estranged from all reality. Moreover, these deductions have drawn attention to a wealth of further relations between entities which can be experienced and observed, while at the same time they have furnished means for a quantitative check on the assumptions involved.

And even though mathematics has no logical safeguard against the occurrence of contradictions it has in recent years undergone developments of a splendor admired just by those who penetrate the deepest logical concerns.

So if today we oppose to Poincaré's celebrated simile of the shepherd another allegory, then because of the insights of the new logic today's picture must be more modest with regard to the claims of mathematics; yet in view of the enormous successes of recent mathematics it may be free of pessimism. I would say: Mathematicians are like men who build houses – homes that are not only pleasant to live in but enable their inhabitants to do many things which a cave-dweller could never accomplish. Mathematicians are like men who build, although they are not sure that an earthquake will not destroy their buildings. Should that happen, then new constructions, if possible such as promise to be more resistant to earthquakes, will be erected. But men will never permanently decide to give up the building of houses with all their conveniences – the more so because absolute security against the effects of earthquakes is not to be obtained even by the inconvenient habit of living in caves. They will continue to elaborate their structures and to erect new ones.

APPENDIX
THREE PASSAGES OF [1937, 6] NOT INCLUDED IN THE PRECEDING TRANSLATION OF THE 1932 LECTURE

1

What seems to me of importance, what I have for years tried to emphasize, and what I have attempted to sum up in the foregoing remarks are the following facts:

(1) All that any intuitionist does is to realize one out of a great variety of possible conditions which can partially be arranged in order of stringency, but which partially overlap one another.

(2) Were the intuitionist to omit the statement that his theory is the one founded on intuition whereas all more far-reaching arguments transcend intuition, he would lose no mathematical advantage. Such dicta are only weak points open to attack, for they introduce nothing but psychological elements, personal attitudes and value judgments.

(3) It is certainly possible to construct mathematical systems more restricted than any one so far developed. The negation of universal propositions might be excluded; or, after elementary steps of control have been defined, only conclusions open to control by a limited number (say a million) steps might be admitted. Perhaps some one will some day succeed

in stating precisely what is meant by a finitistic mathematics; for this task, in spite of much talk, has not yet, so far as I can see, been accomplished. Another possibility would be the development, on the basis of modern logic, of a mathematics suitable for physics, a problem of great importance for science.

2

There are also relations between certain hypotheses and modes of inference on the one hand and certain rather vague psychological phenomena like inner conviction or intuition on the other. It is important though usually overlooked that such phenomena change with time, and depend upon history and practice. Or is it to be believed that the methods of inference and construction which are self-evident to an intuitionist in 1930 are identical with the methods of inference and construction which were self-evident to Pythagoras, to Archimedes or to Euler, or which will be self-evident to a mathematician in 2930? Again, the relation between certain parts of mathematics and the feeling of inner conviction is important both from an historical and from a psychological point of view. I only question whether it be practical to deal with such problems within mathematics itself. At any rate, I would insist that the mathematician sharply separate them from the mathematical theories in which they play no role.

But mathematics (and particularly mathematics in the sense of the transformation of propositions) besides its aesthetic qualities and the importance of its relations and applications to many sciences has another highly important function. It clarifies discussion on any subject. The theme may be philosophical problems, methodological theories or sociological questions – although this last aspect of mathematics has not yet been extensively developed.[12a] This clarifying function may yet turn out to be of considerable importance for mankind. Mathematics might be the source of a new insight into human intellectual affairs as well as into practical matters insofar as they involve intellectual elements.

3

Formalism and metamathematics have made it possible to deal generally, systematically and accurately with the question which in my opinion is of supreme interest to mathematicians and logicians: What can be inferred from given hypotheses (precisely formulated systems of propositions) by given methods of inference (precisely formulated systems of rules for the transformation of propositions)? Under this heading come also the intuitionist systems at any given moment, although some intuitionists are not fond of formulating precisely what hypotheses and methods of inference they do at the moment actually use. It cannot be repeated too often that *various* intuitionist systems have been constructed, and that still other and more restricted ones might be. Almost every one of these theories so far invented has claimed to be the one and only intuitive mathematics and has ruled out farther-reaching systems. But, in my opinion, this attitude is a philosophical interpretation unessential for mathematics, since it plays no role at all in the system itself. I should treat this interpretation as a physiologist might treat the different diets preached by various philosophical schools. He would be interested; he would discuss and study the consequences of each one of these diets for bodily and mental well-being; he would compare the consequences of different diets, etc. But he would leave to the individual the choice of which diet to adopt. In mathematics the essential fact is that various systems of hypotheses and various systems of rules of inference (rules for the transformation of hypotheses) may be adopted. Everyone is free to choose as he pleases. Precisely what is interesting is to observe the development (that is the investigation of the consequences) of different systems, and to compare them one with another.

NOTES

[1] See Boole, *The Mathematical Analysis of Logic*, Cambridge, 1847, and *An Investigation of the Laws of Thought*, London 1854 and Chicago 1916; Schröder, *Vorlesungen über die Algebra der Logik*, 3 vols. Leipzig 1890–1910, and *Abriss der Logik*, Leipzig 1909/10.

[2] According to Ladd-Franklin, these can moreover be combined into a single statement, namely, the theorem that for three classes *A*, *B*, *C* it cannot be simultaneously true that the intersection of *A* and *B* is empty, the intersection of *not-B* and *C* is empty, and the intersection of *A* and *C* is not empty. In the case of mode Barbara, the premises are: All *M* is *P* and All *S* is *M*; The intersection of the classes *M* and *not-P* is empty, and the intersection of *S* and *not-M* is empty. Then, by Ladd-Franklin's theorem it is impossible that the intersection of *S* and *not-P* should not be empty; in other words, the intersection of *S* and *not-P* is empty; that is, All *S* is *P* – the assertion of mode Barbara.

[3] The calculus of classes is not only methodologically the same as an axiomatic geometry. It can, in fact, as I have found, be brought formally under one heading with elementary geometrical theories. That is, there exists a theory including elementary geometry and the calculus of classes. From this theory, each of the two more special theories may be obtained by specific additional axioms (see Chapter 19).

[4] More detailed expositions of the calculus of propositions as well as of the calculus of functions are to be found in Frege, *Begriffsschrift*, Halle 1879; Whitehead-Russell,

Principia Mathematica, Cambridge 1925; Hilbert-Ackermann, *Grundzüge der Theoretischen Logik*, Berlin 1928; Carnap, *Abriss der Logistik*, Vienna 1929.

[5] An easily comprehensible exposition of the logical foundations of mathematics is Russell's *Introduction to Mathematical Philosophy*, London 1920.

[5a] [Russell later gave up this view.]

[5b] [An excellent exposition of this entire material can be found in F. Waismann's book, [1936, 8]].

[5c] [Tarski, "*Der Wahrheitsbegriff in den formalistischen Sprachen*," *Studia Philosophica*, Lwów 1935. This important paper contains the first successful treatment of a problem in philosophy by means of modern logic.]

[6] Cf. particularly the Introduction to Whitehead-Russell's work and Carnap's book mentioned in [4].

[7] Easily comprehended expositions: Bernays, *Blätter f. deutsche Philosophie*, vol. 4, 1930; Herbrand, *Revue de Metaphysique et de Morale*, vol. 37, 1930; Hilbert *Die Grundlagen der Mathematik*, *Hamburger Mathematische Einzelschriften*, vol. 5, Leipzig 1928; [von Neumann, *Erkenntnis*, vol. 2, 1931].

[8] An exposition of the results of these and other not easily accessible papers will appear in Heyting-Gödel, *Mathematische Grundlagenforschung* in the series *Ergebnisse der Mathematik*, Springer, Berlin. [Post, *Amer. J. Math.*, vol. 43, 1921, Łukasiewicz, *Comptes Rendus Soc. Sci. et Lettres*, Warsaw, vol. 23, 1930; Łukasiewicz and Tarski, ibid. See also the older literature there quoted. *Addition to the present version of this paper:* Instead of the book by Heyting-Gödel, announced in the original 1933 version of this paper, there appeared Heyting, *Mathematische Grundlagenforschung. Intuitionismus. Beweistheorie* in *Ergebnisse der Mathematik*, vol. 3, 1934.]

[9] A start in this direction has been made by Reichenbach, *Sitzungsber. Preuss. Akad. d. Wissensch.*, 1932.

[9a] [Cf. Lewis and Langford, *Symbolic Logic*, New York 1932.]

[10] Gödel, *Monatschefte f. Math. u. Phys.*, vol. 38, 1930 and *Erkenntnis*, vol. 2, 1931.

[10a] [Gentzen, *Math. Annalen*, vol. 112, 1936. Transfinite methods in the theory of proofs are also used by Church.]

[10b] [Cf. Gödel, *Ergebnisse e. math. Kolloquiums*, no. 7, 1936. Cf. also Church, *Amer. J. of Math.*, vol. 58, 1936. *Addition to the present version of this paper:* More precisely what Gödel demonstrates to be undecidable is the question as to the existence of integer solutions $x_1, x_2, \ldots, x_m$ for certain polynomial equations $P(x_1, x_2, \ldots, x_m, a_1, a_2, \ldots, a_k) = 0$ with integer coefficients *for all k-tuples of integers* ("*parameters*"), $a_1, a_2, \ldots, a_k$.]

[10c] [Cf. Gödel's paper quoted in [10]. Skolem, *Norsk. Matem. Forenings Skrifter*, 1933, proved that it is impossible to characterize the sequence of natural numbers by a finite number of axioms if the law of the excluded middle is one of them.]

[11] I gave an account of the historical development of this school of thought with many literature references in an article *Der Intuitionismus* (hierin Chapter 2). An account of the discussions of foundation questions together with many literature references is to be found in Fraenkel's *Mengenlehre*, 3rd ed., Berlin, 1928. [More recent results are summed up in Heyting, loc. cit.[8]]

[11a] [Heyting in two papers in *Sitzungsber. Preuss. Akad. d. Wissensch.*, 1930.]

[12] Gödel, *Ergebnisse e. math. Kolloquiums*, no. 4, 1933.

[12a] [An attempt in this direction was made in my paper [1937, 7].]

CHAPTER 2

ON INTUITIONISM

In 1929, the editors of the *Blätter für Deutsche Philosophie* (the journal of the German Philosophical Society) decided to devote a double issue to the *Philosophical Foundation of Mathematics*. They consulted with the set-theoretician A. Fraenkel about the selection of topics and the contributors. I was asked to write about intuitionism. The double issue of the journal appeared in the fall of 1930. The following translation of [1930, 19] is the work of Mr. Robert Kowalski.

INTUITIONISM
(1930)

1. Intuitionism in Philosophy and in Mathematics

In philosophy, Intuitionism is the view that besides the knowledge we derive from scientific inquiry we possess another, different kind of knowledge, whose source is called intuition. According to Bergson, intuition is concerned with our Ego, especially with its duration, as well as with time in general, and thus forms the basis of metaphysics. In contrast, the analytic method reduces new concepts to previously known ones, and thus forms the basis of the positive sciences. According to Husserl we possess besides the factual knowledge of experimental science another type of knowledge derived from what he calls "intuition of essences" (*Wesenserschauung*).

Whereas in philosophy Intuitionism strives to add to ordinary knowledge other, controversial insights by appeal to intuition, in mathematics it seeks, likewise by an appeal to intuition, to restrict the scope of commonly accepted truth. Mathematical Intuitionism regards a special type of cognition, called intuition, as the sole source of mathematical truth. It holds that the currently accepted mathematics is not entirely within the boundaries of this intuition, and that insofar as mathematics transcends these boundaries it is nothing more than a game with empty words devoid of intuitive meaning.

2. The Founding of Mathematical Intuitionism by Kronecker

Mathematical Intuitionism goes back to the number theorist and algebraist Kronecker, who propounded the following two theses[1] to his contemporaries of the 1880's who were trying to establish a rigorous theory of real numbers and functions:

1) The whole numbers and calculations with them are based on intuition. Using only the whole numbers one can unobjectionably define fractions (as pairs of whole numbers) as well as algebraic numbers – numbers such as $\sqrt{2}$ or $\sqrt[4]{3} + \sqrt[3]{5}$, which are the roots of polynomials with integer coefficients (by introducing the number and operations with it in terms of the system of coefficients of the defining polynomial). But arbitrary real numbers are not expressible in this way. Kronecker's first thesis is summed up by his often-quoted dictum: "The whole numbers were created by God, all else is the work of man".

2) Purely logical deduction does not necessarily yield mathematical truths, that is, statements with intuitive meaning. In particular, to prove a certain mathematical theorem, it is not sufficient to show that its negation is contradictory. Rather, one must have a procedure which leads to the desired conclusion in a finite number of arithmetic operations, and this procedure is the essence of the proof. In the number-theoretical work of Kronecker and of similarly oriented algebraists it was naturally the existence theorems for which such procedures were worked out. For example, the theorem "there is a whole number with a certain property" is not provable in Kronecker's scheme by merely deriving a contradiction from the assumption "no whole number has the stated property". Instead the exhibition of an actual number with the stated property is required or, failing that, a procedure which allows one to produce such a number after a finite number of arithmetic operations.

3. Poincaré

The second representative of mathematical Intuitionism is Poincaré.[2] He sees the source of mathematical truth in specifically mathematical procedures, especially complete induction. He calls these methods synthetic a priori judgments. He is further of the opinion that mathematical objects do not exist unless they are being thought of, and requires (for the purpose of conceivability) that all mathematical concepts be definable with a finite number of words and that all assertions be verifiable in a finite number of operations. Other concepts and assertions are held to be meaningless. For example, he rejects Zermelo's axiom of choice, which states that given a system of pairwise disjoint sets, there exists a "choice set" which has exactly one element in common with each set of the system. He does not attribute any meaning to this axiom, since it does not specify any procedure for determining the choice set (determining which element in each set of the system is to be chosen), thereby introducing ideas not defined in a finite number of words, leading to conclusions not verifiable in a finite number of operations.

In particular, Poincaré opposes the so-called impredicative methods, which entail defining an object by a relation to all elements of a class of which the object to be defined is itself an element. He detects this method in the formation of sets which contain themselves as elements and holds it responsible for the well-known paradoxes that result. Poincaré considers impredicative methods circular, because in using them to define an object, one must know all the elements in a class to which the object to be defined itself belongs; in particular, one must also know the object to be defined. Poincaré rejects most of the concepts and deductions of Cantor's set theory as impredicative.

4. The Function Theorists of the Paris School

The next proponents of intuitionistic thought, the function theorists of the Paris School at the turn of the century, stood in their fundamental work on real functions and measure theory on the ground of Cantor's point set theory. They took pains, however, to make their assertions and proofs verifiable whenever possible.[3] Borel repeatedly emphasized that only the study of entities defined legitimately (with finitely many words) is meaningful. The totality of these entities is countable. For from a finite number of symbols (letters) one can form at most a countable set of finite combinations. On the other hand, this totality, or even the totality of legitimately definable real numbers, is not effectively denumerable; that is to say, it is not possible to give in a finite number of words a procedure which will assign to each legitimately definable real number a unique integer label. The confusion between countability and effective denumerability is held to be the reason for the set-theoretical paradoxes. Lebesgue expressly took Kronecker's position of requiring constructive procedures with regard to the axiom of choice. He noted that one cannot always exhibit an infinite sequence of elements in a non-finite set; that one must perhaps distinguish between non-finiteness and infiniteness; and that the impossibility of a set which is neither finite nor infinite has by no means been proven. Baire not only rejected the axiom of choice, he further disputed that if an infinite set is given, all its subsets are given.

Also of importance (even for foundational questions) are the positive set-theoretical results of the Paris School which are directed towards the definition of classes of special sets for which problems such as choice sets and the formation of subsets present less difficulty than for arbitrary sets. (Originally these special sets were always sets of real numbers or subsets of the n-dimensional Cartesian spaces.) For example, Borel sought to extend the elementary geometrical concept of length to a more general measure of linear sets.[4] He began with intervals, whose measure he defined to be their length. He then defined measure for open sets (which are representable as countable unions of open intervals), and for closed sets (whose complements are open sets). Next he defined measure for countable unions of closed sets, for the complements of such sets, and so on. In this way, he arrived at a definition of measure for what are today called Borel sets.

Following upon the investigations of Lebesgue, sets of values of special functions were studied, among them those of the sums of countably many polynomials. These are sets which, so to speak, can be defined analytically. The systematic theory of these *analytic sets* (developed primarily by Lusin and Sierpiński) goes back to the Russian mathematician Suslin, who distinguished these sets by certain properties that they have in common with the set of all real numbers. A set S of real numbers is called analytic if a countable system $\mathfrak{J}$ of intervals exists, such that

1) $\mathfrak{J}$ contains: countably many intervals called first-step intervals; associated with each of these intervals countably many subintervals called second-step intervals; associated with each of these second-step intervals countably many third-step subintervals, and so on.

2) $\mathfrak{J}$ determines S in the following sense: a real number r is in S if and only if r belongs to at least one first-step interval, at least one of its associated second-step intervals, at least one of the latter's associated third-step intervals, and so on.

Every Borel set is analytic, but the converse is not true. An analytic set is a Borel set if and only if its complement is analytic. The class of sets which are complements of analytic sets is also an important extension of the class of Borel sets. Recently other fruitful generalizations have been made which allow certain sets to be treated constructively (although with different degrees of constructivity).

5. *Weyl's "Continuum"*

In his booklet 'The Continuum' (1918), Weyl went beyond the French intuitionists in a number of points. Poincaré believed he was attacking nothing more than set theory in his rejection of impredicative definitions. In defense of such definitions, however, Zermelo showed that the concept of the maximum of a continuous function on a closed interval, basic in analysis, is also impredicative in the sense of Poincaré. For the maximum value of a function is defined to be that value of the function which is greater than all other values of the function. This involves a relation of the maximum value of the function to the elements of the set of all its values – a set to which the maximum value obviously belongs itself. Weyl was of the opinion that modern analysis is indeed built upon circular foundations. If one defines real numbers in the usual manner as cuts in the domain of rational numbers, and then defines the supremum of a bounded set of real numbers to be a cut in the domain of real numbers, one obtains a real number of a second level. According to Weyl, the ubiquitous confusion of these two levels of real numbers in modern analysis is a vicious circle. Since an analysis with numbers of different levels would be impracticable, one must sacrifice numerous classical theorems (for example, that every bounded set of real numbers has a supremum, and that every continuous function on a closed interval has a maximum) and stay in the realm of cuts in the rational numbers. Weyl then went on to give construction principles with the aid of which one could build a noncircular foundation for a part of today's real analysis.

6. *Brouwer*

In a long series of publications,[5] Brouwer has developed intuitionistic thought in a critical as well as in a constructive sense.

The basis of Brouwer's critique is Kronecker's thesis that purely logical deduction does not necessarily yield mathematical truth. Brouwer regards the laws of logic, such as the laws of contradiction and of the excluded middle and the syllogism, as rules of the language of mathematical deduction with respect to finite systems. Their thoughtless extension to infinite systems is inadmissible, and leads in the case of the law of contradiction and the syllogism to correct results, but in the case of the excluded middle to incorrect consequences. For example, if a finite set of numbers is given, then we can test by finitely many operations whether or not a number in the set has a certain property. Even if we do not know which of these two cases obtains, we may nevertheless say with certainty that either there is a number in the set with the stated property or that all numbers in the set fail to have the property. If, however, we pose the corresponding question for the set of all natural numbers then the situation is quite different, since an examination of infinitely many entities is not possible. Here, the question can be settled only if either a number with the stated property or a finite procedure leading to such a number is actually exhibited, or if it is proved that no natural number can have the stated property. As long as neither of these possibilities materializes, not only do we not know *which* of the two cases holds, but we are not even justified in asserting *that* one of them occurs. In other words, as long as a problem concerning infinity is unsolved we not only do not know its solution, but we do not even know whether the problem is solvable. If one nonetheless assumes always that one of the two propositions "there is a number with a certain property" and "no number has that property" holds, then one is unjustifiably applying the law of the excluded middle (which demonstrably holds for finite sets) to infinite sets, for which it is not proven and is presumably false. One application of this assumption, repugnant to Kronecker, is the indirect proof, especially of existential propositions. As when, for instance, one asserts that "there is a number with a certain property" when one has merely derived a contradiction from the assumption "no number has that property", without giving a procedure which will actually produce the desired number. For some theorems, in particular for the fundamental theorem of algebra (that every polynomial assumes the value zero somewhere), Weyl as well as Brouwer and de Loor have given new proofs which dispense with purely existential propositions.

Taking any currently unsolved problem, we can construct a real number for which we cannot currently decide whether it equals zero or not. This fact was noted by Sierpiński and construed by Brouwer to mean that there are real numbers which are neither equal zero nor unequal zero. With the aid of such numbers, Brouwer subjects the current theory of real functions to a critique which prompts him to deny the validity of most theorems of analysis to an even greater extent than did Weyl in *The Continuum*. In set theory, he refutes along with Poincaré the notion that all elements of a set which possess a certain property can be combined into a set. He further rejects as meaningless cardinalities greater than that of the continuum as well as a large part of the theory of ordinal numbers.

With regard to the positive development of mathematics, Brouwer assumes, like Bergson, a time-related primordial intuition (*Urintuition* concerning, in his phrase, *Zweieinigkeit*), which justifies the theory of whole numbers and, in particular, complete induction. In addition, Brouwer develops a set theory independent of the law of the excluded middle. It begins with an, at first sight, extremely complicated definition of the set concept. Brouwer then introduces two special kinds of sets called *individualized* and *finitistic* (*finite Mengen*, not to be confused with finite sets) and then, *species*. Species are any conceptually well-defined properties of set elements (*begriffllich fertig definierte Eigenschaften von Mengenelementen*) such that within a given set all elements with a certain property are combined into a subspecies of the set. The definitions are followed by an excluded-middle-free treatment of these ideas, including species, in which the usual concepts of ordinary set theory, such as identity of sets, countability, ordering, well-ordering, closure of point sets, and so forth are stated for arbitrary species. The rejection of the excluded middle naturally engenders some concept-splitting. Thus, identity and nonidentity of real numbers is replaced by the three concepts of coincidence (identity), deviation (nonidentity) and distance (demonstrability of a positive lower bound for the difference). Distance implies deviation, but not conversely.

7. *The Positive Content of Brouwer's Intuitionism*

First of all, Brouwer's rules for logical inference were stated in concise axiomatic form by Heyting.[6] They make use, as far as the propositional calculus is concerned, of the four customary logical concepts: 'from *a* follows *b*', '*a* and *b*', '*a* or *b*' and '*not-a*', none of which, in contrast to the classical logic, can be defined in terms of any of the others. The four concepts are connected by eleven (provably independent) axioms: From *a* follows *a* and *a*. From *a* and *b* follows *b* and *a*. If *b* follows from *a*, and *c* follows from *b*, then *c* follows from *a*. From *b* follows: from *a* follows *b*. If *a*, and if *b* follows from *a*, then *b*. From *a* follows *a* or *b*. From *a* or *b* follows *b* or *a*. If *c* follows from *a*, and *c* follows from *b*, then *c* follows from *a* or *b*. From *not-a* follows: *b* follows *a*. If *b* follows from *a*, and *not-b* follows from *a*, then it follows: *not-a*.

From these axioms one can prove the absurdity of the absurdity of the law of the excluded middle; that is, that the negation of the alternative *a* or *not-a* does not hold. But one cannot prove that from the double negation of a statement that statement follows. For example, the law of the excluded middle cannot be proved, but the absurdity of its denial can be proved. If the law of the excluded middle is added to the eleven axioms the usual logic results, and the twelve axioms can be reduced.

The predicate and functional calculi used by Brouwer include complete induction. The negation of a universal proposition is, in contrast to the traditional logic, not an existential proposition.

The author has shown[7] that Brouwer's definition of a set is nothing other than a highly complicated definition of an analytic set, which, as was mentioned before, has recently become the starting point for a broad theory within the framework of the usual set theory. What is more, individualized sets are seen to be the same as Borel sets, and finitistic sets none other than compact sets – all of them concepts resting on Cantorian set theory. But as to species (which Weyl, who in the early 1920's followed Brouwer in set theory,[8] has repudiated), they are nothing but all the ordinary subsets of analytic sets. Brouwer did not draw any noteworthy conclusions from his definitions of set, individualized set and finitistic set (i.e., analytic, Borelian and compact set), but rather developed the theory of ordering and well-ordering for any species, that is, subsets of analytic sets. Since the continuum of real numbers and the *n*-dimensional Cartesian spaces are sets in Brouwer's sense (analytic sets), it follows from the translation of Brouwer's terminology into the usual one that with regard to Cartesian spaces Brouwer has developed the usual set theory, with only the restriction to his principles of logical inference, that is, without the law of the excluded middle. If one disregards the fact that Brouwer rejects set-theoretical concepts transcending analytic sets and their arbitrary subsets as meaningless, then the difference between his set theory and the classical set theory is the same as that between his (Kroneckerian) arithmetic and the usual arithmetic. Unhappily, no progress whatever is achieved by all this in the matter of foundations of set theory. For Brouwer's definition of the set concept or, as he calls it, the species concept includes the unexplained phrase "conceptually well-defined property", which embodies all the logical difficulties attendant to the phrase "definite property" in Zermelo's axiomatics of set theory.

8. *Terminological Discussions*

About 1925, intuitionistic thought became generally known, mostly through the work of Weyl and Brouwer, and has led to discussions over terminology. Against Brouwer it has been objected that the statements 'one can exhibit a natural number with a certain property' and 'no natural number has that property' are not really contradictory; so that even if they should in some particular instance both be false the law of the excluded middle, which deals with contradictory statements, is not involved. Brouwer has further noted that on the basis of a presently unsolved problem P a real number r_P can be constructed for which in his parlance the proposition "neither $r_P = 0$ nor $r_P \neq 0$" is true today, whereas tomorrow, if P is solved, that proposition will be false and "either $r_P = 0$ or $r_P \neq 0$" valid. Brouwer thus uses the words "true" and "false" in a time-dependent, subjective manner. Also, the use of the designation "false" for any theorem which has not been proven is analogous to the use of the words "there is" for certain numbers which have not been exhibited – a usage decried by Brouwer.

But it would of course be easy to avoid these paradoxical formulations altogether. The major languages have two currently synonymous phrases: there is, there exists; French: *il y a, il existe*; German: *es gibt, es existiert*; Polish: *istnieje, egzystuje*. One could therefore express the fact – interesting to the majority of mathematicians despite Kronecker and Brouwer – that the assumption 'no number has a certain property' leads to a contradiction by the sentence 'there exists a number with that property', and the further circumstance that one can exhibit the number by the sentence 'there is a number with that property'. Similarly, one could designate any number for which the assumption that it be equal to zero leads to a contradiction as 'unequal zero', and a number which differs from zero by a specifiable amount as 'different from zero'.

9. *The Discussion between Formalism and Intuitionism*

About 1920 intuitionists initiated an important discussion with Hilbert, who set himself the goal to prove that analysis and the fruitful parts of set theory are free from contradiction. The main objective is to show that statements about real numbers are consistent, although in the language of the intuitionists they are impredicatively, non-effectively or circularly defined. It seems that this proof cannot be carried out without the principle of complete induction. Once it is completed, the difference between Formalism and Intuitionism will reduce to the following: both start from certain basic principles, such as complete induction, but the intuitionist confines himself to theorems provable with these principles, whereas the formalist engages in analysis and set theory and proves, with the aid of these same principles, that his creations are free from contradiction. [For the change in the situation due to Gödel's 1930 discovery, see pp. 32–42.]

The intuitionists state in objection to Hilbert that a consistency proof for mathematics is meaningless, that mathematical theorems derive their validity solely from their intuitive proofs, and that metamathematical considerations cannot increase our collection of meaningful theorems in the least. Hilbert's response is that the only criterion of acceptability of a mathematical concept or system is its freedom from contradiction (a view sometimes also expressed by Poincaré); that when consistency has been demonstrated for analysis and set theory, the intuitionistic mutilation of these disciplines by insistence on constructive proofs is unnecessary and unjustified, and that on the contrary, the intuitionistic torso of mathematics would receive its true foundation only by way of metamathematical proofs of consistency. To appreciate the significance of Hilbert's remark one should bear in mind that Brouwer's version of intuitionistic mathematics introduces the continuum, the justification of which causes the formalists the greatest difficulty, on purely intuitive grounds. To put it another way, if the intuitionists consider Brouwer's theory of the continuum constructive, then their constructivity principles (never explicitly formulated in their entirety) transcend the principles used in metamathematical consistency proofs.

Brouwer further objects that Hilbert's identification of the consistency with the correctness of a mathematical theorem presupposes the law of the excluded middle. A proof of the law of the excluded middle by an appeal to consistency is therefore circular. In other words, Hilbert's attempt at a foundation of analysis ultimately presupposes the unjustified belief in the law of the excluded middle. To this Hilbert might reply that he has no use for Brouwer's idea of correctness, but only for consistency, and that he accordingly proves the law of the excluded middle to be consistent in his system, whereas Brouwer, on the other hand, considers his (quite far-reaching) constructions to be consistent without good reason, and, since he rejects consistency proofs of his system on principle, in a way bases his propositions on an unfounded belief in the law of contradiction. That this law and also the syllogism can be carried over from finite to infinite systems is tacitly accepted by Brouwer. Were one to attempt a construction of infinite mathematics without the law of contradiction, the results would be scanty indeed. This aspect of the debate (not always a model of communication) between Formalism and Intuitionism can be summarized thus: From an intuitionistic standpoint, Formalism assumes the law of the excluded middle, and from a formalistic standpoint, Intuitionism assumes the law of contradiction.

10. General Epistemological Remarks

At the beginning it was mentioned that in mathematics Intuitionism at first sight appears to use an appeal to intuition for the achievement of greater rigor and a restriction of the permissible statements, whereas in philosophy it seeks to base on intuition otherwise unjustifiable propositions and thereby to broaden cognition. If we consider the positive intuitionistic investigations to date, we find that they all begin with at least the natural numbers (complete induction), introduced by an appeal to intuition. Now, as is known, some foundation theorists, especially logicists, are trying to analyze and give a basis for the natural numbers too. The modern intuitionists repeatedly label as superfluous, impossible and simply confusing any attempt to provide logical foundations for what they themselves accept as intuitive starting points. Hilbert, on the other hand, had already labelled Kronecker's position as dogmatic because the latter had accepted as a starting point, on intuitive grounds alone, the theory of the whole numbers, which could and should be logically founded. [See, however, Skolem's important result, p. 45, [10c]]

But one can also accuse Intuitionism in both its critical and its positive forms of dogmatism on altogether different grounds: it recognizes constructively proved theorems but rejects all other results of mathematics. Before we discuss the dogmatism inherent in this scheme, we should note that the fundamental concept of constructivity has never been adequately defined. All that is certain is that the constructivity requirements of various mathematicians differ substantially from one another. For instance, Brouwer rejects the law of the excluded middle but admits uncountable sets, whereas Borel does not object to the law of the excluded middle but rejects uncountable sets as inadequately defined. The author has repeatedly expressed the opinion that the heretofore undefined concept of constructivity could be made precise in various ways and degrees. Even in the most intuitive parts of geometry there is no word or idea which would inevitably demand a particular definition and could not be made precise in several different ways. No doubt this is true to an even greater extent of the nebulous idea of constructivity. For each of the various versions of constructivity one could develop a corresponding deductive mathematics. For example, it is perhaps possible to give a constructivity principle so strict that it would allow only finite sets, or a somewhat weaker one which would include countable sets, or a weaker one still which would admit analytic sets, or a very general one which would allow arbitrary sets of real numbers.

The requirement of consistency may in this sense be considered the weakest possible constructivity principle.

What the intuitionistic attempts to date have done is to attach themselves dogmatically to some particular notion of constructivity (in most cases not clearly circumscribed), to accept only the resulting developments as meaningful, and to reject any others as meaningless. In the opinion of the author such a position is totally devoid of cognitive content. For what matters in mathematics and logic is not which axioms and rules of inference are chosen, but rather what is derived from them. That mathematician A declares the axiom of choice to be "valid" and "believes" in it while mathematician B rejects it as "nonconstructive" or because he cannot "associate any meaning with it" – these facts are of interest for the biographies of the mathematicians A and B and perhaps for the history of mathematics, but they are not relevant for mathematics and logic. These are concerned solely with the consequences of the axiom of choice.[9] All that matters is into which statements certain others can be transformed by the use of given transformation rules. Attempts to found the acceptance or rejection of propositions or transformation rules on intuition are ultimately empty words.

Without entering into the question (doubtless full of pseudo-problems) of 'founding' finite mathematics and of its applicability to 'reality' we wish to make one observation concerning deductive mathematics of infinity. Even if the latter should never be capable of application and never provide us with knowledge in some restricted sense of the word, it would still find justification on its overwhelming aesthetic merits, just as does music, which certainly does not provide us with knowledge. We claim, in effect, that in the practice of mathematics of infinity what counts is not value judgments such as acceptance or rejection, but only implications.

According to this – let us call it the *implicationist* – position, we must reject any intuitionistic program, as described above, concerned with deductive mathematics of infinity, but must also regard the absolute fixation on consistency as arbitrary. As soon as some sharper constructivity principles are made precise, however, it will be possible and appropriate to develop the mathematics corresponding to them. The relations of these various systems of implications to each other and especially to the consistency mathematics will surely be of interest, as are today the relationships between various classes of sets (which may well correspond to various types of constructivity), without anyone finding it necessary to designate one of these classes of sets as admissible and all larger classes as meaningless. Only in one respect, as far as one can tell today, will the mere demand for consistency be distinguished from all the stricter constructivity demands: by the aesthetic appeal of the corresponding mathematics. But this criterion is no more cogent than appeals to intuition.

NOTES

[1] See Pringsheim and Molk in the *Encyclopédie des sc. math.* I, 1904, 3 p. 158 ff.

[2] Cf. the mathematical and logical parts of his books *Science and Hypothesis*, *Science and Method*, *The Value of Science* and *Final Thoughts*.

[3] Cf. in particular the collection of essays and letters in note IV of Borel, *Leçons sur la théorie des fonctions*. (Gauthier-Villars, Paris, 3rd ed. 1928.)

[4] Nonmathematicians may skip the remainder of section 4 on the first reading.

[5] Cf. in particular the pamphlet *Wiskunde*, *Waarheid*, *Werkelijkheid* (Noordhoff, Groningen 1919) and the papers "Zur Begründung der intuitionistischen Mathematik" (*Mathem. Annalen* 93 ff.). Also the notes in the *Jahresbericht d. D. M. V.* 33, p. 251, and in *Crelles Journal* 154.

[6] *Sitzungsber. d. Preuss. Akad. d. Wiss.* 1930.

[7] [1928, 1], herein Chapter 5.

[8] See in particular *Mathem. Zeitschr.* 10, p. 39.

[9] Instructive in this respect is in particular Sierpiński, *L'axiome de M. Zermelo et son rôle dans la théorie des ensembles* (*Bull. Ac. Crac.*, 16.5.1918) or the papers in the *Fundamenta Mathem.* 2, p. 112; 5, p. 177.

PART II

OPUSCULA LOGICA

CHAPTER 3

MEANINGFULNESS AND STRUCTURE

During a visit to Poland in the summer of 1929, I learned about the parenthesis-free notation that Łukasiewicz had just devised for logic – a simple but fascinating idea. Today this symbolism is widely known and its variations are even utilized in mass-produced calculators; but at that time, probably no other Western mathematician was acquainted with it. So upon my return to Vienna, I spoke about it before my Mathematics Colloquium as well as in the Circle. The latter, I felt, should be informed of Łukasiewicz' idea since in the *Tractatus* Wittgenstein had emphasized how significant was "the apparently unimportant fact that the logical pseudo-relations such as $\vee$ and $\supset$ (for *or* and *implies*) require parentheses – in contrast to the real relations." Łukasiewicz' notation demonstrated that the alleged requirement did not exist.

Traditionally, implication is indicated by a symbol that is *sandwiched* between the letters designating propositions ($\supset$ or $\rightarrow$, the latter being used in the sequel), as in $p \rightarrow q$. Parentheses then are needed in order to distinguish, for example,

(1) $\quad p \rightarrow (q \rightarrow r) \quad$ and $\quad (p \rightarrow q) \rightarrow r$;

since $p \rightarrow q \rightarrow r$ would be ambiguous.[1] Łukasiewicz introduced a *frontal* letter for implication writing *C*pq for p implies q, and thereby obviating the need for parentheses. Instead of (1) he writes, unambiguously,

(1′) $\quad C p C q r \quad$ and $\quad C C p q r$.

Accordingly, the traditional expressions of syllogism and contraposition

(2) $\quad ((p \rightarrow q) \rightarrow (q \rightarrow r)) \rightarrow (p \rightarrow r) \quad$ and $\quad (p \rightarrow q) \rightarrow (-q \rightarrow -p)$

are replaced by

(2′) $\quad C C C p q C q r C p r \quad$ and $\quad C C p q C n q n p$,

where the symbols $-$ and *n* indicate negation.[2]

It soon occurred to me that one might formulate a simple arithmetical criterion to distinguish among strings of symbols those that represent meaningful composite propositions in this notation, such as (1′) and (2′), from those that do not, such as

(3) $\quad C C p q, \quad C C C p q r C p r C C r q, \quad C p q n$.

In the fall of 1930, I presented such a criterion to my Mathematical Colloquium[3] and also spoke about it to members of the Circle, where at that time Carnap was beginning to develop interest in syntax; for my study had established *a precise connection between meaningfulness and structure*.

The following note is a translation of [1931, 10]. Further remarks about the frontal notation, some of them excerpted from later publications, are collected in an Appendix.

AN ELEMENTARY REMARK ABOUT THE STRUCTURE OF LOGICAL FORMULAE (1930)

In Łukasiewicz parentheses-free notation using

*C*pq for p implies q and *n*p for not p,

each proposition of the calculus of propositions can be represented by a string consisting of the italics *C* and *n* and lower case letters in roman type. For examples, see (1′) and (2′) of the Preface.

THEOREM. *In order that a string* $\mathfrak{S}$ *of C's, n's and roman type letters represent a proposition the following conditions are necessary and sufficient:*

1. *In* $\mathfrak{S}$ *the number of roman type letters exceeds that of C's by exactly* 1.
2. *Each initial segment of* $\mathfrak{S}$ *contains at least as many C's as roman type letters.*
3. *The last element of* $\mathfrak{S}$ *is not n.* (From Conditions 1 and 2 it follows that the last element is not *C*; hence it is a roman type letter.)[4]

Necessity is proved by complete induction on the number of elements in $\mathfrak{S}$. If $\mathfrak{S}$ contains only one element, then this must be a roman type letter and the conditions 1, 2, 3 are satisfied. If for some natural number m exceeding 1 the conditions are necessary for all strings including less than m elements, let $\mathfrak{S}$ be a string representing a proposition and including exactly m elements. $\mathfrak{S}$ begins either with *n* followed by a string $\mathfrak{R}$, or with *C* followed by two strings $\mathfrak{P}$ and $\mathfrak{Q}$ according as that proposition is the negation of a proposition or an implication. The strings $\mathfrak{R}$, $\mathfrak{P}$, and $\mathfrak{Q}$ represent propositions and include less than m elements each, whence they satisfy the conditions. Therefore, as one readily sees, so does $\mathfrak{S}$. That the conditions are sufficient is also seen by induction on the number of elements in the strings, namely, by translating a string satisfying the conditions into a proposition.[5]

COROLLARY. If α stands for an element of a string $\mathfrak{S}$ representing a proposition, then $\mathfrak{S}$ includes one and only one segment beginning with α that represents a proposition. We denote this segment, which does not necessarily extend to the end of $\mathfrak{S}$, by $\mathfrak{S}_\alpha$. If $\mathfrak{S}$ is the string (2′) of the Preface expressing the syllogism and α stands for the second or the third *C*, then $\mathfrak{S}_\alpha = CC\mathrm{pq}C\mathrm{qr}$; or $\mathfrak{S}_\alpha = C\mathrm{pq}$; if α stands for a roman type letter, then $\mathfrak{S}_\alpha$ consists of that single letter. There need not exist any segment of

$\mathfrak{S}$ terminating in α and representing a proposition (there certainly is none if α stands for an italic letter); on the other hand, there may exist several such segments (for example, two segments terminate with the eighth element of (2′), namely *CC*pq*C*qr and *C*qr).

The segment of a string beginning at a certain element and representing a proposition is called the *scope* of that element in the string.[6]

APPENDIX

REMARKS ON THE FRONTAL NOTATION IN ARITHMETIC, ALGEBRA AND ANALYSIS

Łukasiewicz of course extended his frontal notation to the other binary connectives of logic: *or*, *and*, *incompatible* etc. and applied it moreover to n-valued logic (See p. 30). Clearly, the idea is also applicable to arithmetic, algebra and analysis. For example, ever since the Renaissance, the sum and the product of two numbers have been symbolized by sandwiched + and ×, but may be expressed by frontal *S* and *P*, as in

*S*23 for $2 + 3$ and *P*34 for 3×4.

Parentheses then become superfluous. Instead of

(4) $2 + (3 \times 4)$ and $(2 + 3) \times 4$

one can write

(4′) *S*2*P*34 and *PS*234

without ambiguity.[7] The distributive law of arithmetic and algebra, $a \times (b + c) = (a \times b) + (a \times c)$, then reads

(5) *P*a*S*bc = *SP*ab*P*ac.

Actually, it would have been easier for the Renaissance arithmeticians to devise *S*23 than it was for Łukasiewicz to invent *C*pq; all they would have had to do was to follow the words *Sum of* 2 *and* 3, whereas no one says *Implication* p *to* q.

In a later study of the totally erratic symbolism that the Renaissance mathematicians have bequeathed to us, I distinguished four types of symbols[8] (to which logic and more advanced mathematics have added a fifth):

a) *sandwiched*, as in $2 + 3$;

b) *frontal*, as in $\sqrt{2}$ and *log* 2;

c) *rear*, as the raised 3 for *cube* in 2^3;
d) *pincers*, as the two bars for *absolute value* in $|-2|$;
e) *interlaced*, as in *neither* . . . *nor* . . . and $\int \ldots \mathbf{d} \ldots$.

Interestingly enough, all these types of arithmetical symbols have analogues in descriptions of family relations (cf. *loc. cit.*[8]) in English, though only expressions of type b) occur in analytic languages such as French:

a) *sandwiched*, as in Henry's daughter by Anne;
b) *frontal*, as in the daughter of Henry;
c) *rear*, as in Henry's daughter;
d) *pincer*, as in the oldest of Henry's children;
e) *interlaced*, as in the daughter of Henry by Anne.

In common language, there seems to exist a healthy trend toward frontal expressions. In mathematics, combinations of frontal and nonfrontal symbols create further needs for parentheses,[9] for example, in order to distinguish

(6) $(log\ 2)^3$ from $log(2^3)$ and $(log\ 2) + 3$ from $log(2 + 3)$.

Systematic use of frontal symbols such as *cu* for cube (as in *cu* 2 for 2^3) would obviate that need. Instead of the traditional expressions (6) one then could write, respectively,

(6′) *cu log* 2, *log cu* 2 and *S log* 23, *log S*23.

Conditions 1, 2, 3 in the Theorem on p. 62 of course characterize all meaningful logical strings consisting of:

(α) capital italics for binary connectives (*C* and/or others);

(β) lower case italics for unitary connectives (*n* and/or others, especially in multi-valued logics);

(γ) roman type letters for propositions and proposition variables.

(In Conditions 1 and 2, instead of the *C*'s one has to count the capital italics; in Condition 3, replace *n* by 'lower case italic'.)

In this form, Conditions 1, 2, 3 are also necessary for the meaningfulness of arithmetical strings consisting of

(α) capital italics for binary operations and 2-place functions (traditionally called functions of two variables) such as *S*, *P*, *D* and *Q* (for difference and quotient, *Q*34 = 3/4);

(β) lower case italics for (1-place) functions such as *log*, *cu*, *sq*, *rec* and *neg* (the last for square, reciprocal and negative, *rec* 3 = 1/3, *neg* 3 = −3);

(γ) numerals and roman type letter for number variables.

For example, the strings in formula (5) include 2 (and 3) capital italics and 3 (and 4) roman type letters, while no initial segment includes fewer capital italics than roman type letters and neither expression terminates in a lower case italic.

Conditions 1, 2, 3 are both necessary and sufficient for the meaningfulness of arithmetical strings in which all italics designate operations or functions that are *universal*, that is, applicable to *each* number or pair of numbers as are, for example, *sq*, *cu*, *neg*, *S*, *P*, *D*. They are only necessary but not sufficient if designations of nonuniversal operations or functions are included, such as *rec*, *log*, *Q*. Indeed, *rec* 0, *log* 0, *Q*10, *Q*00 are expressions satisfying the three conditions without being meaningful[10].

A concluding remark may be devoted to the relation between logical and functional notation. The expression *P*(a, *S*(b, c)) is the traditional denotation of the value of the function *P* for the pair (a, *S*(b, c)), whose second member is the value of the function *S* for the pair (b, c). If in this expression the parentheses and commas are deleted, then what remains is the expression on the left side of (5). In the same way, one obtains what is on the right side of (5) from the traditional expression *S*(*P*(a, b), *P*(a, c)). One readily sees that, more generally, *all traditional expressions of function values can be transformed into correct parentheses-free expressions* (corresponding to Łukasiewicz' logical strings) *simply by deleting the parentheses and commas but without any other alteration*. Conversely, *meaningful logical strings in Łukasiewicz' notation can be converted into the traditional denotation of function values by the insertion of parentheses and commas but without any other alteration*. For example, the strings in (2′) are converted into

(2″) $C(C(C(\mathrm{p},\mathrm{q}), C(\mathrm{q},\mathrm{r})), C(\mathrm{p},\mathrm{r}))$ and $C(C(\mathrm{p},\mathrm{q}), C(n(\mathrm{q}), n(\mathrm{p})))$.

These are traditional expressions in functional notation, except that for the sake of greater transparency one often writes

$$C\big(C\big(C(\mathrm{p},\mathrm{q}), C(\mathrm{q},\mathrm{r}\big), C(\mathrm{p},\mathrm{r})\big) \quad \text{and} \quad C\big(C(\mathrm{p},\mathrm{q}), C\big(n(\mathrm{q}), n(\mathrm{p})\big)\big).$$

In logic, however, the functional notation is not normally used.

NOTES

[1] Actually one can dispense with the parentheses in *one* of the two expressions (1) by adopting the rule: *read all formulae from the left*; for example, read $\mathrm{p} \to \mathrm{q} \to \mathrm{r}$ and $\mathrm{p} \to \mathrm{q} \to \mathrm{r} \to \mathrm{s}$ as $(\mathrm{p} \to \mathrm{q}) \to \mathrm{r}$ and $((\mathrm{p} \to \mathrm{q}) \to \mathrm{r}) \to \mathrm{s}$ and so on (or all formulae from the right). But even then parentheses would be indispensable in the *other* formula (1).

[2] Łukasiewicz himself used N for negation. I write *n* and reserve capital italics for binary connectives conforming to the typography explained on p. 70 ff.

[3] Cf. [1931, 10]. During a later visit to Poland, Łukasiewicz told me that his student Jaskowski had independently found the same criterion. A corresponding criterion might of course also be given for strings including the traditional sandwiched arrows and encumbered with superfluous parentheses of two kinds, (and). But that criterion would be more complicated and harder to apply.

[4] [*Addition in the Translation.*] The three strings in (3) in the Preface are examples of strings violating just one of the three conditions: the first, the second and the third, respectively, while satisfying the other two.

[5] [*Addition in the Translation.*] Assume that each string satisfying Conditions 1, 2, 3 and including less than s elements represents a proposition, and that $\mathfrak{S}$ is a string of s elements satisfying the three conditions. Let $\mathfrak{T}$ be the terminal segment of $\mathfrak{S}$ following the first letter. If $\mathfrak{S}$ begins with *n*, then $\mathfrak{T}$ satisfies the three conditions. By assumption, $\mathfrak{T}$ represents a proposition, and $\mathfrak{S}$ represents its negation. If $\mathfrak{S}$ begins with *C*, and includes, say, m *C*'s and consequently $\mathrm{m} + 1$ roman type letters, then the number of *C*'s in $\mathfrak{T}$ is $\mathrm{m} - 1$. Let $\mathfrak{T}'$ be the first initial segment of $\mathfrak{T}$ that includes more roman type letters than *C*'s. If the number of *C*'s in $\mathfrak{T}'$ is k, then $\mathfrak{T}'$ clearly includes $\mathrm{k} + 1$ roman type letters and also satisfies Conditions 2 and 3. In the terminal segment, $\mathfrak{U}$, of $\mathfrak{T}$ after $\mathfrak{T}'$, then umber of *C*'s is $\mathrm{m} - 1 - \mathrm{k}$, that of roman type letters is $(\mathrm{m} + 1) - (\mathrm{k} + 1)$. Thus $\mathfrak{U}$ satisfies Condition 1. No initial segment of $\mathfrak{U}$ includes fewer *C*'s than roman type letters, since no initial segment of $\mathfrak{S}$ does so. Hence $\mathfrak{U}$ satisfies all three conditions as does $\mathfrak{T}'$. By assumption, both $\mathfrak{T}'$ and $\mathfrak{U}$ represent propositions, and $\mathfrak{S}$ represents the proposition that the former implies the latter.

[6] The concept of the scope of an element in a string provides a means of making precise the postulate that "the *Wirkungsbereiche* of quantifiers belonging to variables denoted by the same letter do not overlap." (Hilbert-Ackermann, *Grundzüge d. Theoretischen Logik*, 1928).

[7] Even though by introducing sandwiched symbols mathematicians have made it necessary to use parentheses they regard them as nuisances and, wherever possible, eliminate them by special rules. Just as in one of the two logical expressions (1), parentheses are traditionally omitted in one of the two expressions (4), namely in $2 + (3 \times 4)$ by the rule, *first multiply, then add*; but then they are indispensable in $(2 + 3) \times 4$. If, in another culture, mathematicians were to stipulate *first add, then multiply*, then $2 + 3 \times 4$ would stand for $(2 + 3) \times 4$ and one would have to write $2 + (3 \times 4)$. The arithmetic and algebraic formulae would look quite different from ours while expressing the same content. Of course, even an expression such as *S* 2 3 4 7 is ambiguous unless either separators (commas or spaces) are used in order to distinguish

$$S\,2{,}347,\quad S\,23{,}47,\quad S\,234{,}7 \quad \text{for} \quad 2 + 347,\ 23 + 47,\ 234 + 7$$

or each numeral is terminated with an asterisk, as in *S* 2*347* and *S* 23*47*. Sums and products of more than 2 numbers, say of k terms, are traditionally expressed by k − 1 symbols, + and ×, alternating with the k numerals. Instead, frontal letters $S^{(k)}$ and $P^{(k)}$ might be used unambiguously without parentheses, as in

$$S^{(3)}2\,3\,4 = S\,2\,S\,3\,4 = S\,S\,2\,3\,4\,,$$

$P^{(4)}$a b c d $= P^{(3)}$a b P c d $= P^{(3)}$a P b c d $= P^{(3)}P$ a b c d $= P$ a $P^{(3)}$b c d $= PP^{(3)}$a b c d $= P\,P$ a b P c d. Here, $S = S^{(2)}$ and $P = P^{(2)}$.

[8] Cf. [1957, 1].

[9] Again parentheses-saving rules, such as *unitary operations have precedence over binary operations*, make it possible to write, for example, log 2 + 3 for (log 2) + 3 while the parentheses in log(2 + 3) must be retained.

[10] A generalization of the Conditions 1, 2, 3 to expressions including also symbols for ternary, quaternary, . . . *n*-ary operations or *n*-place functions (usually called functions of *n* variables) has been given in D. C. Gerneth's paper, 'Generalization of Menger's Result on the Structure of Logical Formulas', *Bull. Amer. Math. Soc.* **54** (1948).

CHAPTER 4

A NEW POINT OF VIEW ON THE LOGICAL CONNECTIVES

"The following remarks are very elementary. But . . . logicians may be interested to see that a basic field as elementary and as old as the calculus of propositions still has room for a new point of view. Such a viewpoint is indeed suggested by the algebra of functions," which I have developed in several papers[1] and which can also be applied to many-valued logics. Passages concerning the logical connectives *not*, *or*, *implies*, etc. from these papers will be quoted or excerpted in this chapter. While the results may be of no immediate concern to logicians they do reveal previously unexplored interrelations in the realm of logical connectives and operations. "Classical results and methods are resolved into layers of theorems and procedures"[2] which will be presented in point D below. Points A, B, C first describe (in a not quite traditional manner) three ways in which the connectives arise in the classical calculus of propositions:

A. In the study of compound propositions.

They are composed of simpler propositions and connectives (and expressed in strings of symbols as described in Chapter 3). Each binary connective associates a proposition to each pair of propositions, p, q; for example, $p \rightarrow q$ (p *implies* q), $p \vee q$ (p *or* q *or both*), $p \mid q$ (*not both* p *and* q), $p \equiv q$ (p *equivalent* q) – in frontal notation *C*pq, *A*pq, *D*pq, *E*pq, respectively. A unary connective associates a proposition to each proposition, for example, *not*-p or *n* p to p.

B. In the study of "truth values".

One such value is associated to each proposition. The 2-valued calculus of propositions is based on two assumptions:

ASSUMPTION I. *There are exactly two truth values: true and false – briefly,* t *and* f, *or* 1 *and* 0 (only these *numerical* values are used in what follows).

The truth value of any proposition r will herein be denoted by v(r), the argument always between parentheses, whereas all strings describing propositions will be parentheses-free in frontal notation (See Chapter 3). According to Assumption I, for each proposition p, either $v(\mathrm{p}) = 1$ or $v(\mathrm{p}) = 0$. Hence for a pair of propositions, there are four possible pairs of truth values:

$$(1) \qquad (0,0), \quad (0,1), \quad (1,0), \quad (1,1).$$

If p, q and p′, q′ are two pairs of propositions with the same pairs of truth values, then also the compounds Cpq and Cp′q′ have the same truth value and so do np and np′.

More generally,

ASSUMPTION II. *For every binary connective F and every unary u,*

$$\textit{if } v(\mathrm{p}) = v(\mathrm{p}') \textit{ and } v(\mathrm{q}) = v(\mathrm{q}'),$$

$$\textit{then } v(F\mathrm{pq}) = v(F\mathrm{p}'\mathrm{q}') \textit{ and } v(u\mathrm{p}) = v(u\mathrm{p}').$$

C. In the definition of numerical functions corresponding to connectives.

According to Assumption II, if F is any binary connective, then for every pair p, q of propositions having truth values (0, 0), the truth value $v(F\mathrm{pq})$ is the same and, therefore, may be denoted by $F(0,0)$. Similarly, one defines $F(0,1)$ as the truth value $v(F\mathrm{pq})$ for any pair p, q having the truth values (0, 1); and one can introduce $F(1,0)$ and $F(1,1)$. This pairing of a number to each of the four pairs in (1) defines what is called *a function on the domain* (1). Clearly, to each binary connective F there corresponds such a function. The number $F(\mathrm{x},\mathrm{y})$ is called the *value* of that function for (not *of*!) the pair (x, y) – to be distinguished from the *truth value* of (not *for*!) a proposition Fpq.

In the literature, these functions have often been confused with the connectives to which they correspond, even though they are of a totally different nature – the connectives associating propositions with pairs of propositions, the functions pairing numbers to pairs of numbers. It therefore would be desirable that at least slightly different symbols, say F, u and $*F$, $*u$, be used for connectives and functions, respectively. But in view of a parallelism between them (to be discussed presently), a sufficient symbolic differentiation is achieved by denoting the function values in the usual way, that is, by symbols such as $F(\mathrm{x},\mathrm{y})$, $C(1,0)$, $n(1)$, with parentheses, while describing compound propositions by parentheses-free strings[3] such as Fpq, Crs, nt. In what follows, the functions *themselves* (as distinguished from their values) will be denoted by $F(,)$, $C(,)$ and $n()$ where they might be confused with the connectives F, C and n. The functions corresponding to unary, binary (m-ary) connectives will be called 1-*place*, 2-*place* (m-*place*) *functions*, the idea being that they have one place, two places (m places) for what will be called the *entering of arguments* or *input*. The simplest inputs are, respectively, numbers and pairs (m-tuples) of numbers, resulting in the *values* of the functions for the entered arguments. The parallelism mentioned earlier, is described in the following

BASIC LAW. *The truth-value of a compound proposition is the value of the corresponding function for the truth values of the components of the compound;* in formulae:

$$v(F\mathrm{pq}) = F(v(\mathrm{p}), v(\mathrm{q})) \qquad \textit{and} \qquad v(u\mathrm{p}) = u(v(\mathrm{p}))$$

for every binary connective F, every unary u, and any two propositions, p *and* q.

In the case of n, one sees that the compound np is true or false and $v(n\mathrm{p})$ is 1 or 0 according as p is false or true and $v(\mathrm{p})$ is 0 or 1. Consequently, $v(n\mathrm{p}) = 1 - v(\mathrm{p})$. The function with the value $n(\mathrm{x}) = 1 - \mathrm{x}$ for x corresponds[4] to the connective n; and clearly, $v(n\mathrm{p}) = n(v(\mathrm{p}))$. As for implication, the traditional calculus of proposition assumes that every proposition is implied by each false proposition and implies each true proposition (See p. 21), and that Cpq is false if p is true and q is false. For the function corresponding to implication, these assumptions imply that

$$C(0,0) = C(0,1) = C(1,1) = 1 \qquad \text{and} \qquad C(1,0) = 0.$$

The respective values of the function $C(,)$ for the four pairs in (1) or, as we say, on (1) are 1, 1, 0, 1. The function may also be described by associating to any pair of numbers (x, y) the number $C(\mathrm{x},\mathrm{y}) = 1 - \mathrm{x} + \mathrm{x}\cdot\mathrm{y}$.

In the 2-valued logic there are

1) four unary connectives and 1-place functions: n (negation) and three that are hardly ever mentioned, namely j (*identity*) assuming for x the value $j(\mathrm{x}) = \mathrm{x}$, and two that are *constant: o* and i, assuming for each x the values 0 and 1, respectively.[5]

2) sixteen binary connectives and 2-place functions: C and the already mentioned A (*inclusive disjunction*), D (*incompatibility*) and E (*equivalence*)[6]

the corresponding functions assuming on (1) the values

$$0, 1, 1, 1; \qquad 1, 1, 1, 0; \qquad 1, 0, 0, 1,$$

respectively, whence for any pair (x, y)

$$(2) \qquad A(x, y) = x + y - x \cdot y, \quad D(x, y) = 1 - x \cdot y,$$
$$E(x, y) = 1 - x - y + 2x \cdot y.$$

Furthermore, a less important B such that Bpq stands for p *is implied by* q and $B(x, y) = 1 - y + x \cdot y$; two important, though usually neglected connectives and functions that I call *selectors*, I the *first* and J the *second* selector, because they select from any pair (of propositions or numbers) the first and the second member, respectively:

$$I\text{pq} = \text{p}, J\text{pq} = \text{q} \qquad \text{and} \qquad I(x, y) = x, J(x\,y) = y;$$

the eighth and ninth are *constant functions*, *0* and *1*, such that $0(x, y) = 0$ and $1(x, y) = 1$ for each x, y. The 2-place and 1-place constant functions (for which alternative symbols would be C_0, C_1 and c_0, c_1) must not be confused with each other or with the numbers 0 and 1. Finally, if F is A, B, C, D, E, I or J, then there exists an *opposite* connective F' such that for any p, q the propositions F'pq and Fpq have opposite truth-values (one is 0, the other is 1) and $F'(x, y) \neq F(x, y)$ for each x, y. Examples include:

A' (*joint rejection*); A'pq for *neither* p *nor* q;

D' (*conjunction*); D'p, q for p *and* q

E' (*exclusive disjunction*); E'pq for p *or* q *but not both*;

the corresponding functions assuming on (1) the values

$$1, 0, 0, 0, \qquad 0, 0, 0, 1, \qquad 0, 1, 1, 0,$$

respectively, whence for any pair (x, y)

$$(2') \qquad A'(x, y) = 1 - x - y + x \cdot y, \quad D'(x, y) = x \cdot y,$$
$$E'(x, y) = x + y - 2x \cdot y.$$

0 and *1* are opposites of one another: $1 = 0'$ and $0 = 1'$.

The traditional calculus of propositions demonstrates that either of the following pairs (among others) can be used as a basis: C and n (Frege, Łukasiewicz) or A and n (Russell); and most famous is the result (Peirce, Sheffer) that D can serve as a basis by itself; that is, each proposition is equivalent to a compound including no connectives other than *D*. This result is often expressed by saying: *All connectives of the 2-valued logic can be expressed in terms of D alone (or in terms of C and n or A and n)*. In particular,

(3) $Cpq = DpDpq.$

(3a) $Apq = DDppDqq$ and $np = Dpp,$

whence $Apq = Dnpnq.$

I now turn to a fourth and new approach to the connectives.

D. Operating with the corresponding functions themselves (rather than with their values). This procedure reveals various deeper *interrelations between the functions and between the operations.* The operation *par excellence* in the realm of functions is what is often referred to as *composition* and for the purposes of this paper[7] more conveniently called *the entering of an input* into a function – the inputs themselves consisting of functions: of a single function G into a 1-place function, $f()$, or a pair of functions (G, H) into a 2-place function, $F(,)$. The results are 2-place functions denoted by $f(G)$ and $F(G, H)$. Their values for (x, y) I denote[8] by $\langle f(G)\rangle(x, y)$ and $\langle F(G, H)\rangle(x, y)$ using $\langle\ \rangle$ as a mere (and, by the way, dispensable) indication of *grouping*, whereas ( ) and (,) preceded by the symbol for a function in this chapter always indicate an *operation*: the entering of an input. These values are defined as follows:

(4) $\langle f(G)\rangle(x, y) = f(G(x, y))$ and

$$\langle F(G, H)\rangle(x, y) = F(G(x, y), H(x, y))$$

for each pair (x, y) in (1) or, briefly, on (1).

Example 1. If $f = n$ and $G = D$, then

$$\langle n(D)\rangle(x, y) = n(D(x, y)) = 1 - D(x, y) = D'(x, y) \text{ on } (1),$$

the last equality resulting from a comparison of formulae (2) ad (2′). More generally, $n(G) = G'$ and $n(G') = G$ for $G = A, B, C, D, E, I, J, O$.

Example 2. If $F = G = D$ and $H = I$, then

$$\langle D(D, I)\rangle(x, y) = D(D(x, y), I(x, y)) = D(D(x, y), 1).$$

This number is 0 or 1 according as $D(\mathrm{x}, \mathrm{y})$ is 1 or 0. But the same is true for $D'(\mathrm{x}, \mathrm{y})$. Hence

$$\langle D(D, 1)\rangle(\mathrm{x}, \mathrm{y}) = D'(\mathrm{x}, \mathrm{y}) \text{ on (1), whence } D(D, 1) = D'$$

according to the

BASIC EQUALITY DEFINITION FOR FUNCTIONS.

(5) $F_1 = F_2$ *if and only if* $F_1(\mathrm{x}, \mathrm{y}) = F_2(\mathrm{x}, \mathrm{y})$ *on* (1)

and $f_1 = f_2$ *if and only if* $f_1(\mathrm{x}) = f_2(\mathrm{x})$ *for* $\mathrm{x} = 0, 1$.

In the same way one establishes

(6) $D(D, D) = D', \quad D(D, D') = 1, \quad D(D', 1) = D, \quad D(1, 1) = 0$

and, using Example 1,

(7) $D(F, F) = n(F)$ for each F.

One obtains the complete output of F with respect to the operation (,) by listing (beginning with F itself) all functions expressible in terms of F along the left and upper edge of a table and putting $F(X, Y)$ in the row headed by X and the column headed by Y. The table is complete if all functions that occur in it also appear on the edge.

$F(\ ,\ )$	F . .	Y . .
F	$F(F, F)$	
:		:
X	. . .	$F(X, Y)$ . .
:		:

$A(\ ,\ )$	A
A	A

$C(\ ,\ )$	C	1
C	1	1
1	C	1

$D(\ ,\ )$	D	D'	1	0
D	D'	1	D'	1
D'	1	D	D	1
1	D'	D	0	1
0	1	1	1	1

The (,)-output table for A displays the single entry A since $A(A, A) = A$, whence no other function can be expressed in terms of A alone. The complete table for C includes 1 besides C. For D, it follows from (7) that the edge must include at least D, D', 1 and 0; and by completing the table one sees that these four functions do not produce any others[9]. Hence

THEOREM I. *The only functions that can be expressed in terms of D alone are*

(8) $D,\ D' = D(D, D),\ 1 = D(D, D') = D(D, D(D, D)),$

$0 = D(1, 1) = D(D(D, D(D, D)), D(D, D(D, D))).$

How can this statement be reconciled with the famous traditional result mentioned above about all connectives being expressible in terms of *D* alone? Actually, for example formula (3) represents, for any p and q, a definition of *the proposition* Cpq *in terms of D*, p *and* q. And for the function corresponding to implication, one obtains from (3) an analogous formula expressing *the value of C for* (x, y) *in terms of D*, x *and* y, namely[10]

(9) $C(\mathrm{x}, \mathrm{y}) = D(\mathrm{x}, D(\mathrm{x}, \mathrm{y}))$ on (1).

This is an, as it were, *extensional* definition of *C*, but not an expression of *C* in terms of *D* the way (6) expresses *D'*, *1* and *0*. There is not on the right side of (9) the value for (x, y) of an immediately identifiable function, since the term x is not presented as a function value for (x, y). Actually, however, x *is* such a value, namely of the (traditionally neglected) first selector, *I*. Introducing *I*(x, y) into (9) one obtains by (4)

$$C(\mathrm{x}, \mathrm{y}) = D(I(\mathrm{x}, \mathrm{y}), D(\mathrm{x}, \mathrm{y})) = \langle D(I, D)\rangle(\mathrm{x}, \mathrm{y});$$

thus

(10) $C = D(I, D).$

(10) expresses *C* – the function *itself*, not its value – with the help of *D*, but not in terms of *D alone*: it expresses *C* in terms of *D* and *I*. Theorem I states that *C*, *A*, *E* and the other functions except *D'*, *1* and *0* *cannot* be expressed in terms of *D* alone. The main results concerning the (,)-operation, which can be readily verified, are summarized in

THEOREM II. *In the 2-valued logic, no single function F has an* (,)-*output of more than four functions (including F). The functions expressible in terms of D and I besides D, D', 1, 0 are C, C', I, I'. No pair of functions has a joint* (,)-*output of more than eight functions. At least three functions are necessary in order to express all sixteen of the functions. Examples of triples generating all functions include, among others, D, I, J and D, B, C.*[10a]

Definition (4) only covers 2-place inputs. But one can also enter a 1-place function *g* into *f*, and a pair (*g*, *h*) of 1-place functions into *F*, both operations resulting in 1-place functions, $f(g)$ and $F(g, h)$, whose values are defined by

(11) $\langle f(g)\rangle(\mathrm{z}) = f(g(\mathrm{z}))$ and $\langle F(g, h)\rangle(\mathrm{z}) = F(g(\mathrm{z}), h(\mathrm{z}))$ for $\mathrm{z} = 0, 1$.

Example 3. If $f = g = n$, then $\langle n(n)\rangle(\mathrm{z}) = n(n(\mathrm{z})) = 1 - (1 - \mathrm{z}) = \mathrm{z} = j(\mathrm{z})$, whence $n(n) = j$. Similarly[11], $n(o) = i$, $n(i) = o$.

Example 4. If $F = D$, $g = n$, $h = j$, then because $D(0, 1) = D(1, 0) = 1$

$\langle D(n, j)\rangle(\mathrm{z}) = D(n(\mathrm{z}), j(\mathrm{z})) = 1 = i(\mathrm{z})$ for $\mathrm{z} = 0, 1$, whence $D(n, j) = i$.

By similar reasoning, one readily completes the first of the following tables

$D(,)$	n	j	o	i
n	j	i	i	o
j	i	n	i	i
o	i	i	i	i
i	o	i	i	o

$D[,]$	n	j	o	i
n	A	B	1	I
j	C	D	1	I'
o	1	1	1	1
i	J	J'	1	0

There is yet another way to enter g and h into F, the result being a 2-place function that I denote by $F[g, h]$. Its value is defined by

(12) $\langle F[g, h]\rangle(\mathrm{x}, \mathrm{y}) = F(g(\mathrm{x}), h(\mathrm{y}))$ for each (x, y).

Because of the symbols used, I refer to (g, h) as *parenthesis input* and to $[g, h]$ as *bracket input*[12]. Obviously, 1-place functions admit only one kind of input which I will denote, synonymously, by parentheses or brackets, whichever is more convenient: $n(n) = n[n]$ and $f(G) = f[G]$.

Example 5. $\langle D[j, n]\rangle(\mathrm{x}, \mathrm{y}) = D(j(\mathrm{x}), n(\mathrm{y})) = D(\mathrm{x}, 1 - \mathrm{y}) = 1 - \mathrm{x} \cdot (1 - \mathrm{y}) = C(\mathrm{x}, \mathrm{y})$.

Consequently, $D[j, n] = C$. Thus by means of bracket input C is obtainable from D and n; hence so is $C' = n[C]$. But the second table above, which displays the bracket output of D for all four 1-place functions,[13] shows that E and E' cannot be thus obtained.

The bracket output tables for A, B, C are easily seen to include the same ten functions that appear in the table for D while those for A', B', C', D'

include in the left upper corner the four functions A', B', C', D' and elsewhere the same six other functions that occur in the D-table. This proves the first two assertions of

THEOREM III. *The functions expressible in terms of n and D by bracket inputs are*

$$n, j, A, B, C, D, A', B', C', D'.$$

These are also the functions expressible in terms of n and any of the other seven functions A, . . . , D' while all that can be [,]-expressed in terms of n and either E or E' is j and both E and E'.

The last assertion follows from the fact that the left upper corners of the bracket tables of both E and E' include only the two functions E and E'.

An operation altogether different from any form of composition is the *diagonalization*, δ. It transforms 2-place into 1-place functions. If F is 2-place, then the 1-place function δF is defined by

(13) $\langle \delta F \rangle(\mathrm{t}) = F(\mathrm{t}, \mathrm{t})$ for $\mathrm{t} = 0, 1$.

Example 6. $\delta D = n$, $\delta D' = j$, $\delta C = i$, $\delta C' = o$.

THEOREM IV. *The functions expressible in terms of D by bracket inputs and diagonalization are the four 1-place functions and fourteen of the sixteen 2-place functions – all except E and E'. The same is true for A' while each of the other functions generates at most three functions different from itself*[14].

Indeed, $\delta D = n$, whence according to Theorem III the ten functions listed in that theorem are expressible in terms of D and bracket inputs and δ. For example,

$$C = D[j, n] = D[n[n], n] = D[\delta D[\delta D], \delta D].$$

Diagonalization of C and C' yields i and o. Hence also the six remaining functions listed in the input table of D, that is, $I, J, I', J', 0, 1$ are expressible by such inputs.

In order to express E and E' along with the eighteen functions mentioned in Theorem IV one must also use parenthesis inputs, as in $E = D(D, A) = D(D, D[\delta D, \delta D])$. Hence

THEOREM V. *Only by the use of parenthesis as well as bracket inputs and diagonalization can E and E′ and thereby all sixteen 2-place functions and four 1-place functions be expressed in terms of D.*

NOTES

[1] [1961, 5], [1962, 1], [1962, 3], [1963, 2], [1964, 1], [1964, 2], [1966, 2]. The preceding is a quote from [1962, 3].

[2] Cf. [1964, 1], p. 86.

[3] Moreover, I use p, q, r, s, ... as proposition variables, and x, y, z, ... as number variables, that is, symbols standing respectively for any proposition and any number – in the 2-valued logic just for either 0 or 1. The interrelations of the functions corresponding to the connectives, however, will be presented without the use of *any* number variables – a characteristic feature of the entire algebra of functions. This theory does, however, employ function variables where their use is necessary or convenient, e.g. the letters *F*, *G*, *H* in note[9] and *f* in note[11] below.

[4] So do of course all functions with the values 1 for 0, and 0 for 1, for example, those whose values for x are $1 - x^2$, $(1 - x)^4$ or even $\cos \pi x/2$.

[5] The constant connective *o* is related to what is introduced in some developments of the calculus of propositions as the primitive constant, *f*, designating a false proposition.

[6] I am using here the letters *A*, *B*, *C*, *D*, *E* as introduced by Łukasiewicz for the corresponding connectives. In the papers referred to in[1], written at a time when unfortunately I only knew Łukasiewicz' symbol *C* for implication, I used *B*, *D*, *A* instead of his *A*, *B*, *D*.

[7] In the papers referred to in[1], I have called the operation "*substitution*" occasionally pointing out that it is a substitution of functions *into* a function and not a substitution of functions *for* something. Yet the term has led to numerous misunderstandings. I therefore here use "*composition*" which is employed widely, especially in the important work on the algebra of functions and the grammar of functions by B. Schweizer and A. Sklar. Unfortunately, "composition" seems to suggest a nonexisting symmetry between the function into which something is entered and the function or functions which are entered.

[8] Cf. [1964, 1].

[9] All functions expressible in terms of functions that are expressible in terms of *D* are also expressible in terms of *D*. This follows from the fundamental property of composition that I have called *superassociativity* (cf. especially [1964, 2]).

$$\langle F(G_1, G_2)\rangle(H_1, H_2) = F(G_1(H_1, H_2), G_2(H_1, H_2)).$$

[10] According to the Basic Law,

$$v(C\text{pq}) = C(v(\text{p}), v(\text{q})) \text{ and}$$

$$v(D\text{p}D\text{pq}) = D(v(\text{p}), v(D\text{pq})) = D(v(\text{p}), D(v(\text{p}), v(\text{q}))),$$

where $(v(\text{p}), v(\text{q}))$ is any pair (x, y) in (1). Hence (3) implies (9).

[10a] In [1966,2], A. I. Whitlock and I developed a general theory of p-place functions in n-valued logic. In this theory, the fact that *three* functions are needed to generate the sixteen 2-place functions in 2-valued logic represents a singular exception. For in the 3-, 4-, and more-valued logics, the generation of all 2-place functions (although their numbers are close to 20 000, 4000 millions, and astronomical, respectively) requires only *two* of them. And even in 2-valued logic not more than three 3-place functions are needed to generate all two hundred and fifty-six 3-place functions. In fact, for all integers p and n exeeding 1, all p-place functions in n-valued logic can be generated by p of them except in the case where both p and n equal 2.

[11] Moreover, for any 1-place function f,

$$j(f) = f(j) = f, \quad o(f) = o, \quad i(f) = i.$$

[12] The distinction between parenthesis and bracket composition is also relevant in analysis (cf. [1944, 3]).

[13] The second table is identical with a parenthesis output table for D with I', I, 0, 1 on the left edge and J', J, 0, 1 above. For example, $D[n,j] = D(I',J)$.

[14] I' generates n by diagonalization, and j and I by bracket inputs. C generates i by diagonalization, and only 1 by bracket inputs. $\delta A = j$, $\delta I = j$ and since $F[j,j] = F$ for every function F, the bracket input tables of A and I each have only one entry, namely A and I, just as do their parenthesis input tables.

CHAPTER 5

AN INTUITIONISTIC-FORMALISTIC DICTIONARY OF SET THEORY

This chapter is a translation of the first of four articles on questions concerning the foundations of mathematics, [1928, 1]. I remember the set-theoretician Hausdorff writing me that after reading this paper he understood for the first time what Brouwer was talking about in his definition of set.

In translating the paper I have expanded Section 1 and shortened Section 7.

ON RAMIFICATION SETS

In this paper there is established a close connection (oddly enough, not heretofore noticed in the literature[0]) between Brouwer's definition of *set* in intuitionism and the concept of *analytic set* in traditional set theory. We start from a more general, comprehensive concept.

1. *Ramification sets.*

Two tools will be used in defining ramification sets relative to a given set $\mathfrak{S}$ of entities of any kind.

a) The set of all natural numbers arranged in the *linear scheme* 1, 2, . . . , n, ... A *sequence* of elements of $\mathfrak{S}$ is the result of pairing an element of $\mathfrak{S}$ to each natural number – not necessarily different elements to different numbers. The result of pairing, for example, *e*, *f*, *e*, . . . to 1, 2, 3, . . . , respectively, is denoted by (*e*, *f*, *e*, . . .); the sequence of the odd numbers, by (1, 3, 5, . . . , 2*n*-1, . . .). Sequences in this sense are called *nonterminating* if one wishes to distinguish them from sequences where the pairing terminates with an initial segment of the natural numbers, say, 1, 2, . . . , *k*. One then denotes, for example, by (2, 2, 1) the association of 2, 2, 1 to 1, 2, 3, respectively, and by (4) the "singleton" sequence associating 4 to 1.

b) The set of all terminating sequences of natural numbers arranged in the following *ramification scheme*:

(1) (2) ...(k) ...
(1,1) (1,2) (1,3) ...(2,1) (2,2) ...(k,1) (k,2) ...
(1,1,1)(1,1,2)...(1,2,1)(1,2,2)...(1,3,1)(1,3,2)...(2,1,1)(2,1,2)...(2,2,1)(2,2,2)...(k,1,1)(k,1,2)...(k,2,1)(k,2,2)...
. . . .

The scheme resembles a sequence of family trees originating in singletons, each family member having a (nonterminating) sequence of descendants. For each nonterminating sequence $\nu = (n_1, n_2, \ldots, n_k, \ldots)$ the ramification scheme includes what will be called the *line of descendency* corresponding to ν, that is, the nonterminating sequence of terminating sequences $\Delta_\nu = ((n_1,),(n_1, n_2), (n_1, n_2, n_3), \ldots, (n_1, n_2, n_3, \ldots, n_k), \ldots)$. Each member of Δ_ν is an immediate descendant of the preceding member. (The letters ν and μ without subscripts will hereinafter be reserved for nonterminating sequences of natural numbers.)

A *ramification system* of elements of $\mathfrak{S}$, an analogue to a sequence, is the result of pairing an element of $\mathfrak{S}$ to each element of the ramification scheme – the element $S_{n_1, n_1, \ldots, n_k}$ to $(n_1, n_2, \ldots, n_k)$ – not necessarily different elements of $\mathfrak{S}$ to different sequences. This system will be denoted by the customary (though somewhat incongruous) symbol $\{S_{n_1, n_1, \ldots, n_k}\}$ or, briefly, by Σ.

We can now define *ramification sets* relative to $\mathfrak{S}$. More specifically, the ramification set, R, with the *determining* (ramification) *system* Σ is the set of the (nonterminating) sequences

$$(S_{n_1}, S_{n_1, n_2}, S_{n_1, n_2, n_3}, \ldots, S_{n_1, n_2, n_3, \ldots, n_k}, \ldots)$$

corresponding to the lines of descendency Δ_ν, for all sequences $\nu = (n_1, n_2, n_3, \ldots, n_k, \ldots)$ of natural numbers.

2. *Analytic Sets*

Now let $\mathfrak{S}$ be a set *of sets*. Then the ramification set R with the determinating system Σ is a set of sequences of sets belonging to $\mathfrak{S}$. Next, for each $\nu = (n_1, n_2, \ldots, n_k, \ldots)$, consider the intersection

$$A_\nu = S_{n_1} \cap S_{n_1, n_2} \cap \ldots \cap S_{n_1, n_2, \ldots, n_k} \cap \ldots$$

of the sets corresponding to the line of descendence Δ_ν. The union of the (not necessarily different) sets A_ν for all sequences ν is called an *analytic* set relative[1] to $\mathfrak{S}$ or, more specifically the analytic set with the determining system Σ. Conversely, a set A, is called analytic relative to $\mathfrak{S}$ only if there exists a ramification system of sets of $\mathfrak{S}$ which in this way determines A.

In general, an analytic set has many determining systems. For example, if for each terminating sequence $(n_1, n_2, \ldots, n_k)$ we set

$$S_{n_1} \cap S_{n_1, n_2} \cap \ldots \cap S_{n_1, n_2, \ldots, n_k} = T_{n_1, n_2, \ldots, n_k},$$

then the ramification systems Σ and $\mathsf{T} = \{T_{n_1, n_2, \ldots, n_k}\}$ determine the same analytic set. The system T is *monotonic* in the sense that each set contains all its immediate descendants (and hence all its descendants) as subsets; that is, for every $(n_1, n_2, \ldots, n_k)$

$$T_{n_1, n_2, \ldots, n_k} \supset T_{n_1, n_2, \ldots, n_k, n_{k+1}} \text{ for } n_{k+1} = 1, 2, \ldots$$

Clearly, unless Σ happens to be montonic the system T differs from Σ. It is furthermore evident that each analytic set possesses (among others) determining systems that are monotonic.

3. *Analytic Point Sets*

In a coordinate plane with two perpendicular axes, let P be a ramification system consisting of closed squares with sides parallel to the axes, the empty set being considered as such a square. The analytic set with the determining system P is called an *analytic point set* in the plane. The following remark (used in Section 6) can be proved in an essentially simple, if lengthy way[2] without the use of the law of the excluded middle. Each analytical point set in the plane possesses (among others) a determining system P such that:

1. The corners of all squares belonging to P have rational coordinates.
2. Each square belonging to P contains all its immediate descendants in its interior.
3. For every sequence v, the intersection of the squares corresponding to the line of descendence Δ_v either consists of a single point or is empty.[3]

A point p belongs to the analytic point set with the determining system P if and only if there exists a sequence v such that p constitutes the intersection of the squares associated with Δ_v (see p. 49).

4. *Three Remarks about Ramification Sets*

a) In defining an analytic set A one usually finds sequences of sets with empty intersections uninteresting since they make no contribution to A, the union of such intersections. We say also, synonymously, that the sequences of *numbers* to which such sequences of sets correspond make no contribution to A. For example, if the set $S_{2,1}$ in the determining system of A is empty, then none of the sequences $(2, 1, n_3, n_4, \ldots, n_k, \ldots)$ makes a contribution to A. But even if the elements of $\mathfrak{S}$ are entities other than sets, $\mathfrak{S}$ may include an entity N, called the *null entity*, with the following property: If, in the determining system of a ramification set R, the element $S_{m_1, m_2, \ldots, m_r}$ is N, then each sequence $(m_1, m_2, \ldots, m_r, m_{r+1}, \ldots)$ beginning with $(m_1, m_2, \ldots, m_r)$ is considered to make no contribution to R and can be disregarded. For analytic sets, the empty set clearly is a null entity. For analytic point sets in the plane the empty set is the only null entity. Indeed, each sequence of nonempty closed squares corresponding to a line of descendence is nested (by Property 2), whence according to a theorem of Cantor its intersection is nonempty and does make a contribution to A, namely a single point (by Property 3).

b) Using the law of the excluded middle we now prove: If Σ is a determining system of the ramification set R, then by replacing, if necessary, certain elements of Σ by a null entity, N, one can obtain a determining system Σ' of R with the following property: *Each element of Σ' other than N is a member of at least one sequence that makes a contribution to R.* We shall express this also by saying that, *except for N, each element of Σ' participates in the determination of at least one element of R.* The system Σ' can be obtained from Σ as follows: By the law of the excluded middle, for each terminating sequence $(m_1, m_2, \ldots, m_k)$, the element $S_{m_1, m_2, \ldots, m_k}$ either does or does not belong to a sequence that makes a contribution to R. Accordingly we set

$$S'_{m_1, m_2, \ldots, m_k} = S_{m_1, m_2, \ldots, m_k} \text{ or } = N.$$

Each analytic point set A in the plane therefore possesses a determining system Σ' with the properties 1., 2., 3. and such that

4. Each nonempty square belonging to Σ' contains at least one point of A.

One can, however, define analytic sets for which the actual exhibition of a determining system Σ' presupposes the solution of an as yet unsolved problem and, therefore, is at present (and perhaps forever) impossible.[4] For example, let A be the set determined by the ramification system Σ in which 1) the empty set is paired to each sequence $(n_1, n_2, \ldots, n_k)$ including a number $\neq 1$;

2) the set paired to the sequence $(1, \ldots, 1)$ of k ones is empty if the k-th decimal of the number π is the beginning of a sequence of 10 sevens; otherwise it is the closed interval between $-1/2^k$ and $1/2^k$. Clearly, if the decimal expansion of π includes 10 consecutive sevens, then the set A with the determining system Σ is empty while otherwise A consists of the point 0. But none of the elements of Σ' paired to sequences $(1, \ldots, 1)$ can be exhibited, not even the element paired to (1). For only after it has been

decided whether the decimal expansion of π does or does not include 10 consecutive sevens can one say whether the set paired to (1) is empty or the interval from $-\frac{1}{2}$ to $\frac{1}{2}$.

c) Let $\Sigma = \{S_{n_1, \ldots, n_k}\}$ be a monotonic determining system of the analytic set A and assume that for the sequence $\mu = (m_1, m_2, \ldots, m_k, \ldots)$ all sets corresponding to Δ_μ after a certain one are equal, say,

$$S_{m_1, \ldots, m_k} = S_{m_1, \ldots, m_k, m_{k+1}} = \ldots$$
$$= S_{m_1, \ldots, m_k, m_{k+1}, \ldots, m_{k+r}} = \ldots$$

Then also A_μ, the intersection of all sets corresponding to Δ_μ is equal to $S_{m_1, \ldots, m_k}$; and A_μ may be said to be *determined by the terminating sequence* $\mu_k = (m_1, \ldots, m_k)$ and be denoted by A_{μ_k}. If the set $S_{m_1, \ldots, m_k}$ is equal to *all* its descendants in Σ, then for each sequence ν beginning with μ_k the set A_ν ($= A_{\mu_k}$) may be said to be determined by μ_k.

5. *Brouwer's Definition of Sets and its Explanation*

Let there be given an infinitude of strings – each made up of a finite number of symbols, just as each decimal numeral is a string made up of the symbols 0, 1, . . . , 9. The following is a translation of Brouwer's definition of set which preserves the involved syntax of the original[5]: "A set is a law which, whenever an arbitrary number is being chosen, over and over, each of these choices either produces a definite string with or without the termination of the process or brings about the stoppage of the process with the definitive annihilation of its result, where for each $k > 1$ after each unfinished and unstopped sequence of $k - 1$ choices at least one number can be specified which, when chosen as the k-th number, does not bring about the stoppage of the process. Each sequence of strings produced in this way by an unlimited sequence of choices (*Wahlfolge*) is called an *element* of the set."

What this means is that, given a set $\mathfrak{S}$ of strings, in defining a Brouwer set B one produces a ramification set relative to $\mathfrak{S}$. For if one has chosen ("over and over") natural numbers $n_1, n_2, \ldots, n_{k-1}$ then the law defining B associates to the ("arbitrary") choice of a number n_k either N or a string $\sigma_{n_1, n_2, \ldots, n_{k-1}, n_k}$ other than N. In the former case, the sequences beginning with $n_1, \ldots, n_k$ are to be disregarded as not making any contribution to B (the law brings about a "stoppage of the process with the definitive annihilation of the result"). In the latter case, the strings associated to

$$(n_1), (n_1, n_2), \ldots, (n_1, n_2, \ldots, n_{k-1}), (n_1, n_2, \ldots, n_{k-1}, n_k)$$

may or may not determine an element of B (may or may not "terminate the process"). The element of B corresponding to a line of descendence Δ_v is the *concatenation* of the strings

$$\sigma_{n_1}, \sigma_{n_1, n_2}, \ldots \sigma_{n_1, n_2, \ldots, n_k}, \ldots$$

The condition at the very end of Brouwer's definition of sets means that if a string other than N is associated to $(n_1, n_2, \ldots, n_{k-1})$ ("after an unfinished and unstopped sequence of $k - 1$ choices"), then the law must specify at least one natural number n_k such that a string other than N is also associated to $(n_1, n_2, \ldots, n_{k-1}, n_k)$. This means that each element of the determining system of B other than N *participates in the determination of at least one element* of B. For example, if $\sigma_1 \neq N$, then there is also a number m such that $\sigma_{1,m} \neq N$.

Disregarding some terminological differences we thus can say: *The Brouwerian sets are ramification sets in whose determining systems each element other than the null entity participates in the determination of at least one element.* Making essential use of the law of the excluded middle we have proved (See Section 3b)) for each ramification set the existence of a determining system satisfying that additional condition. Hence, again disregarding terminological differences, we can say that *on the basis of the law of the excluded middle each ramification set is a Brouwerian set.*

The differences which must be disregarded originate especially in the fact that Brouwer first defines the concept of *element* of a set by a law of the type described. Since he does not presuppose the set concept but introduces sets by means of laws that produce elements (in fact identifies them with such laws) he naturally cannot speak of the set of all elements produced by a law of the type described. According to the phrasing of Brouwer's definition one therefore would have to say: A law that is a Brouwerian set also defines a ramification set in such a way that each element other than N participates in the determination of at least one element of the set. Conversely, each association of entities to terminating sequences of natural numbers which satisfies the additional condition follows a law that is identical with a Brouwerian set. Using the law of the excluded middle, and only by its use, one can derive from each association defining a ramification set a definition of the same set satisfying the additional condition. As indicated, the difference between the last formulations and the preceding shorter statements seems to me to be only of a terminological nature.

To clarify the situation further it should be noted that Brouwer also uses a term for the totality of the elements of a set in the sense of his definition (that is, for the totality of the elements defined by a ramification system). But since he reserves the word *set* for certain laws he had to choose another term for that totality and indeed speaks of the *species* of all elements of a set.

6. *Brouwerian Point Sets*

In the plane, a Brouwerian point set is a ramification set relative to the system of squares whose corners have rational coordinates (the empty set being the null element of this system) and satisfying Conditions 2., 3. and 4. of Sections 3 and 4. Using the law of the excluded middle one sees that conversely each analytic point set in the plane is a Brouwerian set. We now exemplify two important classes of Brouwerian plane point sets.

a) *Sets with Finite Generations.*[6] These are ramification sets whose ramification system resembles family trees originating in a finite number of ancestors and in which each member has only a terminating sequence of immediate descendants other than N (that is, of nonempty squares), whence each generation includes only a finite number of members $\neq N$. Using the law of the excluded middle one can show that each set with finite generations is closed and bounded, and that each closed and bounded set is a set with finite generations.

b) *Individualized Sets.* Brouwer calls the set B individualized if, for any two nonidentical sequences $v = (n_1, n_2, \ldots, n_k, \ldots)$ and $v' = (n_1, n_2', \ldots, n_k', \ldots)$ that determine points, these points (that is, the intersections of the squares corresponding to the lines of descendence Δ_v and $\Delta_{v'}$) are nonidentical. According to well-known theorems,[7] analytic point sets with this property are *Borelian* sets (that is, obtainable from closed and open sets by the iterated formation of unions and intersections). They can also be characterized as the analytic sets whose complements are likewise analytic.

Summarizing we thus obtain the following intuitionistic-formalistic dictionary of set theory:

Totalities in Brouwer's Terminology	Subsets of the Plane in Traditional Terminology
Sets with finite generations	Bounded closed sets
Individualized sets	Borelian sets
Sets	Analytic sets
Species	Sets

7. *Methodological Consequences*

Analytic sets have recently not only been extensively studied but are considered by some traditional set-theoreticians as the only sets that in a certain sense are really definable[8] – a viewpoint that represents a rapprochement between the formalist and intuitionist point of view. At the same time it is clear that the differences between the intuitionistic and the formalistic development of some of the most important parts of set theory (including the theories of plane point sets and of sets of real numbers) concern only the non-use or use of the law of the excluded middle in operating with the concepts but not the specific concepts themselves.

NOTES

[0] [*Added in this Translation:*] Even after the publication of this paper in 1928, its main result, the relation between Brouwerian and analytic sets, has been widely ignored, especially by intuitionistic writers. Yet in 1928 Heyting wrote me that he hoped the dictionary would help many people to overcome the difficulties of Brouwerian terminology. See below Chapter 21, p. 246.

[1] Analytic sets, often called simply *(A)-sets* or, after their discoverer, *Suslin sets* (See *Paris Comptes Rendus*, **164** (1917) p. 88) have been extensively treated by Lusin (*Fundamenta Mathematicae* **10** (1926) p. 1) and in the second edition of Hausdorff's book *Mengenlehre* (1927), where for the first time abstract analytic sets (not only analytic point sets) are considered.

An analytic set relative to a set $\mathfrak{S}$ of sets does not necessarily belong to $\mathfrak{S}$. Let $\mathfrak{S}$ be, for example, the set of all closed intervals in a straight line; and consider the ramification system Σ in which, for each natural number k, each member of the k-th generation (each set with k indices) is the interval from $-1/k$ to $1/k$. Then the analytic set with the determining system Σ consists of the single point 0 and hence does not belong to $\mathfrak{S}$.

By definition, an analytic set relative to a set $\mathfrak{S}$ of sets is derived from a ramification set R relative to $\mathfrak{S}$ by a certain operation, namely by forming the union of the intersections of all sequences that are elements of R. By forming the intersection of the unions of all elements of R one obtains a set that is the complement of an analytic set. According to well-known theorems, such a set is not necessarily analytic (see Section 6).

[2] For analogous considerations, cf. Hausdorff, loc. cit.[1] p. 92. For the transformation of the determining system of an analytic point set into one with properties 1., 2., 3., cf. Brouwer, *Verhandelingen Akad. Amsterdam*, **(1)** 12, No. 7, p. 5 ff.

[3] Condition 3. can be satisfied by simply choosing, for each k, all nonempty squares of the k-th generation with sides of the length $1/k$.

[4] For the method of referring to an unsolved problem, cf. Brouwer, *Jahresber. d. Deutschen Mathem. Vereinigung* **33** (1924) p. 242 and *Crelles Journal* **154** (1924) p. 3 and Sierpiński, *Fundam. Math.* **2** (1921) p. 114 ff.

[5] Cf. Brouwer, *Jahresber. d. Deutschen Math. Vereinigung* **23** (1914) p. 79 and **28** (1920) p. 204; *Mathem. Annalen* **93** (1924) p. 244.

[6] Brouwer uses the term *finite Menge*; but in English *finite* might lead to confusion.

[7] Cf. Hausdorff, loc. cit.[1]

[9] In order to make this view precise it would seem to me interesting if one could derive the definition of analytic sets from certain postulates of constructivity for the set concept or if one could at least formulate certain properties of constructivity by which analytic sets are characterized. I regard it as rather likely that such a characterization is possible. I wish to emphasize, however, that I consider *constructivity* as a term that has not as yet been made precise and that can be made precise, if at all, then probably *in various ways* and *in various degrees*.

CHAPTER 6

ULTRASETS AND THE PARADOXES OF SET THEORY

In nontechnical German, rather paradoxically, sets and numbers (*Mengen* and *Zahlen*) that are exceedingly large are called *Unmengen* and *Unzahlen* – literally, non-sets and non-numbers. The closest English analogue that occurs to me is the word *invaluable* for something that is exceedingly valuable. In what follows, I translate *Unmengen* and *Unzahlen* by "ultrasets" and "ultranumbers". Actually I did not take this little note quite seriously in 1927 when writing it. It appeared as the third in the series [1928, 1]. I here include a (condensed) translation of the note because it demonstrates in an utterly elementary fashion the need for the distinction between two kinds of totalities (here called *sets* and *ultrasets*) now generally made in the axiomatics of set theory between *sets* and *classes*.

THE PARADOXES OF SET THEORY

We start from the certainly unparadoxical arithmetical fact that if S is any finite set of natural numbers, then there exists a natural number that is not an element of S – for example, the sum of all numbers belonging to S – a fact expressed in the sentence: The natural numbers constitute an infinite set or, as in this paper it is preferable to say, an *ultrafinite set*. A similar argument applies to the prime numbers. Equally unparadoxical is the fact that if S is any denumerable set of numbers in decimal form, then there exists such a number that is not an element of S, for example, a number obtained from those belonging to S by Cantor's diagonal procedure – a fact expressed by saying: The numbers in decimal form constitute an indenumerable set – here better, an *ultradenumerable set*. A similar argument applies to the ordinal numbers of the second class.

Now let S be any set of sets that do not include themselves as elements (such as the set of all integers, which is not an integer). Then there exists a set not including itself as an element which is not an element of S; for example, as one readily sees, the set S itself. We express this fact in the sentence: The sets that do not include themselves as elements constitute an *ultraset*. It was in anticipation of the analogy of the three cases that in the first paragraph the terms ultrafinite and ultradenumerable were preferred to infinite and indenumerable. (Actually, it would be more precise to speak of ultra-(finite sets) and ultra-(denumerable sets), since a totality that is not a finite or denumerable set need not be a set.)

Speaking of the *set* of all sets that are not elements of themselves is like misspeaking of the finite set of all natural numbers or the denumerable set of all numbers in decimal form. Some applications of theorems on finite sets to the "finite set" of all natural numbers happen to yield correct results; for example, each finite set of natural numbers includes a smallest number; hence so does the "finite set" of all natural numbers. Similarly, each set of sets that are not elements of themselves has the empty set as a subset; hence so does the "set" of all sets that are not elements of themselves. But each finite set of natural numbers includes a largest number; yet the "finite set" of all natural numbers does not. And each set of sets that are not elements of themselves either is or is not an element of itself; yet the "set" of all sets that are not elements of themselves neither is nor is not an element of itself.

Other examples of ultrasets include that of all ordinal numbers. For if S is any set of ordinal numbers, there is an ordinal number that is not an element of S; for example, the ordinal number of the well-ordered set S' consisting of the elements of S and all, if any, of the ordinal numbers surpassed by at least one element of S. Similarly, the cardinal numbers constitute an ultraset. So do sets. For if S is a set of sets, then there exists a set S' that is not an element of S; for example, the set S' of all sets that are subsets of some element of S, since the set S' has a higher cardinality than S.

Clearly, the preceding remarks do not yield a criterion as to whether or not the elements of a given totality, for example, the real numbers, constitute a set or an ultraset since the definition of the ultrasets involves the term set. But in investigating the formalism of ultrasets one finds that the ultrasets do not have cardinalities like those of sets – H. Hahn suggested calling their cardinalities *ultranumbers* – and that the well-ordered ultrasets have no ordinal numbers like those of well-ordered sets. This explains the paradox of Burali-Forti and similar paradoxes, which treat the ordinal or cardinal ultra numbers of ultrasets as ordinal or cardinal numbers of sets.

In dealing with ultrasets of ultrasets it is convenient to speak of *classes* for which term one may substitute, for example, the words finite set, denumerable set or set. A totality including for each given class of a certain kind an element not contained in the given class may be called an ultraclass of the kind under consideration. If classes not including themselves as elements are called *normal* one readily sees that the normal classes constitute an ultraclass. Application of this reasoning to ultraclasses of normal ultraclasses necessitates the introduction of ultra-ultraclasses and, more generally, of a sequence of types of classes, where the jumps from a type to the consecutive type are much greater than those between consecutive types in the sense of Russell.

CHAPTER 7

A LOGIC OF THE DOUBTFUL ON OPTATIVE AND IMPERATIVE LOGIC

The paper [1939, 1], (with minor improvements of the notation) is here included despite its obvious shortcomings, since some of the ideas expounded may be at least of historical interest.

1.

The ordinary 2-valued logic divides propositions into two classes, the class of the true and the class of the false propositions. It furthermore combines propositions into more complex ones by means of the particles ′ (not), & (and), → (implies), etc. According to the ordinary calculus of propositions the class to which a compound proposition belongs is uniquely determined by the classes to which the component propositions belong; e.g. the proposition p&q belongs to the class of true propositions if and only if both p and q belong to this class.

The *n*-valued logics of Post and Łukasiewicz start with a division of propositions into n classes, the 3-valued into 3 classes. They share with the ordinary 2-valued logic the above-mentioned feature that the class of the compound is uniquely determined by the classes of the components.

But the logic of every-day life lacks this feature with regard to doubtful propositions. 'Tomorrow it will rain,' 'Tomorrow it will snow,' 'Tomorrow it will not rain' are three doubtful propositions. So is 'Tomorrow it will rain and it will snow.' But 'Tomorrow it will rain and it will not rain' is a false proposition. Thus p&q may be either doubtful or false if both p and q are doubtful. The proposition (p&q)′&(p′&q′)′ (exactly one of p and q) may even be either true or doubtful or false if both p and q are doubtful. It is true if p stands for 'it will rain,' and q for 'it will not rain.' It is doubtful if p stands for 'it will rain,' and q for 'it will snow.' It is false if p stands for 'it will rain,' and q for 'it will rain.'

In the logic of the doubtful we divide the propositions into three mutually exclusive classes of modality: μ_+ consisting of the *asserted*, μ_0 consisting of the *doubtful*, μ_- consisting of the *negated* propositions. If p is a proposition, then we shall denote by $\mu(p)$ the class of modality to which p belongs. In contrast to the traditional 3-valued logic, *the modality of a compound is not determined by the modalities of the components.*

We make the following assumptions:

I. μ_+ *is a deductive system* in the sense of Tarski, that is to say, a system containing each proposition that follows from any set of propositions of the system. In particular, μ_+ contains

$\mathbf{I_1}$. the tautologies of the ordinary 2-valued calculus of propositions, i.e. the compound propositions which remain true no matter which (true or false) propositions we substitute for the components. Besides these tautologies, μ_+ may contain other propositions, expressing, for example, observations or physical laws. Being a deductive system μ_+ contains

$\mathbf{I_2}$. the proposition p&q if it contains p and q,

$\mathbf{I_3}$. the proposition q if it contains p and p $\rightarrow$ q or (p&q')'.

From this it follows that μ_+ contains

$\mathbf{I_4}$. the propositions p, q and q&p if it contains p&q.

$\mathbf{I_5}$. the proposition q' if it contains (p&q)' and p.

$\mathbf{I_6}$. the proposition (p&q)' if it contains p'.

II. *If* p *belongs to* μ_+ *then* p' *belongs to* μ_- *and conversely.*

We shall call two propositions P and Q *tautologically equivalent* if the propositions P $\rightarrow$ Q and Q $\rightarrow$ P are tautologies in the ordinary 2-valued calculus of propositions. (Tautological equivalence which, obviously, can only subsist between two compound propositions, should not be confused with the equivalence of the ordinary 2-valued logic subsisting between any two true as well as between any two false propositions.)

THEOREM 1. *If P and Q are two tautologically equivalent compound connectives made up of the particles* ', &, $\rightarrow$, $\vee$ *etc., then*

$$\mu(P(\mathrm{p},\mathrm{q},\ldots)) = \mu(Q(\mathrm{p},\mathrm{q},\ldots)) \textit{ for any propositions } \mathrm{p},\mathrm{q},\ldots$$

Proof. Let p_0, q_0, ... be particular propositions, and P_0 and Q_0 the propositions obtained by substituting them into *P* and *Q*. First Case: $\mu(\mathrm{P}_0) = \mu_+$. The proposition $\mathrm{P}_0 \rightarrow \mathrm{Q}_0$ being a tautology is likewise an element of μ_+ by $\mathbf{I_1}$. Hence $\mu(\mathrm{Q}_0) = \mu_+$ by $\mathbf{I_3}$. Second Case: $\mu(\mathrm{P}_0) = \mu_-$. Then $\mu(\mathrm{P}'_0) = \mu_+$. By assumption $\mathrm{Q}_0 \rightarrow \mathrm{P}_0$ is a tautology. So is, therefore, $\mathrm{P}'_0 \rightarrow \mathrm{Q}'_0$. Since $\mathrm{P}'_0 \rightarrow \mathrm{Q}'_0$ belongs to μ_+, by $\mathbf{I_3}$ also $\mu(\mathrm{Q}'_0) = \mu_+$, whence $\mu(\mathrm{Q}_0) = \mu_-$ by **II**. Third Case: $\mu(\mathrm{P}_0) = \mu_0$. Then $\mu(\mathrm{Q}_0) = \mu_0$. For by what has just been proved, from $\mu(\mathrm{Q}_0) = \mu_+$ or $\mu(\mathrm{Q}_0) = \mu_-$ it would follow that $\mu(\mathrm{P}_0) = \mu_+$ or $\mu(\mathrm{P}_0) = \mu_-$, respectively.

THEOREM 2. *In the usual notation, we have the following tables of modality*

p	p′
+	−
0	0
−	+

p	q	p&q
+	+	+
+	0	0
0	+	0
−	*	−
*	−	−
0	0	0 or −

Here (and in what follows) $*$ *stands for any of the symbols* $+, 0, -$.
Proof. The table for p′ follows from **II**. In the table for p&q, the first row follows from $\mathbf{I_2}$. Since $+ \;\&\; 0 = +$ is excluded by $\mathbf{I_4}$, and $+ \;\&\; 0 = -$ is excluded by $\mathbf{I_5}$ and **II**, it follows that $+ \;\&\; 0 = 0$ and, by $\mathbf{I_4}$, that $0 \;\&\; + = 0$. By $\mathbf{I_6}$, $- \;\&\; * = -$ whence $* \;\&\; - = -$ by virtue of $\mathbf{I_4}$, while $0 \;\&\; 0 = +$ is excluded by $\mathbf{I_4}$.

THEOREM 3. *There are at most seven classes of pairs of doubtful propositions. If* $\mu(\mathrm{p}) = \mu(\mathrm{q}) = \mu_0$. *then the modalities of* p&q, p′&q′, p′&q, p&q′ *have one of the seven following quadruples of values:*

p&q	p′&q′	p′&q	p&q′	and we say:	and write:
−	0	0	0	p and q are *incompatible*	p*i*q
−	−	0	0	p and q are *contradictory*	p*c*q
0	−	0	0	p and q are *alternative*	p*a*q
0	0	−	0	p is *weaker* than q	p*w*q
0	0	0	−	p is *stronger* than q	p*s*q
0	0	−	−	p and q are *equivalent*	p*e*q
0	0	0	0	p and q are *unrelated*	p*u*q

Proof. By Theorem 2 $\mu(\mathrm{p}') = \mu(\mathrm{q}') = \mu_0$ and $\mu(\mathrm{p\&q}) = \mu_0$ or μ_-. First assume $\mu(\mathrm{p\&q}) = \mu_-$. Then $\mu(\mathrm{p\&q}') \neq \mu_-$. For the equality would imply $\mu(\mathrm{p\&q}')' = \mu_+$. From the assumption it follows that $\mu(\mathrm{p\&q})' = \mu_+$; and $((\mathrm{p\&q})'\&(\mathrm{p\&q}')') \to \mathrm{p}'$ being a tautology likewise belongs to μ_+, whence p′ would belong to μ_+, whereas $\mu(\mathrm{p}') = \mu_0$. Moreover, $\mu(\mathrm{p\&q}') \neq \mu_+$. For from $\mu(\mathrm{p}) = \mu_0$ and $\mu(\mathrm{q}') = \mu_0$ it follows by virtue of Theorem 2 that $\mu(\mathrm{p\&q}') = \mu_0$ or μ_-. Hence $\mu(\mathrm{p\&q}') = \mu_0$. Similarly, $\mu(\mathrm{p}'\&\mathrm{q}) = \mu_0$. As to $\mu(\mathrm{p}'\&\mathrm{q}')$ there are two possibilities: μ_0 and μ_-. They correspond to the first two of the seven cases mentioned above. Next assume $\mu(\mathrm{p\&q}) = \mu_0$. If $\mu(\mathrm{p}'\&\mathrm{q}') = \mu_-$, then an argument similar to the one used under the first assumption shows that the third case of the theorem obtains. If $\mu(p'\&q') = \mu_0$ then, by the same reasoning, either $\mu(p'\&q) = \mu_-$ and the fourth or the sixth case is present, or $\mu(p'\&q) = \mu_0$ and the fifth or the seventh case obtains.

There follow two sets of examples for the seven cases, one from meteorology and one from the game of dice. Let p be the proposition 'it will rain.' Then an incompatible proposition is 'there will be sunshine, but no rain'; a contradictory proposition 'it will not rain'; an alternative proposition 'there will be no heavy rain'; a stronger proposition 'it will rain heavily'; a weaker proposition 'there will be bad weather'; an equivalent proposition 'it will rain'; an unrelated proposition 'it will snow.' If a die is to be thrown once and p stands for the proposition 'either 1 or 2 will turn up,' then '3 will turn up' is incompatible; '3 or 4 or 5 or 6 will turn up' is contradictory; '2 or 3 or 4 or 5 or 6 will turn up' is alternative; '1 will turn up' is stronger; 'either 1 or 2 will turn up' is equivalent; '1 or 2 or 3 will turn up' is weaker; 'either 2 or 3 will turn up' is unrelated to the proposition p.

Of course, there do not exist pairs of each of these seven types in all examples. If we are to toss a coin just once, then we have two doubtful propositions: 'heads will turn up' and 'tails will turn up.' There are only one pair of contradictory, and two pairs of equivalent propositions.

THEOREM 4. *Set* $P = (p\&q)'\&(p'\&q')'$ *and* $Q = (p'\&q)'\&(p\&q')'$. *The propositions* P *and* Q *are both doubtful if exactly one of the propositions* p *and* q *is doubtful. If* p *and* q *are both doubtful, then we have*

	p*i*q	p*c*q	p*a*q	p*w*q	p*e*q	p*s*q	p*u*q
P	0	+	0	0	–	0	0
Q	0	–	0	0	+	0	0

Proof. If p*i*q then $\mu(p\&q) = \mu_-$ and $\mu(p'\&q') = \mu_0$, thus $\mu(p\&q)' = \mu_+$ and $\mu(p'\&q') = \mu_0$. It follows that $\mu(P) = \mu_0$. By a similar argument one shows that $\mu(P) = \mu_+$ if p*c*q, and that $\mu(P) = \mu_0$ if p*a*q. Analogously, $\mu(Q) = \mu_0$, μ_+, μ_0 if p*w*q, p*e*q, p*s*q, respectively, and that $\mu(P) = \mu(Q) = \mu_0$ if p*u*q. Using the fact that P and Q' are tautologically equivalent one determines $\mu(Q)$ in the first three and $\mu(P)$ in the second three of the seven cases.

In the 2-valued logic each compound of two propositions is tautologically equivalent to one of the following 16 propositions: p, q, p&q,p'&q',

p'&q, p&q', P, p&p', and their negations. By virtue of Theorem 1, each compound of two propositions is thus equivalent in the sense of the logic of the doubtful to one of these 16 propositions whose modalities can be determined by means of Theorems 2, 3, 4.

According to M. H. Stone's results, propositions behave like subsets of a set if we identify any two propositions p and q for which (p'&q)'&(p&q')' is asserted (e.g. any two doubtful propositions p and q for which p*e*q). For instance, we may establish the following correspondence:

Asserted proposition – universal set; negated proposition – empty set (any two asserted and any two negated propositions are identified); doubtful proposition – nonempty proper subset; p&q – intersection of two sets; p' – complement of a set.

Then the seven classes of pairs of doubtful propositions correspond to the following seven classes of pairs of nonempty proper subsets:

- p*i*q – sets whose intersection is empty but whose union is not the universal set;
- p*c*q – sets whose intersection is empty and whose union is the universal set;
- p*a*q – sets whose intersection is nonempty and whose union is the universal set;
- p*w*q – the second set is a proper subset of the first;
- p*s*q – the first set is a proper subset of the second;
- p*e*q – sets that are identical;
- p*u*q – sets with a nonempty intersection neither being a subset of the other and whose union is not the universal set.

For illustrative purposes we may thus use plane diagrams in the calculus of propositions as well as in the calculus of classes, all but two sets representing the doubtful propositions.

The preceding statements about the belonging of a proposition to a class of modality have not been considered as propositions. But instead of writing $\mu(p) = \mu_+, \mu_0, \mu_-$ we might write + p, 0p, − p and consider these symbols as denoting propositions. In this case we should have to make assumptions about the iteration of modality symbols. One possibility would be the assumption of a law of the excluded fourth $+(+p \vee 0p \vee -p)$, assumptions concerning the mutual exclusiveness of the modalities

$+(+p \,\&\, 0p)'$, $+(+p \,\&\, -p)'$, $+(0p \,\&\, -p)'$, and assumptions about some categories of the modalities, like $-0{*}p$ ($*$ standing for $+$, 0, $-$).

Moreover one might develop a logic of the doubtful with regard to the quantifiers "all" and "some" of the calculus of propositional functions.

2

An exact theory of wishes and commands is one of the mathematical approaches to ethical problems. An attempt in this direction is a theory of E. Mally[1] which introduces into the calculus of propositions '!p' symbolizing that p *ought to be*, or that the proposition p *ought to be true*. Mally sets up five postulates for !p. They can be written in the following form:

I. $[(p \rightarrow\, !q)\&(q \rightarrow r)] \rightarrow (p \rightarrow\, !r)$;

II. $[(p \rightarrow (!q\&!r)] \rightarrow [p \rightarrow\, !(q\&r)]$;

III. $(p \rightarrow\, !q) \rightleftarrows\, !(p \rightarrow q)$;

IV. !u; that is, there exists a proposition u for which !u holds.

V. (!v)′; that is, there exists a proposition v for which the negation of !v holds. (In fact, Mally's Postulate V is more complicated but may be replaced by the preceding postulate.)

After deriving 50 theorems from these assumptions Mally arrives at the conclusion:

$$p \rightleftarrows\, !p;$$

that is to say, if p is true then p ought to be true, and conversely. On the basis of Mally's assumptions, this result can be made plausible in a few lines.

According to the calculus of propositions, $(!q \rightarrow\, !q)$ is implied by any proposition, $[(!q)' \rightarrow\, !q]$ by !q and $(q \rightarrow p)$ by p. Consequently,

$$(p\&!q) \rightarrow [(!q \rightarrow\, !q)\&((!q)' \rightarrow\, !q)\&(q \rightarrow p)].$$

Applying Postulate I to $(!q \rightarrow\, !q)\&(q \rightarrow p)$, and to $((!q)' \rightarrow\, !q)\&(q \rightarrow p)$ one obtains (by virtue of the before mentioned formula)

(a) $(p\&!q) \rightarrow\, !p$ (Mally's Theorem 7).

According to III, $(q \rightarrow !p) \rightleftarrows !(q \rightarrow p)$ and $!(p' \rightarrow q') \rightleftarrows (p' \rightarrow !q')$. Since $(q \rightarrow p) \rightleftarrows (p' \rightarrow q')$, it follows, if $!q'$ stands for $!(q')$, that

$$\text{(b)} \quad (q \rightarrow !p) \rightleftarrows (p' \rightarrow !q') \text{ (Mally's Theorem 14).}$$

According to the calculus of propositions, $!p \rightarrow [(q \rightarrow !p)\&(q' \rightarrow !p)]$. Because of (b),

$$[(q \rightarrow !p)\&(q' \rightarrow !p)] \rightarrow [(p' \rightarrow !q')\&(p' \rightarrow !q)].$$

Hence $p' \rightarrow (!q\&!q')$ and, by Postulate II, $p' \rightarrow !(q\&q')$. Clearly also $[p' \rightarrow !(q\&q')]\&[(q\&q') \rightarrow !r)]$, for any r, whence by Postulate I, $p' \rightarrow !r$ or, equivalently, $p \vee !r$. Consequently we have $!p \rightarrow (p \vee !r)$ for any r. Since $(!v)'$ by Postulate V, it follows that $!p \rightarrow p$. Since $!u$ by Postulate IV, it follows that $p \rightarrow (p\&!u)$, whence $p \rightarrow !p$ by (a). Hence $p \rightleftarrows !p$.

This result seems to me, however, to be detrimental to Mally's theory. It indicates that the introduction of the sign ! is superfluous in the sense that it may be cancelled or inserted in any formula at any place we please; and this (in spite of Mally's philosophical justification) clearly contradicts not only the common use of the word 'ought' but also some of Mally's own correct remarks about this concept, for example, his comparison of $p \rightarrow (!q \text{ or } !r)$ and $p \rightarrow !(q \text{ or } r)$. Mally is quite right that these two propositions are not equivalent according to the ordinary use of the word "ought". But by virtue of the equivalence of p and !p, they are equivalent in his theory.

3

In what follows I wish to make a few points that seem to be fundamental for the logic of wishes and commands.

1. No normal person wishes that tomorrow it may either rain or not rain, nor that tomorrow it may both rain and not rain. It is likewise unreasonable to command to another man that he should either pay or not pay a certain sum, as well as that he should both pay and not pay a sum. What normal people wish is that there may be fine weather tomorrow, that the dice may fall in their favor and the like, just because it is neither certain nor impossible that the weather will be fine tomorrow or that the dice will fall in their favor. In general one can say: *The objects of human wishes and commands are neither necessities nor impossibilities.* In building up a logic of wishes and commands we therefore shall merely deal with propositions which, in the sense of the logic of the doubtful, are neither

asserted nor negated, that is, with doubtful propositions. One of the reasons for the failure of Mally's interesting attempt is that it was founded on the 2-valued calculus of propositions.

2. In everyday language, *the formal properties of the words "command" and "wish" differ from each other*. If I command p&q, then, in particular, I command p and I command q. But if I wish p&q, then, at least according to one of the ways in which the word "wish" is used, it does not necessarily follow that I wish p. If p and q are what economists call complementary goods (e.g., a cigarette and a match), then it is quite possible that I wish p&q without wishing either one of the goods by itself.

Another way of pointing out differences between the formalisms of wishes and commands is by what may be called *implicational interpretations of attitudes*. The statement 'I command p' – briefly, Cp – may be interpreted as 'Unless p, something unpleasant will happen (e.g., you will be punished).' Denoting by 'a' the statement that the unpleasant thing will happen one may write $Cp \rightleftarrows (p' \rightarrow a)$. Applying the equivalence

$$[(p\&q)' \rightarrow r] \rightleftarrows [(p' \rightarrow r)\&(q' \rightarrow r)]$$

for any three propositions p, q, r

to r = a yields $C(p\&q) \rightleftarrows (Cp\&Cq)$, which expresses the feature of commands mentioned above: If I command p&q, then I command p and I command q, and vice versa.

The statement 'I wish p' – briefly, Dp, where D for desire corresponds to C for command – is frequently interpreted as 'If p, then I shall be glad.' Denoting by 'b' the statement that I shall be glad one can write $Dp \rightleftarrows (p \rightarrow b)$. Applying the implication

$$[(p \rightarrow r)\&(q \rightarrow r)] \rightarrow [(p\&q) \rightarrow r]$$

for any three propositions p, q, r

to r = b yields $(Dp\&Dq) \rightarrow D(p\&q)$, while the converse, which would be the analogue of $C(p\&q) \rightarrow (Cp\&Cq)$ cannot be inferred.

The implicational interpretation of commands and wishes reaffirms the necessity of founding the imperative and optative logics on a logic of the doubtful. Application of the formula

$$[(p' \rightarrow r)\&(p \rightarrow q)] \rightarrow [q' \rightarrow r] \text{ for any three propositions p, q, r}$$

to r = a yields $[Cp\&(p \rightarrow q)] \rightarrow Cq$. On the basis of the 2-valued calculus of propositions this formula would have the following absurd consequences: If I command one true proposition p, then I command every true proposition q (since p → q if both p and q are true), and if I command one false proposition p, then I command every proposition q (since p → q for any q if p is false). The paradox arises from the further assumption that each proposition is either true or false so that if I command any proposition it would have to be either a false or a true proposition, whence every proposition or at least every true proposition would be commanded. On the basis of the logic of the doubtful, however, p, q, a are three doubtful propositions [and only restricted, but more realistic deductions are possible]. Cp means that

(1) p′ → a belongs to the class μ_+ of all asserted propositions.

The same fact might also be expressed by saying that p′&a′ belongs to the class μ_- of all negated propositions. Hence, in the terminology of the first part of this paper, the two doubtful propositions p and a are either alternative or contradictory. Now assume furthermore that p → q. Then

(2) (p → q) belongs to μ_+.

(3) [(p → q) → (q′ → p′)] being a tautology belongs to μ_+.

Since μ_+ is a deductive system it follows from (2) and (3) that

(4) (q′ → p′) belongs to μ_+;

and from (1) and (4), that

(5) [(q′ → p′)&(p′ → a)] belongs to μ_+.

(6) [(q′ → p′)&(p′ → a)] → (q′ → a) being a tautology belongs to μ_+.

Since μ_+ is a deductive system it follows from (5) and (6) that (q′ → a) belongs to μ_+, that is, that Cq belongs to μ_+.

If the implicational interpretation of the statement Cp is adopted, then the system of all commanded propositions forms what may be called a *restricted deductive system*, that is, a system which contains each *doubtful* proposition that is a consequence of any set of propositions of the system (without containing the asserted propositions that may follow, as for instance tautologies, since no asserted proposition can be the object of a command).

3. *The expression "I wish" of the everyday language is incomplete* and therefore ambiguous. The statement "I wish a dog" fails to make clear whether I wish a dog under all conditions (e.g. even if I get a cat) or merely

under certain conditions (e.g. provided that I shall not get a cat). If Dp means desire under all conditions, then clearly $(Dp\&Dq) \rightarrow D(p\&q)$. If Dp means desire under certain conditions only, then this implication obviously does not generally hold. On the other hand, if Dp means desire under all conditions then the implication $D(p\&q) \rightarrow Dp$ does not always hold. If I want to smoke I wish a cigarette and a match; but this does not imply that I wish a cigarette under all conditions (e.g. without a match). According to another justifiable interpretation of Dp, the implication $D(p\&q) \rightarrow Dp$ may always hold. The same criticism applies to the statement $p \rightarrow b$ (if p, then I shall be glad), mentioned above as a possible interpretation of Dp.

4. A theory of wishes should take into account some results of economic theory, in particular the following two:

(1) The distinction of certain classes of pairs of goods, e.g., *complementary goods* wanted only in combination (like a cigarette and a match) or *goods that can be substituted for each other* and each of which is wanted without, possibly even only without, the other (like a dog and a cat as pets). Similar distinctions hold for triples and n-tuples of goods.

(2) For each individual, $\mathscr{I}$, the combinations of goods constitute a set that is partially ordered by a preference relation such that absence of preference is transitive; that is, if M and N are two combinations of goods, then either N is preferred to M, or M is preferred to N, or neither is preferred to the other. In the last case (the absence of preference) we say that $\mathscr{I}$ is indifferent to M or N. If these three cases be expressed by $M < N$, $N < M$, $M \parallel N$, then not only the asymmetric $<$-relation of being preferred but also the symmetric $\parallel$-relation of being indifferent is transitive. For if $\mathscr{I}$ is indifferent to M and N as well as to L and M, then $\mathscr{I}$ is in general also indifferent to L and N. Moreover, in general, $L < M$ and $M \parallel N$ imply $L < N$.

In conclusion it may be briefly indicated how these remarks apply to the simplest cases of optative logic. Analyzing concrete wishes one finds that few of them, if any, are desires under all conditions. Even the most elementary wishes presuppose at least normal conditions in some respects (e.g. that no major earthquake or hurricane interferes). While the number of conditions is practically unlimited, a theory must start with simplifying assumptions; for instance, with the restriction to definite finite sets of doubtful propositions (as the objects of wishes) and of conditions.

The following assumptions cover the simplest case, that of one proposition, p, and its negation, denoted by p':

I. If Dp, then p belongs to μ_0. (Only doubtful propositions are objects of $\mathscr{I}$'s wishes.)

II. Dp does not belong to μ_0. (There shall be no doubt as to whether or not $\mathscr{I}$ wishes p.)

III. (Dp&Dp′) belongs to μ_-. ($\mathscr{I}$ does not both wish p and wish p′ – a law of optative consistency.)

Writing D′p for (Dp)′, and D′p′ for (Dp′)′ one immediately derives from these assumptions that one of the following three possibilities holds for each doubtful proposition p:

(1) Dp&D′p′ – briefly, Dp ($\mathscr{I}$ wishes p).
(2) D′p&Dp′ – briefly, Dp′ or, synonymously, Ep ($\mathscr{I}$ wishes p′).
(3) D′p&D′p′ – briefly, Ip or, synonymously, Ip′.

The fourth possibility, Dp&Dp′, is ruled out by the law of optative consistency. The three cases correspond to $p' < p$, $p < p'$ and $p \parallel p'$, respectively.

The next case is that of one condition, q, affecting $\mathscr{I}$'s attitude toward p. We distinguish the attitudes Ap(q) and Ap(q′) toward p under the conditions q and q′, respectively. Here, A stands for one of the three symbols D, E, I. Altogether there are nine possibilities, each of which is characterized by a pair of attitudes. We obtain examples of these cases if a die is to be thrown twice and we consider the following conditions:

q: six turns up the first time,

q′: a number other than six turns up the first time;

and the following objects of $\mathscr{I}$'s wishes:

p: six turns up the second time,

p′: a number other than six turns up the second time.

Then $\mathscr{I}$'s attitudes toward p and p′ will be

Ap(q),	Ap(q′) and	Ap′(q),	Ap′(q′)	if $\mathscr{I}$ bets on
D	I	E	I	two sixes
I	E	I	D	no six
D	E	E	D	two sixes or no six
D	D	E	E	a six on the 2nd throw
I	I	I	I	a six on the 1st throw.

The attitudes Dx(y), Ex(y), Ix(y) may be interpreted in terms of preferences between the pairs (x, y) and (x′, y), namely respectively by

$$(x', y) < (x, y), \quad (x, y) < (x', y), \quad (x, y) \parallel (x', y).$$

The preceding table expresses (in improved form) an example mentioned in the original paper; what follows is an abbreviated and corrected version of the final paragraphs.

We finally consider two possible objects of desires, p and q (with their negations, p′ and q′), such that either one affects the optative attitude toward the other. As before, Ax(y) denotes the attitude toward x under the condition y (where A is D or E or I, while A* will denote E, D or I, respectively, and A** = A). Two plausible, though not cogent assumptions, which might be weakened, read:

I. If Ax(y) then Ay(x);

II. If Ax(y) then A*x′(y).

There are eight attitudes which can be cyclically arranged:

$$\begin{array}{cccc} & A_1p(q) & A_2q(p) & \\ A_8p'(q) & & & A_3q'(p) \\ A_7q(p') & & & A_4p(q') \\ & A_6q'(p') & A_5p'(q') & \end{array}$$

From Assumption **I** it follows that $A_{i+1} = A_i$ for i = 1, 3, 5, 7;
From Assumption **II** it follows that $A_{i+1} = A_i^*$ for i = 2, 4, 6, 8.
Jointly, the assumptions imply that

$$A_1 = A_2 = A_5 = A_6 = A_3^* = A_4^* = A_7^* = A_8^*.$$

Complementary objects (such as a cigarette and a match) illustrate the case $A_1 = D$; incompatible objects that can substitute for one another (such as a dog and a cat) as pets correspond to the case where $A_1 = E$.

Attitudes can again be interpreted in terms of preferences between (unordered) pairs; for example, Dp(q) by

$$(p', q) < (p, q), (p, q') < (p, q), (p', q) < (p', q'), (p, q') < (p', q').$$

NOTE

1 *Grundgesetze des Sollens*. Graz, 1926.

PART III

FUNDAMENTAL CONCEPTS IN PURE AND APPLIED MATHEMATICS

Nonmathematicians are advised to begin the reading of Part III with Chapter 10.

CHAPTER 8

A COUNTERPART OF OCCAM'S RAZOR

This chapter combines two papers on a counterpart of the famous methodological tool developed by Occam in the 14th century: [1960, 6], on the use of the counterpart in ontology, and [1961, 1], on its semantical applications. By combining the two papers repetitions have been avoided and the introductions merged. Other changes are minor.

INTRODUCTION

Occam's celebrated razor is the methodological maxim that

entities must not be multiplied beyond necessity.

Among the most famous sentences that call for the application of this law is a response of M. Argan, the imaginary invalid, to his examiners in Molière's satire upon 17th century medicine. When asked *quare opium facit dormire* (why opium causes sleep), the candidate answers *quia est in eo virtus dormitiva cujus est natura sensus assoupire* (because there is in it a soporific force whose nature it is to lull the senses). Occam's razor, which is primarily concerned with ontology, would shave off *virtus* and *natura* as superfluous entities. In a more general form, often called the *Law of Parsimony*, the principle states that

it is vain to do with more what can be done with less.

This law may also be construed as a maxim opposing synonyms.

Thanks to Occam and Molière, superfluous entities no longer present a danger. Science and mathematics rarely consider unnecessary entities. But the past three centuries have witnessed the development of the opposite menace. In many cases, too few entities are being considered. Necessary entities have been suppressed or allowed to merge in misconceptions that are insufficient to account for the variety of ideas. Perhaps the most important example is the medley of ideas called *variables* to which for the past 200 years mathematicians and scientists have reduced many totally different types of entities. To mention only two such types: algebraic symbols such as x and c in the assertions

$x + 1 = 1 + x$, where x stands for any number,

$\log 2c = \log 2 + \log c$, where c stands for any positive number, and physical concepts such as time, position, velocity, which Newton called *fluents*. Actually, fluents constitute an all-important category *sui generis*, which as such has been neglected – in fact ignored – in the traditional literature. This serious lacuna in the ontology of applied mathematics can only be filled by recognizing that countless scientific concepts including time, position and velocity constitute a category that is *toto caelo* different from the category (or categories) constituted by algebraic symbols.

Thus what is needed is a counterpart of Occam's law – a law stipulating that

entities must not be suppressed below sufficiency.

This principle used as a methodological tool would separate compounds of various constituents passing for homogeneous into their diverse components. It might be called a *prism* resolving conceptual mixtures into the spectrum of their meanings or, if one wishes to remain in the tonsorial domain of the razor, a *comb* disentangling and straightening out the various threads of thought.[1] In a more general form – as it were, as a *Law against Miserliness* – the principle might be stated:

it is vain to try to do with less what requires more.

This law may also be construed as a semantic maxim opposing equivocations. And while synonyms are bad, equivocations are worse.

Among the mathematical sentences that call for a semantic application of the Law against Miserliness, there is even an exact counterpart of the imaginary invalid's answer. M. Argan speaks of three "entities" connected with opium

why it causes sleep, *its soporific force*,

and *the lulling nature of that force*,

in order to "explain" one single meaningful predicate of the drug, namely,

opium puts people to sleep.

The mathematical sentence

The function x assumes the value x for any number x,

makes use of a single symbol, namely,

$$x$$

to "denote" three totally unlike ideas:

a function, a value of that function

and *any argument of that function.*

These and other serious inadequacies are the results of withholding proper symbols from entities of paramount importance, in particular, of denying a symbol to the *identity function*, which in the preceding mathematical sentence is misdenoted by x. But such withholding is not the virtue of parsimony; it is the vice of miserliness. Semantic violations of the Law against Miserliness cause gaps in the mathematical symbolism just as its ontological violations cause lacunae in the ontology of mathematics.

The resolution of equivocal uses of terms into various nuances of meaning is, of course, no novelty. This method has led to the clarification and improvement of technical expositions in many a branch of knowledge. But the material that will be resolved by the methodological prism on the following pages is absolutely fundamental and not only of technical nature, and it includes conceptual combinations of such incongruity that the analysis will resolve more than mere nuances.

As a general rule, equivocations are the less dangerous the more unlike the diverse significations and the greater the differences in meaning of one and the same term or symbol. No one has, for example, ever confounded the totally unlike mathematical ideas that have the name 'tangent' – the tangent of a curve and the trigonometric tangent of an angle. Mathematico-scientific methodology, however, presents odd exceptions to this general rule. Time, position and velocity are as different from the algebraic symbols x and c mentioned before as the tangent of a curve is from the tangent of an angle; and yet for centuries they have been confounded under the name of variables. Nor is the identity function akin to a number any more than the tangent line is to the trigonometric tangent number; and yet in the sentence that is a counterpart of Molière's, both are denoted by the same symbol, the letter x.

In ontology as well as in semantics, the counterpart of Occam's principle has as extended and as ramified a use as had the medieval principle. In fact, just as the razor characterized currents in the late scholasticism and the post-scholastic period, so the prism or comb is representative of tendencies in modern philosophy.

Part I of the present paper is devoted to ontology, Part II to semantics. The methodological tool will be referred to as *prism* in Part I, and as *comb* in Part II. The gaps will be filled with parsimony but without miserliness.

PART I. THE PRISM IN ONTOLOGY

1. The Notion of Variable

By a variable logicians mean a symbol that stands for any element of a certain class, which I will call the *scope* of the variable. The first articulate definition of certain variables seems to have been given by the mathematician Weierstrass. He defined a number variable, that is, a variable whose scope is a class of numbers.[2]

Variables in the sense of the logicians are hardly ever used outside pure mathematics, while they are ubiquitous in this field. There is a sound though purely practical reason for this state of affairs. It is that most nonmathematical assertions can be well understood without variables, whereas complicated mathematical assertions in purely verbal form are almost incomprehensible. The *basic* situation is as follows:

(a) *Variables may be used in nonmathematical* (*as well as mathematical*) *assertions.* Consider, for example, the proposition:

> The (paternal) grandfather of the father of any person is the father of the grandfather of that person.

This assertion could be expressed in terms of the abbreviations g and f, and by the use of a variable x – one might say by the use of a *person variable* – as follows:

$$g(f(x)) = f(g(x)) \text{ for any person } x.$$

(b) *Variables may be dispensed with in mathematical* (*as well as in nonmathematical*) *assertions.* Not a single variable occurs in Alkhowarizmi's *Al Jebr*, the famous book whose title is at the root of the word algebra. Nor are purely verbal assertions by any means inferior to formulations in terms of variables. Compare, for example, the assertion:

> The cube of the square of any number
> equals the square of the cube of that number

with the formulations

$$cu(sq(x)) = sq(cu(x)) \text{ or } (x^2)^3 = (x^3)^2 \text{ for any number } x.$$

The latter are shorter but not otherwise superior. (Throughout this paper, 'number' stands for '*real* number'.)

But if there is no basic difference, with regard to variables in the sense of logic, between mathematical and nonmathematical assertions, why has Russell[3] called variables "perhaps the most distinctly mathematical of all notions"? What probably prompted Russell's dictum is the fact that pure mathematics and especially applied mathematics and science overstep the use of variables in the sense of logic and abound in totally discrepant uses. In the traditional literature, 'variable' is used as a reference to x, y, s, and t in each of the following examples including six variations on the idea of something being the square of something else:

(1) *Comprehensive assertions about numbers*, such as $x^2/x = x$ for any number $x \neq 0$; $(x^2)^2 = x^4$ for any number x; or, for any two numbers x and y, if $y = x^2$, then $y^2 = x^4$. And *descriptions of classes of pairs of numbers*; for example, the class of all pairs (x, x^2) for any number x; or of all pairs (x, y) such that $y = x^2$.

(2) *Assertions about functions*; for example, that the function x^2 (or, as some mathematicians say, the function $y = x^2$) is nonconstant.

(3) *Physical laws*, such as Galileo's Law $s = \frac{1}{2}gt^2$ or, in proper units, $s = t^2$.

(4) *Discussions in the analytic geometry of curves*, such as the parabola $y = x^2$.

(5) *Laws of analysis*; for example, if $y = x^2$, then $dy/dx = 2x$.

(6) *Algebraic formulae about polynomials* such as $x^2 = x \cdot x$ or $x^2/x = x$.

These examples, among which only (1) will be seen to include number variables in the sense of the logicians, also clarify why, in the continuation of the passage quoted above, Russell has called variable "certainly one of the most difficult notions to understand". A full understanding presupposes that the melange illustrated in Examples (1) to (6) be resolved in the prism.

2. *The Resolution of Variables in the Prism*

The prism separates each two of the six listed types of variables.

I. *The Separation of* (2) *from* (1). Each of the three assertions in (1) synthesizes countless specific assertions such as $3^2/3 = 3$, $e^2/e = e$ (where e designates the base of the natural logarithms); $(0^2)^2 = 0^4$, $(3^2)^2 = 3^4$, $(e^2)^2 = e^4$; and if $0 = 0^2$, then $0^2 = 0^4$, and if $9 = 3^2$, then $9^2 = 3^4$. Similarly, the definitions in (1) combine countless specific pairs of numbers such as $(0, 0^2)$, $(3, 3^2)$, (e, e^2). The specific assertions or pairs of numbers are obtained by replacing the variables with designations of elements of their scopes. On the other hand, the statement (2) that the function x^2 is nonconstant is not the synthesis of specific assertions such as the function e^2 is nonconstant or the function $9 = 3^2$ is nonconstant. These assertions are not even true. What (2) does assert is a property of a specific entity – the entity obtained by pairing the number x^2 to any number x, for example 3^2 to 3. The most concise definition of that entity is as the *result* of the said pairing; the function x^2 is the class of all ordered pairs (x, x^2) for any number x. (This, incidentally, is the example given in (1) of a definition of a class in terms of a number variable.)

II. *The Separation of* (3) *and* (4) *from* (1). In contrast to the number variables in (1), Galileo's s and t and Descartes' coordinates designate specific entities. One symptom of this profound contrast is the fact that, without changing the meaning of (1), one may replace the number variables with other letters as in

$$(c^2)^2 = c^4 \quad \text{for any number } c,$$

or even interchange them as in

(1′) the class of all pairs (y, x) such that $x = y^2$.

In Galileo's law, on the other hand, s and t must not be interchanged. It is not true that $t = s^2$. Nor is the parabola $x = y^2$ identical with the parabola $y = x^2$.

But, one might ask, does not Galileo's law imply that $t = s^2$ *provided one lets 's' denote the time, and 't' the distance*? Indeed it does. But, similarly, $9 = 3^2$ implies that $3 = 9^2$ *provided one lets* '3' *denote the number of the Muses and* '9' *the number of the Graces*. Yet no one will, on this ground, call 3 and 9 number variables. The point is that the definitions (1) and (1′) are tantamount *without anything being renamed*, since, in contrast to the letters in Galileo's law (3), the number variables in (1) do not name or designate anything specific, to begin with.

That the letters in (3) cannot be interpreted as number variables (whose scopes consist of observed distance and time values) appears also from an attempt to use them as such, as in

(3′) $s = t^2$ for any two values s and t of distance and time.

Obviously (3′) is false[4] except in insignificant cases of miniature scopes, for example, if the scope of t consists only of two numbers, 3 and -3, and that of s includes only 9.

III. *The Separation of* (5) *from* (1). The prism separates the concepts in (5) from the number variables in (1). Indeed, if the letters in (5) are replaced with '3' and '9', the result is the sentence,

$$\text{if } 9 = 3^2, \text{ then } \mathrm{d}9/\mathrm{d}3 = 2 \cdot 3.$$

But this implication is false since the antecedent is true, whereas the consequent, being nonsensical, is not.

IV. *The Separation of* (5) *from* (2), (3), *and* (4). While the letters in assertion (5) are not number variables, they are variables of some kind. The law (5) synthesizes specific assertions that can be obtained by replacing the letters, e.g., with designations of Galileo's s and t or Descartes' x and y, the results being

$$\text{if } s = t^2, \text{ then } \mathrm{d}s/\mathrm{d}t = 2t \qquad \text{and} \qquad \text{if } y = x^2, \text{ then } \mathrm{d}y/\mathrm{d}x = 2x$$

Thus the concepts in (5) have been separated from Galileo's and Descartes'. The latter are, so to speak, special cases of the concepts (5). Being variables of some kind, the concepts (5) are also separated from the specific function in (2).

V. *The Separation of* (6) *from* (1) *and* (5). In contrast to the variables in (1) and (5), the polynomials x and x^2 in the algebraic assertions (6) are not meant to be replaced with designations of either numbers or entities related by Galileo's laws.

VI. *The Separation of* (2) *from* (6). The function in (2) is separated from the polynomials in (6), since the latter are not classes of pairs of numbers. Dividing the polynomial x^2 by the polynomial x yields the polynomial x, just as $3^2/3 = 3$ and $e^2/e = e$. The function x^2, on the other hand, divided by the function x yields only a restriction of the function x, namely, the class of all pairs (x, x) for any number $x \neq 0$, and not the full function x, since $x^2/x = x$ only for $x \neq 0$.

VII. *The Separation of* (3) *from* (2) *and* (6). Galileo's s and t refer to the physical universe, whereas the assertions (2) and (6) belong to pure mathematics.

Only (4) has not yet been separated from all the other examples. This separation, too, will be achieved, but only after the so-called variables in (4) themselves have been treated with the prism and resolved into two discrepant types. The separation of all the other types of so-called variables has been completed.

In the medley of entities indiscriminately called variables, the prism thus separates mathematical and nonmathematical entities; it separates classes and elements; it separates designations of entities and symbols not designating anything specific; and it separates particular and general ideas. It should be difficult to find among objects of intellectual studies another group that is both as prominent and as ill-assorted as the lot of so-called variables.

3. Filling two Lacunae in the Traditional, neo-Pythagorean Ontology. The Definition of Fluents and Fluent Variables

The prism has revealed that variables in the sense of something common to the examples (1) to (6) represent suppression of ideas to the point of inadequacy. What this means is much more, and much more serious, than the mere terminological shortcoming of equivocations – of references to unlike categories by the same term. It appears that several categories that are absolutely basic for mathematical science are not being articulately defined and studied, as such, at all – a defect that in the Introduction has been somewhat loosely described as lacunae in the traditional ontology. These gaps must be filled by definitions of those suppressed categories rather than covered up by equivocal uses of the term variable. The first step in this direction is the ascertainment of the exact nature of Descartes' coordinates x and y and of Galileo's s and t.

In 1637, Descartes introduced coordinates in a physical plane, such as a sheet of paper or a blackboard, where points are pencil marks or chalk dots. Choose a frame of reference consisting of three noncollinear points, say, O, X, and Y. For any point P in the plane (See the figure where O, X and Y have been chosen so as to form a right isosceles triangle) let P_X denote the projection of P on the X-axis (i.e., the line joining O and X).

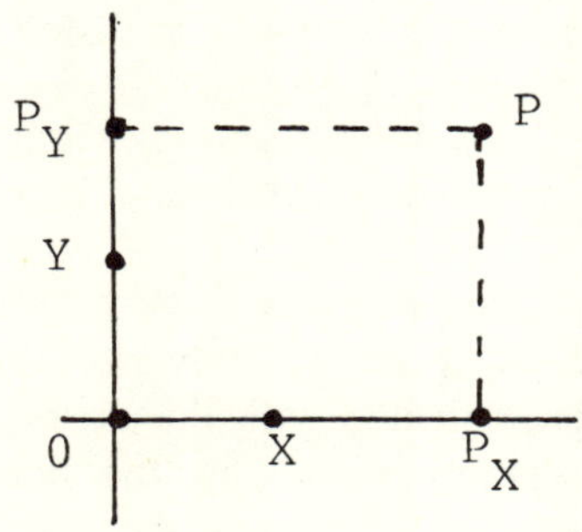

Measure the distance from O to P_X using the segment from O to X as unit and let OP_X be the number thus obtained. Define P_Y and the number OP_Y analogously. Then, in a crucial passage of *La Géométrie*, Descartes says verbatim[5]: *pourceque* OP_X *et* OP_Y *sont deux quantités indeterminées & inconnuës, ie les nomme l'vne x et l'autre y.* Descartes here takes it for granted that special letters are reserved for special purposes – an idea that goes back to the great French algebraist Vieta[6], who used capitalized vowels as unknowns. In algebra, Descartes replaced Vieta's vowels by the last letters of the lower case alphabet and, accordingly, introduced also in geometry 'x' and 'y' for his so-called *quantités indeterminées et inconnuës*.

What with Descartes' epochal contributions to mathematics and what with his celebrated *clarté*, few tasks are as difficult as propounding that one basic aspect of mathematics and science has been obfuscated by his legacy. But it actually is the quoted passage in *La Géométrie* to which some defects of the traditional mathematico-scientific ontology can be traced.

To begin with, Descartes' x and y are not really indeterminate and unknown quantities. As actual examples of unknowns, consider the letters in the equations

$$\text{find all numbers } x \text{ such that } 1 - x^2 = 0;$$

$$\text{find all numbers } x \text{ and } y \text{ such that } x + y = 10 \text{ and } x - y = 2.$$

These letters are tentative designations of numbers satisfying the conditions expressed in the formulae, and the word *find* stipulates that the reader produce numerals designating those numbers without reference to those conditions.

The situation in geometry is basically different. A physical plane with a frame of reference is given. Before a point P has been chosen in that plane, there are no unknown numbers x and y; and after P has been chosen, Descartes' procedure determines two numbers, called the abscissa of P and the ordinate of P, which I will denote by the symbols $x(\mathrm{P})$ and $y(\mathrm{P})$, respectively. For example, the numbers associated with the afore-mentioned points O and X are:

$$x(\mathrm{O}) = 0,\ y(\mathrm{O}) = 0 \text{ and } x(\mathrm{X}) = 1,\ y(\mathrm{X}) = 0.$$

Even someone who knows Descartes' procedure for determining the numbers $x(\mathrm{P})$ and $y(\mathrm{P})$ for any point P still may ask "What is x?" or "What is the abscissa?" His situation is comparable to that of someone who knows the procedure for determining the number $\log x$ for any positive number x but who still may ask "What is log?" or "What is the logarithmic function?" He may be told that the logarithmic function is the result of pairing, to each number x, the said number $\log x$; in other words, that it is the class of all pairs $(x, \log x)$ for any number $x > 0$. Similarly, the question concerning the abscissa may be answered by saying that it is the result of pairing, à la Descartes, the number $x(\mathrm{P})$ to each point P; in other words, it is the class of all pairs $(\mathrm{P}, x(\mathrm{P}))$ for any point P of the plane. An analogous definition is possible for the ordinate y.

Galileo's *t* likewise is the result of pairing a number *t*(T) with each act T of reading a timer that was started when the object under observation was dropped; and *s* is the class of all pairs (S, *s*(S)) for any act S of reading the mark opposite to the falling object on a vertical rod.[7]

As a collective reference to Galileo's *t* and *s*, Descartes' *x* and *y*, and the like, I have revived a term coined by Newton and used by him in reference to the said objects studied in physics and geometry, namely, the term *fluents*. Newton, the supreme virtuoso in operating with fluents, sketched his general idea of their nature in the sentence: *fluentes vocabo quantitates has quas considero tamquam gradatim et indefinité crescentes*[8], but he did not define specific fluents such as *s* or *t* at all. Two centuries later, the term fluent was all but forgotten, even in the British literature. Of course, abandoning Newton's term and replacing it with another word would have been, in itself, harmless; but 'fluent' was almost universally replaced with 'variable'[9] – a particular terminological change that created the danger of confusion. And confusion has actually developed. Fluents such as time and gas pressure have been explicitly identified with *number variables* by some mathematicians[10] and with *classes of numbers* by others.[11] These are procedures of the kind denounced in the Introduction. They are attempts to do with fewer what requires more. Accordingly, these identifications leave a serious gap in the mathematico-scientific ontology.

Such procedures seem to indicate an implicit overestimation of the scientific role of numbers by themselves – something of a neo-Pythagorean philosophy. Mathematicians and scientists often take numbers as quantities. But a quantity is not a mere number. A *quantity* is a number paired to what I will call the *object* of the quantity – something extra-mathematical such as a physical point, a falling object, an observer's act of reading a clock, or even a class of observables.[12] The number paired to the object is called the *value* of the quantity.

For two quantities to be *equal* it is necessary (and of course sufficient) that, besides equal values, they have the same object. For instance, if Peter is the oldest man in his village and is 172 cm tall, then

(Peter, his height in cm) and (the oldest man in the village, 172)

are equal quantities, whereas

(a certain tree, its height in cm),

even if that height is 172, is a different quantity. If two quantities are either equal or have different objects, then I will call them *consistent*. Thus two quantities are *inconsistent* if they have the same object and unequal values.

In this sense, each of Galileo's and Descartes' fluents t, s, x, and y is a *class of consistent quantities*. The class of all extra-mathematical objects of the quantities belonging to a fluent will be called the *domain* of the fluent. The class of all values of those quantities is referred to as the *range* of the fluent. While the range, being a class of numbers, is a concept of pure mathematics, the domain is not.

After the fluents t and s, connected with a moving particle, have been defined separately, they may be interrelated. In the case of a falling particle, they are connected by the law $s = t^2$. In discovering this law Galileo did not primarily pair the ranges of s and t or, more precisely, values of s to values of t. Pairing numbers to numbers is the activity of a mathematician defining a function and not that of a physicist discovering a law of nature. What Galileo actually paired were the domains of s and t or, more precisely, acts of mark reading to acts of clock reading. Specifically, he paired *simultaneous* acts. And he discovered that the corresponding values of the fluents satisfied the condition:

$$s(\mathrm{S}) = (t(\mathrm{T}))^2 \text{ for any two simultaneous acts S and T.}$$

This is what the law $s = t^2$ expresses.

The traditional, neo-Pythagorean ontology ignores objects of quantities, domains of fluents, and those pairings of the domains which are the prerequisite for the connection of fluents by functions.

The theory here outlined fills still another gap in the traditional ontology. It clarifies the role played in (5) by the letters x and y, which definitely are not number variables but appear to be variables of some kind (See IV in Section 2). Since they may be replaced with designations of Galileo's and Descartes' s, t, x and y, each such replacement resulting in an assertion about specific fluents, the letters in (5) are *fluent variables*. The traditional theory fails to distinguish this important category from either number variables or specific fluents by indiscriminately applying the term variable to all three of them.

4. *The Prism and Analytic Geometry*

As far as analytic geometry is concerned, the prism not only analyzes ideas. It separates two altogether unlike theories: the analytic treatment of *physical planes*, where points are ink marks or chalk dots, and the analytic geometry of the *arithmetical plane*, where points are ordered pairs of numbers.

There is no doubt as to which situation Descartes had in mind by dint of his frequent references to compasses and linkages as well as pictorial illustrations in *La Géométrie* of those tools which produce rows of ink marks or chalk dots but certainly not classes of pairs of numbers. What Descartes had in mind when speaking of the parabola $y = x^2$ clearly was a class of points in a physical plane in which a frame of reference had been chosen, namely, the class of all points P such that $y(P) = (x(P))^2$. It was Russell in the course of arithmetizing geometry and Study [13] in clarifying the foundations of nonpostulational geometry who developed the analytic geometry of the arithmetical plane as an algebraic theory of numerically defined objects.

There is – at least on the macroscopic level – a correspondence between the latter objects and the observable points in a physical plane. This correspondence is of course not coincidental but purposely achieved by formulating arithmetical definitions that reflect physical facts. The analogue of the points O, X, and Y constituting Descartes' frame of reference are the points $(0,0)$, $(1,0)$ and $(0,1)$ respectively of the arithmetical plane, (even though the latter points can hardly be called a frame of reference in the arithmetical plane). The analogue of the parabola consisting of all points P such that $y(P) = (x(P))^2$ is the class of all arithmetical points (x,y) such that $y = x^2$. This class, therefore, is called a parabola in the arithmetical plane. In fact, the much-discussed *isomorphism* between the physical model and the arithmetical construct appears to be so perfect that there seems to be no point in separating the two.

Actually, however, dangers lurk below the surface. The prism separates the formula $y = x^2$ in Descartes' description of a physical parabola from the apparently identical formula in the definition of the parabola in the arithmetical plane as the class of all pairs (x,y) such that $y = x^2$. The latter x and y are number variables, whereas Descartes' x and y are fluents. Not only have the two pairs of symbols altogether unlike meanings but they follow incompatible rules. For instance, the variables x and y may be interchanged; the fluents x and y must not be interchanged. A mere juxtaposition of the two results manifests the

DOUBLE-THINK IN ANALYTIC GEOMETRY. *In Descartes' plane,* $y = x^2$ *and* $x = y^2$ *are different parabolae; in the arithmetical plane, the parabolae consisting of all points* (x, y) *such that* $y = x^2$ *and of all points* (y, x) *such that* $x = y^2$ *are identical.*

The best way to avoid paradoxes in analytic geometry will be by keeping apart what the prism has separated: number variables in the theory of the arithmetical plane, and fluents in Descartes' physical plane – concepts that are *toto caelo* different.

5. *Further uses of the prism*

The use of the methodological prism is by no means confined to the analysis of the notion of variables. By using this tool one can, for example, separate the derivative of a function from the rate of change of one fluent with regard to another fluent[14] as well as the integral of a function from the cumulation of one fluent with regard to another fluent.

The prism, moreover, analyzes the mixture of ideas traditionally called *random variables*. It separates Kolmogoroff's purely mathematical concept of a measurable function in whose *domain* a probability measure is given – a concept of basic importance for the theory of probability – from the scientific concept of a random fluent with an extramathematical domain and a frequency distribution or probability measure in its *range* –a concept that seems to be the main object of statistical studies.[15]

PART II. THE COMB IN SEMANTICS

In the Introduction, a comb has been mentioned as a tool for separating tangled ideas in mathematico-scientific semantics. In Part II six *Rules of Dissociation* will be formulated as results of the use of that comb in pure and applied mathematics. They will be supplemented by some positive stipulations, designed to fill the lacunae created by disentangling the misassociations in the traditional symbolism – positive *Rules of Supplementation.*

6. *Paradoxes in Analytic Geometry*

Some of the situations described in the Introduction, even though their obnoxious character seems to have remained unnoticed for centuries, are outright paradoxical. In fact, a semantic paradox arises from the equations of curves in the arithmetically defined plane of analytic geometry. As an example, consider again certain parabolae and the conjunction of the following three propositions.

(α): *The parabola* $y = x^2$ *is the class of all pairs of numbers* (x, y) *such that* $y = x^2$.

(β): *The class of all pairs of numbers* (x, y) *such that* $y = x^2$ *is identical with the class of all pairs of numbers* (y, x) *such that* $x = y^2$.

(γ): *The parabola* $x = y^2$ *is different from the parabola* $y = x^2$.

Proposition (α) may be considered as a definition. The definiendum is the parabola $y = x^2$. According to (β), the letters x and y may be interchanged in the definiens, whereas, according to (γ), these same letters must not be interchanged in the definiendum. It is clear that something is seriously wrong.

This paradox is related to, but not identical with, the double-think in analytic geometry discussed in Section 4. There, the difficulties arose from confusing the fluents in Descartes' treatment of the physical plane with the number variables used in the theory of the plane defined as the set of all pairs of numbers. Section 6 deals exclusively with this arithmetical plane[16].

Discussion of the paradox in the conjunction of (α), (β), (γ) should be preceded by a warning against two approaches that may seem to lead to a solution of the paradox but actually don't.

One might say: "Throughout analytic geometry, it is tacitly understood that the letter x refers to the first coordinate or abscissa, and the letter y to the second or ordinate. In particular, this is true in (α) and (γ). The paradox arises because statement (β) violates this rule." But the said rule, unless it be supplemented by a complete remodeling of the traditional use of number variables, is untenable. How could one adhering to that rule express that the distance between the points (x, y) and (y, x) is $\sqrt{2}|x - y|$? Here, x can be kept out of the place of the ordinate only if x is also avoided as a reference to the abscissa. The rule thus would imply that in certain situations (which one could not easily characterize!) x and y must not be used as number variables at all.

Again, one might say: "A well-known rule of logic forbids the replacement of variables bound by a quantifier with letters that occur in the domain of the quantifier as free variables. In violation of this rule, in (β) the variables x and y are replaced with y and x, which also occur as free variables in the definiendum. The paradox thus is due to the confusion of bound variables in the definiens and free variables in the definiendum." It must be retorted that first of all, whatever the nature of x and y in the definiendum in (α) may be, certainly these letters are not within the domain of the quantifiers occurring in the definiens, while statement (β), where x and y are interchanged, includes only bound variables. Moreover, in Section 9 it will become apparent that the letters in the definiendum in (α) are neither bound nor free variables but no variables at all.

A valid solution of the paradox must take its start from a clarification of the meaning of the definiens in (α). For this purpose I will resort to a much maligned tool – the common language. In some theoretical respects, common language is definitely superior to the traditional mathematical symbolism created during the Renaissance. *Until that symbolism has been reformed, the use of common language in clarifying certain paradoxical situations cannot be recommended too strongly.*

A possible (if not optimal) verbal description of the definiens is the phrase

> (δ): *The class of all pairs consisting of any number and any second number such that the second is the equal to the square of the first.*

The Renaissance symbolism is based on the use of letters that refer to numbers but, in contrast to numerals, are nondesignative[17]. In order to translate (δ) into the traditional symbolism, put any nondesignative letter in the place of the words 'any number' and 'the first (number)'; put any unlike nondesignative letter in place of the words 'any second number' and 'the second (number)'; and append an explanation as to what those letters stand for. Since the choice of the letters is arbitrary, the one verbal phrase (δ) has many (mutually synonymous) symbolic translations. The sentence (β) expresses the synonymity of two translations that may be denoted by ($\delta_{x,y}$) and ($\delta_{y,x}$). In a context where the first letters of the alphabet are nondesignative, (δ) may also be translated into

> ($\delta_{a,b}$): *The class of all pairs* (a, b), *where a and b stand for any numbers such that* $b = a^2$.

The synonymity of ($\delta_{a,b}$) and ($\delta_{x,y}$) in conjunction with (α) would seem to imply the equality of the parabolas $y = x^2$ and $b = a^2$. The words 'the parabola $b = a^2$', however, are incomprehensible. This variant of the paradox has two features not possessed by the first formulation. It has an analogue even for curves whose Cartesian equations are symmetric in x and y; for example,

(α'): *The curve* $x^2 + y^2 = 1$ *is the class of all* (x, y) *such that* $x^2 + y^2 = 1$.

(β'): *The class of all* (x, y) *such that* $x^2 + y^2 = 1$ *is identical with the class of all* (a, b) *such that* $a^2 + b^2 = 1$.

(γ'): *There is no curve* $a^2 + b^2 = 1$.

Secondly, this formulation of the paradox makes it perfectly clear that the letters x and y, which are nondesignative in the definiens, designate something in the definiendum – something that a and b do not designate – and this will explain the paradox.

In order to complete the clarification of the definiens one may describe the use of the letters a and b in the phrase $(\delta_{a,b})$ as follows: replacing them in the symbol (a, b) with numerals (such as 3 and 9) that make the formula $b = a^2$ true yields the designation of a pair (such as $(3, 9)$) belonging to the class defined by $(\delta_{a,b})$.

A nondesignative reference to any element of a class $\mathcal{S}$ is called a *variable*; more specifically, we shall call it a variable with the *scope* $\mathcal{S}$. If $\mathcal{S}$ consists of numbers, as it does in the case of a and b in $(\delta_{a,b})$ or of x and y in $(\delta_{x,y})$ or $(\delta_{y,x})$, then we speak of *number variables* (See Section 1). Number variables are also used in comprehensive assertions such as

(ε_x): $x + 1 = 1 + x$, *where* x *stands for any positive number*,

which is a translation of the verbal sentence

(ε): *any positive number increased by* 1 *equals* 1 *increased by that number.*

Replacement of x in the formal part of (ε_x) by the letter '3' or 'e' yields formulas asserted by the proponent of (ε_x). (He does not assert that $-2 + 1 = 1 + -2$, even though this is true.)

While verbal formulations are self-explanatory, *a formula including a nondesignative letter is a meaningless string of symbols, which may become meaningful only when amplified by a legend explaining what that letter stands for* – a point that cannot be overemphasized, especially since in the traditional literature such amplifications are often omitted.[18]

Incidentally, as a corollary of the use of a variable in the translation of (ε), one immediately derives a rule of dissociation – a rule so trivial that at first glance it hardly seems to warrant being formulated.

RULE I: DISSOCIATING DESIGNATIONS AND VARIABLES. *In a given context, the same letter should not serve both as the designation of an individual and as a variable – least of all, a variable standing for any element of a class including that individual.*

Indeed no one would violate Rule I by writing the assertion (ε) in the form

> (ε_e): $e + 1 = 1 + e$, *where e stands for any number, especially for the base e of the natural logarithms.*

But in some cases, Rule I, notwithstanding its primitive character, has been violated (See Section 11).

The clarification of the *definiendum* in (α), the parabola $y = x^2$, requires further preparation. As indicated in (γ) and (γ'), the letters x and y cannot be interchanged or replaced with other letters without changing the meaning of the phrase, while replacing them with numerals yields phrases such as 'the parabola $9 = 3^2$' or 'the parabola $5 = 3^2$', each of which is nonsensical though each includes a meaningful (true or false) formula. No such formula, however, is in any way comprehended as a particular case in the phrase 'the parabola $y = x^2$'. Thus *the letters x and y in this phrase lack all the characteristics of variables.* Moreover, it has been established above that, in contrast to variables, they designate something. The full elucidation of the paradox presupposes of course an answer to the question: *What is it that x and y in the definiendum of* (α) *designate*? Obviously, further and more penetrating rules of dissociation are indispensable before this question can be answered in Section 9.

7. *The Dissociation of Functions and Numbers*

The style of the verbal phrase (δ) can be improved by writing

> (δ'): *The class of all pairs consisting of any number followed by the square of that number.*

Translating (δ') term by term, just as (δ) has been translated into ($\delta_{a,b}$) and ($\delta_{x,y}$), one obtains the following improvement over the latter translations

> (δ_x): *The class of all pairs* (x, x^2), *where x stands for any number.*

Replacing the number variable with numerals yields designations of elements of the class (δ') such as $(3, 3^2)$ and $(\mathrm{e}, \mathrm{e}^2)$.

The class (δ') or (δ), which in pure analytical geometry is called a parabola, may also be regarded as the result of pairing x^2 to x, where x stands for any number. Since the process of pairing exactly one number to any number of a certain kind is known as *defining a function*, the result of the above procedure, i.e., the class (δ') of all pairs obtained, is a *function*. (The number x^2 paired with x is also called the *value* of that function for x). Hence the definiens in (α) is a function.

Another example of the process of defining a function is the pairing of the number 1 to each number. The result is

(ζ)*: The class of all pairs of numbers whose second members are* 1

or in terms of a variable,

(ζ_x)*: The class of all pairs* (x, 1), *where x stands for any number.*

This class is often called 'the constant function of value 1'; but it is also referred to as 'the function 1'. Yet example (ζ) as well as (δ) shows that a function and a number are basically unlike – comparable, as I have said elsewhere,[19] to a party of married couples and a spinster. Designating them by the same symbol, as one does in writing that the function 1 assumes the value 1 for any number, is utterly inappropriate and calls for

RULE II: DISSOCIATING SYMBOLS FOR FUNCTIONS AND FOR NUMBERS. *In a given context, the same symbol must not be used as a reference both to functions and to numbers. In particular, symbols for constant functions must be dissimilar from the numerals designating the values of those functions.*

One way to implement this rule is by a

TYPOGRAPHICAL CONVENTION. *Symbols referring to functions and symbols referring to numbers ought to be printed in different types.*[20]

For instance – and this is the rule adhered to (except for quotations from the traditional literature) in the remainder of this paper – functions may be referred to in *italics* while references to numbers (numerals and number variables) such as e, x, y, a, and b will be in roman type. Thus, in the remainder of this paper, the number variables in symbolic translations of verbal definitions or assertions such as (δ), (δ'), (ε), and (ζ) will be in roman type. On the other hand, the symbols for the definienda in (α), (δ), (ζ) will be in italics.

In particular, the lack of traditional symbols for constant functions suggests the following

FIRST RULE OF SUPPLEMENTATION. *Symbols for constant functions must be created dissimilar from the numerals designating the values of these functions.*

One possibility is the use of numerals in italic type, as in the sentences

> *the function 1 assumes the value* 1 *for any number* or *1*(x) = 1, *where* x *stands for any number.*

(The typographical convention makes it unnecessary to use parentheses in designating values of functions, as in *1*(x) or *f*(x). One can unequivocally write *1* x, *f* x and *log* x just as one traditionally writes log x and not log (x). In the sequel, the superfluous parentheses will sometimes be retained.)

But it is difficult to distinguish numerals in italic type and ordinary numerals, especially in writing. Even more important, in order to express certain properties of constant functions, especially their parallelism to numbers, it is necessary to introduce symbols such as c_1, c_2, c_b, . . . for the constant functions with the values 1, 2, b, . . . (Here c is in italic type, 1, 2, b are not; in fact, b is a number variable; and c by itself may be used as a variable that stands for any constant function – *a constant-function variable.*) Only by the use of symbols of this kind[21] can one express, for example, that

$$c_2 + c_3 = c_{2+3} = c_5,$$

whereas italic numerals merely permit one to write *2* + *3* = *5*. (It is a general rule that certain analogies cannot be expressed by the mere use of dissimilar types for the designation of analogous elements.)

8. *The Identity Function*

Classical analysis lacks a symbol for what may be called *the identity function*, which pairs any number to itself, that is, (η): *The class of all pairs* (x, x), *where* x *stands for any number*. Traditionally (η) is called 'the function x' is violation of both rules of dissociation so far discussed. The same letter designates an individual and is used as a variable; and it refers to a function as well as to a value of that function – a number. This is an example of what in the Introduction has been called *reduction of symbols to the point of inadequacy, caused by vain attempts to do with less what requires more.* And it is this example that leads to the sentence that is a counterpart of the Imaginary Invalid's response (See p. 106). The obvious remedy is suggested in the

SECOND RULE OF SUPPLEMENTATION. *The lacuna caused by the lack of a traditional symbol for the identity function must be filled by creating such a symbol.*

I will denote the identity function by j, its value for x, by j (x) or jx. Then the traditional rigmarole of the function x assuming the value x for any number x may be replaced by the following meaningful sentence:

> *j is the function whose value* j(x) *is equal to* x, *or* j x = x, *where* x *stands for any number.*

Then, furthermore, one may denote the class (δ'_x), which traditionally is called 'the function x^{2}'', by j^2. Then, finally, the logarithmic and tangent functions, for which symbols have always existed, may at long last be referred to by those symbols as *log* and *tan*, whereas in the traditional literature they, too, have been designated by their values for x as 'the functions log x and tan x' – clearly for the sake of uniformity with the identity function, which was referred to, because no symbol for it existed, as 'the function x'. But a phrase such as 'the function log x' for the function that possesses the symbol *log* is a truly singular oddity since it is, conceptually, a *pars pro toto* and at the same time, semantically, a *totum pro parte*. It is *elliptic as to the content* and, at the same time, *hyperbolic in form*.

Of great importance for mathematics and science is the class of all numbers for which two functions assume equal values. Let $\{j^2 = c_1\}$ denote the class of all numbers x such that $\{j^2(\mathrm{x}) = c_1(\mathrm{x})\}$ or $\mathrm{x}^2 = 1$. Clearly, $\{j^2 = c_1\}$ is the class consisting of the numbers 1 and -1. Similarly, $\{f = c_0\}$ is the class of all solutions of the problem: Find all numbers x such that $f(\mathrm{x}) = 0$. Here, f stands for any function and thus serves as a *function variable*. Replacing f in $\{f = c_0\}$ with the designation of the function (δ') one obtains the class $\{j^2 = c_0\}$, which consists of the single number 0.

9. The dissociation of functions with unequal place numbers. The solution of the paradoxes of analytic geometry

By pairing exactly one number to each element of a certain class of ordered pairs of numbers one defines what I will call a *2-place function*,[22] in contrast to 1-place functions such as *log*, *tan*, c_1, j, and j^2. More generally, the result of pairing a number to each element of a class of ordered n-tuples of numbers is called an n-*place function*, where n stands for any positive integer.

In the traditional literature, a 2-place function is referred to by its value for the pair (x, y). For example, one speaks of 'the function $x^3 + xy^2 + y^3$'. Thus the function whose value for (x, y) is x is called 'the function x', as is the totally unlike identity function j – a situation that calls for

RULE III: DISSOCIATING FUNCTIONS OF UNEQUAL PLACE NUMBERS. *The same symbol should not refer both to an* n-*place and an* m-*place function if* m *and* n *stand for unequal integers.* (If numbers be considered as 0-place functions, then Rule III includes Rule II.)

One way to implement Rule III is by indicating n-place functions for $n > 1$ by superscripts, e.g., by denoting the constant n-place function of value 1 by $c_1^{(n)}$ or $C_1^{(n)}$. Here, where I consider only the cases $n = 1$ and 2, I will denote nonconstant 2-place functions by italic capital letters and 1-place functions by lower case italics. A verbal description of the 2-place function traditionally called "x" reads

> (θ): *The class of all pairs whose first member is itself a pair of numbers and whose second member is the first member of the said pair.*

Translating (θ) in terms of the variables x and y one obtains

> ($\theta_{x,y}$): *The class of all pairs* ((x, y), x), *where* x *and* y *stand for any numbers,*

or

> ($\theta_{y,x}$): *The class of all pairs* ((y, x), y), *where* x *and* y *stand for any numbers.*

In contexts where the first letters of the alphabet are nondesignative, ($\theta_{a,b}$) and ($\theta_{b,a}$) are further synonymous descriptions. Being the function that selects the first element from each pair of numbers, (θ) may be called the *first 2-place selector*. Similarly, the *second 2-place selector*, traditionally called 'the function *y*' is

($\iota_{x,y}$): *The class of all pairs* ((x, y), y), *where* x *and* y *stand for any numbers.*

Rules I and III necessitate the following

THIRD RULE OF SUPPLEMENTATION. *The selectors ought to receive permanent designations.*

For instance, the m-th n-place selector might be denoted by $I^{(n)}_{m}$. (Clearly, $I^{(1)}_{1} = j$.) In what follows, only the 2-place selectors will be studied and denoted by *I* and *J*. Then

$I(x, y) = x$ *and* $J(x, y) = y$, *where* x *and* y *stand for any numbers.*

In accordance with what has been said at the end of the preceding section, $\{J = I^2\}$ is the class of all pairs of numbers (x, y) such that $J(x, y) = (I(x, y))^2$; in words, the class of all pairs (x, y) such that $y = x^2$. But this is the definiens in (α).

This finally clarifies what in the geometry of the arithmetical plane is meant by the letters *x* and *y* in the phrase 'the parabola $y = x^2$', the definiendum in (α). That, far from being variables, these letters designate something – this has already been established. But now it has also become clear what it is that they designate. Since the definiens in (α) is the class $\{J = I^2\}$, the letters *x* and *y* designate the selectors *I* and *J*, respectively. This explains why they are not interchangeable. $\{I = J^2\}$ differs from $\{J = I^2\}$. This clarification removes the paradox mentioned in Section 6. Consistency is restored by replacing the three inconsistent propositions (α), (β), (γ) with

(α*): $\{J = I^2\}$ *is the class of all numbers* (x, y) *such that* $y = x^2$; $\{I = J^2\}$ *is the class of all* (x, y) *such that* $x = y^2$.

(β*) = (β): *The class of all pairs* (x, y) *such that* $y = x^2$ *is identical with the class of all pairs* (y, x) *such that* $x = y^2$.

(γ*): *The classes* $\{J = I^2\}$ *and* $\{I = J^2\}$ *are different.*

10. The dissociation of fluents and variables

In contrast to a point in the arithmetical plane, which is defined as an ordered pair of numbers, a point in a physical plane (such as the page on which these words are printed) is a mark of printer's ink, e.g., a dot over i or j. Descartes' analytic treatment of a physical plane has been summarized in Section 3. In a plane with a Cartesian frame of reference, x is the fluent defined by

(κ): *The class of all pairs* (P, x(P)), *where* P *stands for any point in the physical plane.*

Descartes proved that

(λ): *The class of all physical points* P *such that* $y(\mathrm{P}) = (x(\mathrm{P}))^2$ *is a parabola.*

Extending the symbolism introduced in Sections 8 and 9 one can denote the class (λ) by $\{y = x^2\}$.

Other examples of fluents are the radius r and the area a in the formula $a = \pi r^2$ about circles. Here, r is the class of all pairs (C, rC) and the formula is an abbreviation of $a(\mathrm{C}) = \pi(r(\mathrm{C}))^2$, where C stands for any circle.

In view of the use (often in one and the same sentence!) of the letters x, y, a, r, t, s, . . . both as designations of fluents and as number variables standing for any value of the respective fluents, one of the most penetrating rules of dissociation is

RULE IV: DISSOCIATING FLUENTS AND NUMBERS. *In a given context, the same symbol should not serve as a reference both to fluents and to numbers.*

In what precedes, fluents have been designated by lower case italics (also used for 1-place functions), while numbers have been referred to in roman type, but there are of course other ways of implementing Rule IV. Naturally, it is unimportant *how* references to fluents and to numbers are dissociated; but *some* distinction must be maintained if articulateness and clarity are to prevail in the procedures of applied mathematics. In the traditional literature, on the contrary, fluents not only have been denoted by the same *symbols* that stand for any of their values but they have been explicitly referred to by the *same name* as is used for letters standing for any element of a class. The abscissa, the radius, the time and similar scientific entities, each as definite as definite can be, have been called variables – a *lucus a non lucendo.*

11. The dissociation of fluents and fluent variables

The traditional designation of specific functions and fluents (selectors and coordinates) by the letters x and y that also serve as variables violates Rule I. In what precedes, only their use as *number variables* has been discussed. Vague views on fluents in combination with the addiction of post-Cartesian mathematics to the letters x and y, however, have also led to their use as *fluent variables*, as in

$(\mu_{x,y})$: *If* $y = x^2$, *then* $\mathrm{d}y/\mathrm{d}x = 2x$ *where x and y stand for any two fluents, in particular, for the abscissa x and the ordinate y along the parabola* $\{y = x^2\}$.

$(\mu_{x,y})$ is analogous to the universally shunned formulation (ε_e) of the comprehensive assertion (ε) about numbers. According to Rule I of dissociation, *the italic letters x and y, being permanent designations of Descartes' fluents, cannot serve as fluent variables.* In a context where u and w are nondesignative $(\mu_{x,y})$ may be replaced by $(\mu_{u,w})$ or $(\mu_{w,u})$. The above-described semantic confusion has an ontological counterpart: Besides number variables and specific fluents, also fluent variables are referred to, indiscriminately, as variables.

12. The dissociation of functions and fluents

Both fluents and functions are the result of pairing numbers to elements of a certain class. Yet the prism in its ontological use, establishes a fundamental difference between a fluent and a function – a difference whose basic character certainly cannot escape a philosopher of science. A fluent pairs numbers with extramathematical elements whereas a function pairs numbers with numbers or groups of numbers.[23] Thus, while *log*, *cos*, and j^2 are purely *mathematical* concepts, the coordinates x and y in a physical plane, the time t, the distance s, the velocity v, and the temperature are *extramathematical* notions. Accordingly, functions possess connective powers that fluents lack. It cannot be repeated too often that there is a logarithm of the temperature whereas there is no temperature of the logarithm.

But there are also occasions for the semantic use of the comb in separating symbols for fluents and functions. The traditional literature abounds in formulas of the type

$$(v): \quad v = v(t) = v(s).$$

Formula (v) is supposed to indicate that the velocity, say, of a falling body is some function of the time and some function of the distance travelled.

RULE V: DISSOCIATING FLUENTS AND FUNCTIONS. *In a given context, the same letter should not refer both to fluents and to functions, least of all, to a fluent and to a function connecting that very fluent with other fluents.*

For example, for time in seconds, distance in feet, and velocity in ft./sec.,

$$v = 32j(t) = 8\sqrt{s}$$

or, if one prefers not to specify those functions,

$$v = f(t) = g(s) \text{ for some functions } f \text{ and } g.$$

Another possible implementation of Rule V is by the following

FOURTH RULE OF SUPPLEMENTATION. *If the fluent w is a function of the fluent u, the function connecting w with u may be denoted by* w_u *or* $w//u$.

In the preceding example of the velocity,

$$v_t = 32j \text{ and } v_s = 8\sqrt{j} \text{ or } 8j^{\frac{1}{2}}.$$

This symbolism is particularly convenient for designating functions connecting a fluent with several others, for example, for denoting by $E_{v,p}$ and $E_{v,t}$ the functions connecting the energy of an ideal gas with its volume and pressure and with its volume and temperature, respectively.

Misassociations of fluents and functions result in ontological as well as semantic confusions between the following pairs of concepts:[24]

the derivative of a 1-place function;	the rate of change of a fluent with regard to another fluent;
a partial derivative of a 2-place function;	the rate of change of a fluent with regard to another fluent keeping a third fluent constant;
the integral of a function;	the cumulation of a fluent with regard to another fluent.

13. The Dissociation of Indeterminates and Variables. The completion of the x-oclastic Reform

Sums of scalar multiples of power functions, $j^0 = l = c_1$, $j^1 = j, j^2, j^3, \ldots$ are called *polynomial functions*. Examples of polynomial functions include $j + l, j - l$, and $j^2 - l$, traditionally referred to as $x + 1$, $x - 1$, and $x^2 - 1$, respectively.

According to the definition of the product of functions, one has

$$(\pi) \quad j^2 - l = (j + l) \cdot (j - l)$$

$$\text{or } j^2 - c_1 = (j + c_1) \cdot (j - c_1).$$

There is a corresponding comprehensive assertion about real numbers:

$$(\pi_{\mathrm{x}}) \quad \mathrm{x}^2 - 1 = (\mathrm{x} + 1)(\mathrm{x} - 1), \text{ where x stands for any number.}$$

On the other hand, the comprehensive assertion

$$(\rho_{\mathrm{x}}) \quad \frac{\mathrm{x}^2 - 1}{\mathrm{x} - 1} = \mathrm{x} + 1, \text{ where x stands for any number}$$

and the corresponding formula

$$(\rho) \quad \frac{j^2 - l}{j - l} = j + l$$

are false since, after replacement of x with 1, the left side in the formula (ρ_{x}) is meaningless and not equal to $1 + 1$. All that can be asserted is

$$(\rho_{\mathrm{x} \neq 1}) \quad \frac{\mathrm{x}^2 - 1}{\mathrm{x} - 1} = \mathrm{x} + 1, \text{ where x stands for any number } \neq 1$$

and

$$(\rho_{\neq 1}) \quad \frac{j^2 - l}{j - l} = j + l \text{ on the class of all numbers } \neq 1.$$

The function on the right side is a restriction (or subclass) of $j + l$, namely the class of all pairs $(\mathrm{x}, \mathrm{x} + 1)$ where x stands for any number $\neq 1$.

Just as jurisprudents recognize artificial or juridical persons (corporations clubs, and the like) that have many if not all rights and obligations of human persons, algebraists introduce and operate with artificial numbers that have many if not all properties of the ordinary, real numbers. The most important of those artificial numbers are called *polynomials*. They are in a one-to-one correspondence to polynomial functions. Conceptually, however, a polynomial must be clearly distinguished from the corresponding function. A polynomial function results from associating values to numbers, thus is a class of pairs of numbers, and as such has restrictions or subclasses. A polynomial does not associate values with numbers; it is not a class and hence does not possess restrictions or subclasses. It is an individual *sui generis*, best described as an artificial number. For the past decades, outstanding algebraists have avoided the conceptual, if not always the symbolic, confusion of polynomials and polynomial functions – some of them, by referring to the polynomial functions as *polynomials*, and to polynomials as *polynomial forms*. The polynomial corresponding to $j^2 - l$ may be completely characterized by the system of the 'coefficients' 1 and -1 and their ranks within the artificial number, e.g. by the symbol $1^{[2]} - 1^{[0]}$. Similarly, $1^{[1]}$ and $1^{[0]}$ might denote the polynomials corresponding to the functions j and l. Symbols of this kind were actually introduced by Chuquet (and adopted by Kepler and others) in the 16th century. In the 17th century, however, with the rise of what must be called x-omania, these symbols were superseded by $x^2 - 1$, x, and 1, respectively – the same symbols that came into use for the functions $j^2 - l$, j, and l as well as for the numerical values of those functions for x. This use of x in polynomials has lingered on in large parts of the algebraic literature; the only hint as to the different role that x plays in symbols for polynomials is that, x in this role is sometimes (but not consistently) called an *indeterminate*.

RULE VI: DISSOCIATING INDETERMINATES FROM FUNCTIONS AND NUMBER VARIABLES. *In a given context, neither the designation of a function nor a letter serving as a number variable must be used as the indeterminate in polynomials (if such a symbol be introduced at all – as Chuquet's notation demonstrates, symbols for indeterminates are altogether dispensable).*

FIFTH RULE OF SUPPLEMENTATION. *A permanent symbol should be introduced for the indeterminate in polynomials.*

In what follows, α will be the indeterminate; in polynomials with more than one indeterminate other Greek letters could be used.

The definitions of addition and multiplication in the realm of polynomials are inspired by the corresponding laws for polynomial functions which, in turn, reflect comprehensive assertions about real numbers in the same way as formula (π_x). For instance,

$$(\pi^*) \quad \alpha^2 - \alpha^0 = (\alpha^1 + \alpha^0)\cdot(\alpha^1 - \alpha^0).$$

Here, in order to distinguish Chuquet's $1^{[0]}$ from both the constant function l and the number 1, I designate that polynomial by α^0. No explanatory legend in (π^*) is required or even possible. In (π^*), three individual artificial numbers are connected just as are three real numbers in

$$(\pi_e) \quad e^2 - 1 = (e + 1)(e - 1).$$

And just as (π_e) implies

$$(\rho_e) \quad \frac{e^2 - 1}{e - 1} = e + 1,$$

(π^*) implies

$$(\rho^*) \quad \frac{\alpha^2 - \alpha^0}{\alpha^1 - \alpha^0} = \alpha^1 + a^0,$$

whereas (ρ_x) and (ρ) are false, and only the restricted assertions $(\rho_{x \neq 1})$ and $(\rho_{\neq 1})$ are valid. The formulas (ρ_e) and (ρ^*), on the other hand, cannot be restricted nor is there any need to qualify them, since e and α are neither variables nor designations of classes of pairs of numbers. All this is completely jumbled, however, where the same letter x, is used as indeterminate, as number variable, and as designation of j.

Linear polynomials with multiplication subjected to the condition $1^{[2]} = -1^{[0]}$ are called *ordinary complex numbers*. The said condition partly determines the indeterminate which, in this case, traditionally is denoted by i – one of the few cases where neither x nor y is used in reference to an algebraic or analytic concept. Here, I will write i (in roman type). The said condition reads: $i^2 = -1 + 0i$. But those letters x and y have also penetrated into the theory of complex numbers and functions, spreading even there the ambiguities by which their use throughout post-Renaissance mathematics is characterized. In one and the same context (often in one sentence!), they serve both as (real) number variables and as designations of two real-valued functions having the class of all complex numbers as their domains – the functions assuming for any number x + yi the values x and y, respectively. In complex analysis, the so-called function

x thus is the class of all pairs (x + yi, x), where x and y stand for any two real numbers. Clearly, if analysts took this definition seriously (which fortunately they don't), they would run into paradoxes similar to the one in analytic geometry caused by assertion (α) in Section 6.

14. Conclusion

When the need for conceptual clarification and for unequivocal symbols leads to the acceptance of philosophically well-founded rules of dissociation, mathematics – from the beginners' textbooks on algebra to the most sophisticated monographs on pure and applied analysis – will somewhat change in appearance. But this price is not high for the result: the philosophical soundness of the assertions; the comprehensibility of all formulae; the endowment of every single symbol with significance; the typographical visibility of shades of meaning; the automatic nature of substitutions leading from the general to the specific; the analogy of this automatism to the process of machines; the elimination of guessing tacit meanings and connotations from its prominent if unadmitted role in algebra, analysis and analytic geometry; the possibility of taking all definitions and assertions seriously and all formulae literally, just as a machine takes its instructions literally; and the articulateness of all procedures in applying mathematics to scientific material.

Molière's satire shows that, at least in theoretical medicine, superfluous entities were still rampant 300 years after the contrivance of Occam's razor, designed to shave them off. In this faster-moving age it is perhaps not presumptuous to hope that pure and applied mathematics will move at a faster pace in utilizing the counterpart of Occam's razor – a methodological tool designed to resolve the conceptual and symbolic equivocations inherited from the Renaissance and to fill the resulting lacunae.

NOTES

1 The term 'Menger's Comb' was proposed by Dr. Abbott of the Boston Hospital in a discussion following a report about the author's counterpart of Occam's Razor in W. S. McCulloch's Colloquium, at the Massachusetts Institute of Technology in 1961.

2 Cf. Pincherle, *Giorenale di mat.*, 18 (1880), p. 243.

3 *Principles of Mathematics*, Cambridge 1903, p. 5 and p. 89.

4 Ever since the 18th century, mathematicians have tried to reconcile the interpretation of s and t as variables with the fact that, for all significant scopes of the symbols, (3′) is false. Their attempts have centered on the idea of so-called *dependent variables*. In (3′),

only t is regarded as replaceable with any value of the time (and called an independent variable) while s is said to be *dependent* since, after a value of t has been chosen, the value of s is determined. But a variable (i.e. a symbol that may be replaced with the designation of any element of its scope) which is dependent (i.e., cannot be replaced with the designation of every element of its scope) is a plain contradiction in terms.

[5] On p 321 of the 1627 edition. I am retaining Descartes' spelling, but use different letters for the points.

[6] *In artem analyticam isagoge*, Paris 1591.

[7] I have stressed the difference between these concepts and number variables in the Preface to [1952, 1] and have elaborated the theory in Chapter VII of [1953, 2] and [1955, 4] as well as in several papers.

Carnap defined what he called physical quantities as associations of numbers to quadruples of numbers (thus as 4-place functions) in *The Logical Syntax of Language*, London and New York, 1937, pp. 149 ff., and as number variables in "Foundations of Logic and Mathematics", *Encyclopedia of Unified Science*, vol. I, Chicago, 1939. I therefore am glad to see that later he expressed views more similar to those stressed in my quoted publications. Cf. p. 10 of his *Einführung in die symbolische Logik*, Wien 1954 and *Introduction to Symbolic Logic*, New York, 1958, especially p. 168 ff.

[8] *Methodus fluxionum et serierum infinitarum*, 1737. Apparently influenced by Descartes, Newton continues: *Hasque representabo per ultimas alphabeti litteras u, x, y, et z ut discerni possint ab aliis quantitatibus quae in equationibus considerantur tamquam cognitae et determinatae*.

[9] One may compare Newton's quoted description of fluents with de la Vallée Poussin's description of a variable as *une quantité qui passe par une infinité des valeurs distinctes ou non* in his *Cours d'Analyse*, vol. 1. It should be observed that those descriptions were not elaborated either by rigorous explicit definitions or by postulates defining the concept implicitly.

[10] Cf., e.g., Courant, *Differential and Integral Calculus*, vol. 1, p. 16.

[11] Cf. McKinsey, Sugar, and Suppes, "Axiomatic foundations of classical particle mechanics", *Journ. Rat. Mech. Anal.*, 2. 1953, pp. 253–272, and Artin, *Calculus and Analytical Geometry*, 1957, p. 70.

[12] What laws, if any, govern the association of the value of a quantity with its object; whether the association must be relative to some frame of reference; and if so, what changes of the value correspond to changes of the frame – these questions will be discussed elsewhere. But whatever the answer to these questions may be, strong emphasis must be laid on the (usually neglected) *objects* of quantities, which remain permanent in the said changes. Weyl's elaborate definition of a quantity in *The Classical Groups*, Princeton 1939, pp. 16 ff. does not seem to include a clear reference to an object.

[13] Study, *Die realistische Weltansicht und die Lehre vom Raum*, Vieweg, Braunschweig 1914.

[14] Cf. [1958, 2].

[15] Cf. [1955, 1].

[16] The paradox discussed in Section 6 and the double-think mentioned in Section 4 are related because the same formula, $y = x^2$, designates both a parabola in the physical plane by connecting two geometrical fluents and, as an abbreviation, a class of pairs of numbers in the arithmetical plane.

[17] Besides *permanent* numerals, such as '3' 'π', and 'e' each of which always designates the same number, there are numerals *ad hoc*. For example, in a proof of the irrationality of the number $\sqrt{2} + \sqrt{3} + \sqrt{5}$, in order to avoid numerous anticipated references to that number by the cumbersome numeral '$\sqrt{2} + \sqrt{3} + \sqrt{5}$', a mathematician may start out by saying: 'Set $\sqrt{2} + \sqrt{3} + \sqrt{5} = \mathrm{b}$.' Thereafter, he reserves 'b' for this purpose until he has completed the proof, at which point the letter becomes again available for other purposes (Cf. [1957, 1). Each numeral designates a number. A *nondesignative* symbol referring to numbers came first into use in the late Alexandrian period of Greek mathematics; but no ancient mathematician seems to have used more than *one* such symbol. In the Middle Ages, Alkhowarizmi and his successors altogether avoided symbols confining themselves to purely verbal formulations. Nondesignative letters were systematically introduced into algebra during the Renaissance.

[18] By itself, a formula including a nondesignative letter does not even indicate whether an assertion or a problem is intended, which from the point of view of communication and information makes a great difference. Amplified by the imperative legend 'Find, if possible, a number x such that', the formula $x + 1 = 1 + x$, which occurs in the comprehensive assertion (ϵ_x), becomes a task – a problem that has any number as a solution – while its proponent does not make any assertion. Similarly, the problem 'Find all numbers, if any, such that $x^2 - 1 = 0$' must not be confused with the comprehensive of modest assertion that $x^2 - 1 = 0$ where x stands for either 1 or -1. To a nondesignative letter in a problem I refer as an *unknown* rather than as a variable (Cf. [1957, 1]). If one wishes to attribute to a letter one and the same meaning in a row of formulas, then it is of course permissible to explain the meaning in a common preamble. But the tendency to omit such legends altogether has resulted in a sort of *formula fetishism*.

[19] Cf. [1955, 4], Chapter IV.

[20] In the traditional literature, all single-letter symbols are indiscriminately printed in italics (the number e, the number variable x, the coordinates x and y, the function f, and so on), while all multi-letter symbols are in roman type (log, cos, and so on). Thus an excellent means for visual differentiation between unlike categories is wasted on a total irrelevancy.

[21] Another symbolism of this kind is obtained by introducing italic *1* (and possibly *0*) and denoting other constant functions by scalar products, as in $c_2 = 2\,\mathit{1}$, $c_\mathrm{b} = \mathrm{b}\,\mathit{1}$. Then one can write $2\,\mathit{1} + 3\,\mathit{1} = (2 + 3)\,\mathit{1} = 5\,\mathit{1}$. In longhand, italicization may be indicated by underscoring.

[22] I avoid the traditional phrases 'function of two variables' or 'function of one variable' since functions, as here considered, are self-sufficient entities – classes of pairs of numbers – and not functions of anything.

[23] In an interesting paper "A Classification of Mathematical Concepts" (*Synthese*, **14**, 1962, p. 78), Dr. Josephine Mehlberg demonstrated that fluents and functions are concepts of different type in the sense of Russell's type theory. In this book, both fluents and one-place functions are designated by lower case italics. Another way, which is particularly practical where references to functions of any number of places are in lower case italics with upper indices indicating their place numbers, is the designation of fluents by capital italics, as in

$$Y = X^2 \quad \text{and} \quad A = R^2 \quad \text{or} \quad Y = j^2(X) \quad \text{and} \quad A = j^2(R)\,.$$

[24] Cf. [1958, 2].

CHAPTER 9

A THEORY OF THE APPLICATION OF THE FUNCTION CONCEPT TO SCIENCE

This chapter contains the second half of [1970, 2] – a paper presented at a symposium that the American Association for the Advancement of Science arranged in 1966 on the occasion of the fiftieth anniversary of the death of Ernst Mach. It is here reprinted with a few minor revisions. (For the first half, See Chapter 20.)

THE CLARIFICATION OF FUNCTIONAL CONNECTIONS

In dealing with the application of a branch of mathematics to science the methodologist has to clarify three points: the mathematical concepts to be applied; the nature of the scientific material; and the method of application. Mach's proposal to replace causal explanations by functional connections, therefore, raises three questions: (1) what are functions? (2) exactly what is it that functions are supposed to connect? (3) how is that connection performed or achieved? Traditionally, these questions have been answered as follows.

(1′) A function is a *law* or a *rule* associating a number, denoted by $f(x)$, to each number x belonging to a certain class of numbers, or associating a number $F(x_1, x_2, \ldots, x_n)$ to each n-tuple $(x_1, x_2, \ldots, x_n)$ of numbers of a certain kind. The symbols x and $x_1, x_2, \ldots, x_n$ in the preceding definitions are referred to as *variables*; e.g., the function associating x^2 with x and the function associating $x \cdot y$ with (x, y) are said to be functions *of one variable*, more specifically *of* x, and *of the two variables* x *and* y, respectively. They often are called 'the functions x^2 and xy' or denoted by the so-called equations $y = x^2$ and $z = x \cdot y$.

(2′) Functions are applied to *variables* or *variable quantities* (assumed to be known without explicit definitions except that in some cases units are mentioned), which include the most important objects of traditional geometry and physics such as s and a (length of a side and area of squares) and v, p, t (volume, pressure, and temperature of ideal gases).

(3′) The application is expressed in *formulae* such as $a = s^2$, or (in proper units) $t = v \cdot p$. These formulae are interpreted as summarizing statements of the form: if $s = 3$, then $a = 3^2$. The variables s, v, p in the preceding formulae are said to be *independent*; a and t, *dependent*.

These answers call for several critical remarks and for a complete reformulation.

(1″) Laws and rules are metamathematical rather than mathematical concepts. Mathematicians do not add and multiply laws while they do add and multiply functions. Some mathematicians, therefore, replace(1′) by the statement that a function is defined if, with each element x of a certain set of numbers, exactly one number, denoted by $f(x)$, is somehow associated. But, strictly speaking, this is a definition of the *definition* of function, rather than a definition of *function*. The objects, moreover, of mathematical theorems are not acts of somehow associating or pairing numbers to numbers, but rather the results of such acts, that is, sets of pairs of numbers. This (extensional) point of view in mathematics can be traced, if not further, certainly to Osgood, who in 1906 mentioned it in a footnote on p. 1 of his *Lehrbuch der Funktionentheorie* without, however, utilizing the idea. It was widely adopted in the 1930's. Ever since, mathematicians have considered a binary relation to be a set of ordered pairs of elements, and a function to be a set of ordered pairs of numbers not including two pairs with equal first and unequal second members. If f is such a set, then the set of all numbers that occur as the first members in a pair belonging to f is called the *domain* of f – briefly, dom f. Since, if x belongs to dom f, there is only one pair in f whose first member is x, the second member of that pair, being determined by f and x, may be unequivocally called $f(x)$ or the value of f for x.

In the presence of the clear and simple symbol f for the function, it seems absurd to refer to it rather as *the function* $f(x)$ or *the function* $y = f(x)$ or, for that matter, to refer to the function log as the function log x. The only excuse is the fact that in case of the function referred to as x^2 or $y = x^2$ a clear and simple symbol for the function is lacking – a lacuna in the symbolism of analysis that can be traced to the 17th-century fathers of this branch of mathematics, who left the most important of functions, the *identity function* (i.e., the set of the pairs (x, x) for all x) as well as the constant functions, anonymous. But, as I have repeatedly emphasized[1], this can be helped by adopting a permanent symbol for the identity function, say j. This is actually being done in an increasing number of recently published textbooks and, I venture to predict, will be a very common practice one or two generations hence. For, in possession of such a symbol, one can speak of the function j^2 just as one occasionally speaks of the function f and might speak of the functions log and cos. One then can dispense with the equivocal symbol x^2 for the function j^2 (reserving it for the value that j^2 assumes for x) and can eliminate those so-called equations which actually are not treated as such: the sum of the functions misnamed $y = x^2$ and $y = x$ is the function $y = x^2 + x$ while the addition of the two equations yields $2y = x^2 + x$. However, in the form of f, log and j^2 the function symbols are ready for their role as connectives which, as will be seen in (3″), they play in science.

In defining j^2 to be the set of the pairs (x, x^2) for all numbers x one uses x as a variable, more specifically as a *number variable* in the sense first clearly defined by Weierstrass in his lectures in the 1870's, that is, as a symbol that stands for any element of a certain set of numbers. If in the definition of j^2 the variable is replaced by the designation of a number, this replacement results in the designation of an element of j^2 such as $(3, 3^2)$. At the same time one realizes that it would be absurd to call j^2 (or, for that matter, log or f) a function *of* x or *of* a variable or, in fact, of *anything*.

(2″) The most important task of the clarification of applied analysis is to overcome the century old misidentification of what are called variables, or variable quantities in science, for example, s, a, v, p, t, with variables used in mathematics and logic, as first defined by Weierstrass. As has been mentioned in (2′), traditionally it is assumed that the so-called variable quantities are known and that the symbols for them such as v, p, t are understood without either definitions or postulates concerning them. The closest approximation to definitions are stipulations as to how one should operate with them in pairs or triples appearing in formulae such as $t = v \cdot p$; for example, the rule that if v assumes the value 3, and p assumes the value 4, then t assumes the value $3 \cdot 4$. But what are the individual variable quantities v, p, t? Being denoted by separate symbols they ought to be separately defined. Yet all that is said about them individually is the obscure hint that they are *variable numbers* assuming different values in different cases.

In what misunderstandings these confused indications have resulted is illustrated by the following remark in the first section of Tarski's *Introduction to Logic*: " . . . it is said that the symbols 'x', 'y', . . . also denote so-called 'variable numbers' or rather 'variable quantities' . . . The properties of such a number would change from case to case: that is to say, the number would sometimes be positive, sometimes negative, and sometimes equal to 0. But entities of such a kind we do not find in our world at all; their existence would contradict fundamental laws of thought." While of course a number cannot be sometimes 3 and sometimes 4, Tarski seems to be quite unaware of the fact that the words variable quantity in the literature do not really refer to numbers or symbols for numbers but rather are references – of course obscure and confused references – to a totally different category of objects, including v, p, and t, each of which indeed can sometimes have the value 3 and sometimes the value 4. I have revived Newton's term *fluents* for the entities of this category[2].

Preliminary separate definitions of the fluents v, p, and t can be formulated as follows. Given units, we pair with each sample γ of an ideal gas three numbers, $v(\gamma)$, $p(\gamma)$, and $t(\gamma)$. The gas volume v is the result of the first of these pairings, that is, the class of all pairs $(\gamma, v(\gamma))$ for all samples γ of an ideal gas. Similarly, I define the fluents p and t. This definition of fluents resembles that of functions. But there is a fundamental difference. Dom j^2 consists of numbers, which makes j^2 a purely mathematical concept; dom v consists of gas samples, which makes the concept v extra-mathematical. J. Mehlberg[3] has described the difference between fluents and functions from the point of view of the theory of types. Another difference lies in the fact that only functions (and not fluents) can serve as connectives and, therefore, are used throughout science, while fluents belong to specific branches of science and cannot connect either fluents or functions. We speak of the logarithm of the temperature, the logarithm of the mass, and so on; but we cannot, as I have pointed out before, speak of the temperature of the logarithm or of the mass of the temperature.

Some algebraists seem to take the position that since both functions and fluents, as herein defined, are mappings of a domain into a set of numbers, both deserve the name of function in the wider sense now often adopted; and that nothing new has been achieved by the definition of fluents herein propounded. What those mathematicians seem to overlook (apart from the differences in type between fluents and functions) is the very essence of the clarification: that its main point is not the introduction of a new term for variable quantities (although this in itself might be helpful in forestalling the mistaking of fluents for variables); but that its essence is a manageable definition of a category of extremely important scientific objects (namely, of fluents), which previously were ill-defined or altogether lacked any definition. The fact that a gap in the ontology of applied mathematics has been filled may be unimportant to algebraists but, as will be seen in (3″), it certainly brings about a complete reinterpretation of century old formulae such as $a = s^2$ and $t = v \cdot p$.

(3″) Just as a number connects one object with another object, or one number with another number (e.g., as 12 connects, ratiowise, years with months, feet with inches, 36 with 3, and so on) so a function connects fluents with other fluents, or functions with other functions; e.g., the product function assuming the value $x \cdot y$ for (x, y) connects t with v and p in the realm of ideal gases, the area with length and width in the realm of rectangles, the sine with the cosine and the tangent, and so on. If the product function is denoted by P, then one may describe the gas law by

$$(3) \qquad t = v \cdot p, \quad \text{or} \quad t = P(v,p).$$

Formulae (3), which connect the fluents v, p, and t, are, in view of the definitions of fluents, summaries or abbreviations of the following formulae which connect numbers, more specifically, values of those three fluents:

$$(3_\gamma) \qquad t(\gamma) = v(\gamma) \cdot p(\gamma) \quad \text{or} \quad t(\gamma) = P(v(\gamma), p(\gamma)),$$

for each element γ of the domain of v, p, and t. Here γ is a variable, i.e., a symbol that stands for any element of a certain class – the class of all samples of ideal gases. If in (3_γ) the variable is replaced by the designation of the contents of a container at a certain instant and in a certain place, then a statement about that particular gas sample results. There is a perfect parallelism between (3_γ) and the following formulae connecting numbers that are values of functions

$$(4_x) \qquad \sin x = \cos x \cdot \tan x \quad \text{or} \quad \sin x = P(\cos x, \tan x),$$

for each number x belonging to dom tan. The functions cos, tan, and sin correspond to the fluents v, p, and t; the number variable in the mathematical formula is the counterpart of the sample variable in the physical law. The century-old methodological misinterpretation of physical formulae was based on the invalid supposition of a parallelism between the variable x in (4_x) and the fluents v and p in (3). (That the latter were also called variables seems to have been partly a reason, partly a consequence, of the way the formulae were interpreted.) All that has to be done in order to bring out a complete parallelism between the mathematical and the physical formulae is to formulate an abbreviation of (4_x) that connects the functions themselves, summarizes the statements about their values, and thus corresponds to (3), that is,

$$(4) \qquad \sin = \cos \cdot \tan \quad \text{or} \quad \sin = P(\cos, \tan).$$

That in the 17th century mathematicians wrote on (4_x) and not (4) was one of the corollaries of their failure, mentioned in (1″), to introduce symbols such as j and j^2 for the most important functions. Hence they could not write

$$j^3 = j^2 \cdot j, \quad \text{or} \quad j^3 = P(j^2, j),$$

but had to write $x^3 = x^2 \cdot x$. So, for the sake of uniformity, they did not go beyond (4_x). Thus, while physicists and applied mathematicians never write (3_γ), the diehard adherents of the 17th-century mathematical symbolism even to this day eschew formulae such as (4).

From what has been said it follows that one may well speak of a function of one or of several fluents. But it is clear that in so doing one refers to fluents and not to functions; e.g., the product of v and p is the fluent t. Hence this locution in no way contradicts the remark in (1″) that a function is not a function of anything. There exists in this respect a parallelism between functions and numbers. Just as we speak of the logarithm of the temperature we speak of 12 apples. But only young children are confined to the study of numbers of apples or objects of any kind. The adolescent learns to abstract the idea of number and to use numbers as connectives. The arithmetician never mentions numbers of extramathematical objects at all (though he may apply the number concept, e.g., in order to count the number of prime numbers below a certain number). Speaking of functions only in connection with specific fluents or even in terms of fluent variables (i.e. symbols that stand for any fluent of a certain kind) is a remnant of the childhood of analysis. The analyst never makes any reference to extramathematical fluents though he may use a function in order to connect other functions. (Incidentally, if a formula expresses a fluent as a function of one or several other fluents, then these other fluents correspond to what are traditionally called independent variables, the first, to the so-called dependent variable. All this, of course, is relative to the particular formula since there may well exist another formula expressing the second fluent as a function of the first.)

A few further remarks should be devoted here to that part of the theory which is closest to positivistic ideas, that is, the concept of fluents. The definition of some fluents can be modified by variations of their domains. For example, one may associate the values of v, p, and t with states of gases rather than with gas samples. (This modification has nothing to do with the fact that gas samples change in time since, as goes without saying, the term sample refers to what sometimes is called a time slice of a sample, i.e., a sample at some instant.) With regard to the modified definition of the gas fluents, naturally the question arises as to how the state of a gas can be ascertained and described. The traditional answer to this question maintains that the state of a gas is determined by the values of v, p, and t for that gas. This solution of the problem is not necessarily circular as it may seem to be at first glance, But it leads to a complete arithmetization of the material and altogether detaches it from reality. For the definition amounts to an identification of the fluents with functions in the strict mathematical sense of associations of numbers with triples of numbers. More specifically, v, p, and t are in this way defined as what are called the first, the second, and the third *selector* functions, whose values for each triple of numbers are the first, the second, and the third member of the triple, respectively. A very similar procedure is followed in pure analytic geometry, where points are introduced as triples of numbers and, say, the abscissa is defined (if at all!) as the first selector function. On this foundation (which may well be said to be neo-Pythagorean since Pythagoras taught that everything is number) one can of course develop extensive algebraic and analytic theories; but in order to apply the results to reality one must leave this panarithmetical universe and must formulate procedures connecting the coordinates with observable entities; that is to say, one again faces the problem of defining fluents. In a similar way, the quantum-mechanical definitions of observables as operators must be supplemented, and for the sake of applications are being supplemented, by the formulation of procedures connecting the operators (or probabilities computed by means of the operators) with *observable* observables.

The most positivistic domains which, as I have repeatedly emphasized, can and, in some cases, must be considered, consist of acts of observation, especially, acts of meter readings. The value of a fluent for such an act α is the number read as the result of α. If the domains of v, p, and t consist of volumetric acts β, and acts μ and θ of manometer and thermometer readings, then it is clear that (3) cannot be interpreted as

$$t(\theta) = v(\beta) \cdot p(\mu) \text{ for each triple of acts } \beta, \mu, \theta.$$

The validity of the formula presupposes a correlation of the domains of v, p, and t, namely, that β, μ, and θ be acts simultaneously directed to the same gas container. It is only relative to a correlation of their domains that fluents can be connected by functions.

In general, that is, in terms of fluent variables (which, I repeat, are symbols that stand for any element of a certain class of fluents), the simple functional connection of fluents can be described as follows. If u and w

are fluents, if f is a function, and if Π is a set of ordered pairs (α, β) such that α belongs to dom u and β belongs to dom w, then

$$w = f(u) \text{ rel. } \Pi,$$

if and only if $w(\beta) = f(u(\alpha))$ for any pair (α, β) belonging to Π. Since f is a set of pairs of numbers, this formula is equivalent to the statement that $(u(\alpha), w(\beta))$ belongs to f if (α, β) belongs to Π.

Until recently I have found in discussions that explicit references to acts of observation in the formulation of the laws of classical physics have been unacceptable even to applied mathematicians. It certainly was an important step toward positivism when in 1966 Wigner and collaborators of his[4], included just such explicit references to acts of observation in laws of atomic physics.

NOTES

1 See Chapter 8.

2 Cf. [1951, 1], [1959, 4].

3 J. Mehlberg, 'A Classification of Mathematical Concepts', *Synthese* **14** (1962).

4 Houtappel, Van Dam, and Wigner, 'The Conceptual Basis and Use of the Geometric Invariance Principles', *Rev. Mod. Phys.* **37** (1965), especially pp. 598–600. For explicit references to procedures, see also H. Ekstein, 'Presymmetry', *Phys. Rev.* **153** (1967), 1397–1402, and 'Presymmetry II', *Phys. Rev.*, Aug. 25, 1969.

CHAPTER 10

VARIABLES, CONSTANTS, FLUENTS

This chapter includes a slightly abbreviated version of [1961, 5]. The paper, read at a symposium on the *Logic of Variables and Constants* at the University of Minnesota in 1959, was followed by comments of Professor Ernest W. Adams and my rejoinder. Parts of the latter are herein incorporated in Section 6a.

'Variable' undoubtedly is among the most frequently used nouns in the mathematico-scientific jargon and hence one of the most successful words ever created. No wonder, therefore, that it is a rather thankless task to tell mathematicians, scientists, or engineers what is behind the term that they are using with such splendid efficiency.

Actually, few scientists seem to have given any thought to the problem of what variables are, and still fewer of those who use the term formulate clear and satisfactory answers when the question arises. Nor does there seem to exist a satisfactory comprehensive treatment of the question in the philosophical literature.

I. VARIABLES

The number 5 is not variable; neither are the numbers $\sqrt{2}$ and π. In fact, no number is variable. There are no variable numbers – a thesis stressed by Frege at the turn of the century and today one of the few propositions about which logicians as well as mathematicians are in general agreement.

While there are no variable numbers, mathematicians, in formulating comprehensive arithmetical assertions (that is, assertions about many numbers), frequently use letters each of which stands for any element of a certain class of numbers. Such a letter is called a *number variable*, while the said class will be referred to as the *scope* of the variable. For instance, 'x' and 'c' are number variables in the assertions

$(x^2)^4 = (x^4)^2$ for any number x,

$(\sqrt{c})^2 = c$ for any number $c \geq 0$,

$x^2 - 1 = 0$ for any x belonging to the class consisting of 1 and -1.

One may also say that the proponent of these comprehensive assertions authorizes his listener or reader to replace 'x' and 'c' in the respective formulae with designations of numbers belonging to the scopes of the variables as described in explanatory legends, 'for any . . .', appended to the formulas. In the third example, the scope of the variable x is small – a mere pair of numbers – and this comprehensive assertion expresses the conjunction

$$1^2 - 1 = 0 \quad \& \quad (-1)^2 - 1 = 0.$$

II. UNKNOWNS

The proponent of the sentence

find all numbers x such that $x^2 - 1 = 0$.

is not asserting anything at all. He formulates, as the imperative legend preceding the formula indicates, a task for his readers or listeners: to exhibit numerals which, in place of 'x', make the formula true. Problems of this kind are called *equations* and the letter, whose scope (in contrast to that of a variable in a comprehensive assertion) is not described in an explanatory legend but has to be produced by the listener according to an imperative or sometimes interrogative legend, is called an *unknown*.

A full understanding of the material presupposes a recognition of the difference between variables and unknowns. The distinction is not clearly and consistently maintained in the traditional literature, where far too many formulas are unamplified by legends. Yet any formula that includes a nonself-explanatory letter, such as $x^2 - 1 = 0$, is by itself nothing but a string of symbols. It becomes a sentence only in conjunction with a legend; either an *assertion* (a *true* or a *false* assertion according as the legend describes a proper or an improper scope) or a *problem* (a *soluble* problem, if one aks for all numbers x or for one number x satisfying $x^2 - 1 = 0$; an *insoluble problem*, if one asks for three such numbers).

III. CONSTANTS

Logicians often refer to symbols standing for single objects (for example, to the designations of numbers or functions) as *constants*. But since the designation of a number is called a *numeral*, ordinary arithmetic might altogether dispense with the term 'constant', and it actually ought to do so. For, while it would be harmless if 'constant' were used as a synonym of 'numeral' consistently, the traditional use of 'constant' is inconsistent. Indeed, '1/2' and '$\frac{1}{2}$', being designations of the same number but clearly unlike symbols, would have to be called *different* constants, and so would '*e*' (the mathematico-scientific designation of the base of the natural logarithms) and 'ε' (the symbol used by electrical engineers for the same number). Actually, however, in these cases everyone speaks of the *same* constants, thus obviously thinking of the numbers themselves and not of the symbols.

IV. PARAMETERS

One of the most obscure mathematical ideas is that of parameters – special kinds of variables which, however, in some cases are misnamed constants. The lack of a satisfactory definition of parameters is due (as it is in many cases of mathematical notions for which satisfactory definitions are lacking) to the use of the term in a variety of discrepant meanings. These meanings can hardly be synthesized but must be analyzed.

One meaning of parameter is that of *a variable in descriptions of classes of pairs* (*or* n-*tuples*) *of numbers*, as in talking about

the class of all pairs $(1 + 2t, 1 - 3t)$ for any number t.

Any replacement of the parameter t with a numeral in the two expressions yields the designation of a particular pair of numbers belonging to the class.[1]

Another meaning of parameters is that of *variables occurring in conjunction with unknowns or functions*:

(1) In *comprehensive problems*; for example, a and b are parameters in the following comprehensive equation with the unknown x:

for any two numbers a and b, find x such that $x^2 + ax + b = 0$;

replacing a and b, say, with 0 and -1 one obtains the problem: find x such that $x^2 - 1 = 0$, whence such variables a and b may be called *problem parameters* (problem parameters are often referred to as *constants* or *arbitrary constants*);

(2) In *comprehensive conditions*; for example, n is a parameter in

find x such that $\sin x/n = 0$ for each integer n ;

that is, find x subject to all conditions (or, as it is often expressed, to the

simultaneous conditions) $\sin x = 0$, $\sin \frac{1}{2}x = 0$, $\sin \frac{1}{3}x = 0$, and so on – the only solution being $x = 0$. In the following problem, x is a parameter while a and b are unknowns: Find a and b such that $1/2(x^2 - 1) = a/(x + 1) + b/(x - 1)$ for each $x \neq 1, -1$.
This system of infinitely many (simultaneous) equations with the unknowns a and b has the solution $a = -\frac{1}{2}$, $b = \frac{1}{2}$. Variables such as n and x in the preceding examples might be called *condition parameters*.

V. QUANTITIES

This section opens the study of an entirely different topic. Its essence is the pairing of a real number to some (not necessarily material) object. Without any philosophical pretensions and merely for the purpose of introducing a short technical term for an often-used idea, I have called the result of such a pairing a *quantity*. A quantity thus is an ordered pair whose first member, which may be anything, will be called the *object* of the quantity, while the second member, which is a real number, will be referred to as the *value* of the quantity. Two quantities are *equal* if they have the same object and the same value. I will call two quantities *inconsistent* if their objects are equal while their values are unequal. Quantities that are not inconsistent will be called *consistent*. The following examples illustrate these concepts. The quantity

(Mr. Smith, his weight in pounds)

is consistent with the quantity

(Mr. Brown, his weight in pounds),

but inconsistent with the quantities

(Mr. Smith, his height in feet), (Mr. Smith, his weight in kilograms).

If Smith weighs 160 pounds and has a bank account of \$160, then the following two quantities (both having the value 160) are consistent but unequal, because they have different objects:

(the act of reading Smith's weight in pounds on a scale, the number read as the result of this act),

(the act of ascertaining Smith's account in dollars, the number ascertained).

On the other hand, the following three quantities are not only consistent but *equal* because they have the same object and the same value: (Smith, 160), (Smith, his weight in pounds), (Smith, his account in dollars).

VI. CLASSES OF CONSISTENT QUANTITIES

We next consider classes of mutually consistent quantities such as – to mention two significant examples

> the class of all quantities (A, age in years of A) for any inhabitant A of Illinois,
>
> the class of all quantities (B, the height in feet of B) for any inhabitant B of New York,

and – to mention an example without apparent significance – the union of the two preceding classes.

For any class q of consistent quantities or, as we shall occasionally say[2] for brevity's sake, for any *functor q*, one may form the class of all objects of the quantities belonging to q, and that of all values of those quantities. The former class is called the *domain* of q, the latter – a class of numbers– the *range* of q. The class q itself may be considered as the result of mapping the domain of q on its range.

Some mathematicians propose to call every mapping of a class onto a class of numbers (that is, every class of consistent quantities) a *function.* This proposal is, however, at variance with the usage of most scientists, who apply the term function only to a mapping of a class of numbers (or of a class of systems of numbers) onto a class of numbers. Examples of such functions include the logarithmic function or log, which is the class of all ordered pairs of numbers (x, y) such that $x = e^y$, and the product function whose domain is the class of all pairs of numbers, the function itself being the class of all pairs $((x, y), x \cdot y)$ for any pair of numbers (x, y). The same restricted use of the term function prevails in some parts of mathematics, as can be seen from the fact that, following Hadamard, analysts call a mapping of objects other than numbers or systems of numbers (especially of functions, curves or surfaces) onto a class of numbers a *functional* rather than a function.

Scientific investigations actually, if implicitly, deal with classes of consistent quantities or functors whose domains comprise altogether *non-mathematical* objects such as acts of reading clocks or meters, states of gases, or members of populations. The realization of this fact makes precise the traditionally undefined or misdefined ideas of what Newton called *fluents*, such as t (the time in seconds) or p (the gas pressure in atmospheres). In the eighteenth century, fluents were renamed *variable quantities* or even simply *variables*. (It is this equivocal use that makes the latter term ubiquitous in applied as well as in pure mathematics.)

It has been suggested above that, in order to bring out conceptual differences one should refer to time, gas pressure, and the like as *fluents* rather than as *variables* or *functions*. But it would be a complete misunderstanding if one considered this suggestion (or in fact any merely terminological proposal) as essential for the ideas here expounded. What *is* essential here is that t, p, and other scientific fluents (a) are introduced by clear explicit definitions; (b) are introduced as classes of consistent quantities. In the traditional literature either no definitions are attempted at all or t, say, is defined as a class of numbers read on a clock. The time in seconds observed by a person A is the class of all pairs that consist of (1) a certain act performed by A – namely, an act of reading a clock calibrated in seconds; (2) the number read as the result of that act. The analogue to defining time as a class of numbers read on a clock would be defining the logarithmic function as the class of all numbers that are logarithms – in other words, as the class of all numbers. No mathematician introducing a function would propose such an inadequate definition – a definition that confuses the function with its range (the class of the function values). It is universally recognized that function values are paired to elements of the domain of the function and that it matters which values are paired to which elements. Indeed the sine and the cosine functions are different even though they have the same domain and the same range. It is imperative that physical fluents, too, be introduced as classes of consistent quantities and not as mere classes of numbers.

Whereas the elements of the domain of a function, being numbers or systems of numbers, are permanent, reproducible, and easy to describe, the domain of a scientific fluent consists of acts, states, and the like, which are transient, irretrievable, and hard to characterize. But this discrepancy simply reflects the essential difference between mathematics and science.

VIa. *Volume, Observed Volume, Geometric Volume*

In defining a fluent such as the volume (say, in cubic feet) one has a choice between two positions: one may pair a number either to each act of measuring the volume of a solid or to each solid. The classes v^* and v of all ordered pairs (or of all quantities) thus obtained may be referred to as *the observed volume* and *the volume* of physical solids, respectively.[3]

The former point of view has been criticized because of the vagueness of the notion of act of measurement.[4] It has been asked, for example, whether different acts of measuring the same solid correspond to different quantities belonging to the fluent and whether acts of inaccurate measuring are to be included. These questions, however, while well justified, do not at all discredit the idea of observed volume. With regard to v^*, they simply and most decidedly call for the answer Yes. And it might be added that the fluent v^* also includes quantities which correspond to acts σ_1, σ_2, . . that various observers simultaneously direct to the same solid S and which, nonetheless, may have unequal values $v^*\sigma_1$, $v^*\sigma_2$, . . .

A pair:

(act of meter reading, number read as the result of that act)

might be called an *observed quantity* or here, where nonquantitative observations are not being considered, briefly, an *observation.*[5] Because no two acts are equal, any two observations $(\sigma_1, v^*\sigma_1)$ and $(\sigma_2, v^*\sigma_2)$ are *consistent* quantities, whereas, if $v^*\sigma_1 \neq v^*\sigma_2$, the quantities $(S, v^*\sigma_1)$ and $(S, v^*\sigma_2)$ are inconsistent. (Here S is the object of σ_1 and σ_2).

If anything is vague, then it is not the concept of *observed* (or, so to speak, Machian) volume but, on the contrary, the concept of *objective* (or, as it were, Platonic) volume. For how can the value that this fluent assumes for a solid S (in other words, the volume in cubic feet of S) be ascertained? Clearly, vS is derived from the values $v^*\sigma_1$, $v^*\sigma_2$, . . . by averaging processes that are more or less arbitrary.

There is finally a third type of volume, which I will call *geometric* volume and denote by V. On the basis of certain assumptions, V is connected with other geometric fluents, such as the lengths S and R of edges of cubes and of radii of spheres, by formulae such as $V = S^3$ and $V = 4/3\pi R^3$. The domains of these geometric fluents are undefined elements (called cubes, spheres, etc.) which satisfy certain postulates.

VII. CONSTANT AND NONCONSTANT FUNCTORS

In contrast to numbers, the classes of consistent quantities or functors (for example, functions, functionals, and fluents) can be divided into those that are *nonconstant* (that is, whose range includes more than one number) and those that are *constant* (whose range includes only one number – the common value of all quantities belonging to the functor). Here, finally, the concept *constant* becomes mathematically significant.

In the literature, 'constants' (in the sense of constant functors) are often presented as numbers – a corollary of the misrepresentation of 'variable quantities' (in the sense of nonconstant functors) as classes of numbers. And there is a lack of notational as well as conceptual distinction between a constant function or fluent and its value. The same numeral is used in describing the number 3 and the constant function assuming for each number the value 3. (These errors partly account for the inconsistent use of the term *constant* discussed in Section III of this chapter.)

Attempts have been made to define a physical constant as a *denominate* number – for example, as a number of seconds or of centimeters – and, similarly, a variable quantity as a class of denominate numbers. But including a denomination, unit or dimension in the definition fails to provide a sufficient characterization of a physical constant as such. The radius, the diameter, the circumference, and the electrical capacity of a sphere are all numbers of centimeters.

What has to be specified in defining a physical constant besides its value is its *domain* – that is, the class of objects to which a certain number has been paired or the class of procedures all of which result in that number. In defining the gravitational acceleration in cm sec^{-2} at a certain location on earth, a definite class of experimental acts must be specified each of which, in the past, has resulted (and, in the future, is expected to result) in one and the same number.

VIII. FUNCTOR VARIABLES

In formulating comprehensive assertions about fluents or, more generally, functors, it is convenient to use fluent variables or functor variables – symbols each of which stands for any element of a certain class of fluents. Function variables – letters that stand for any element of a certain class of functions or may be replaced with the designation of any element of that scope – are extensively used in comprehensive assertions about functions.[6]

Functor variables are used, for instance, in the following definitions which are applicable to any two functors, u and w. If α belongs to the domain of u, and β belongs to the domain of w, then to the ordered pair (α, β) there corresponds an ordered pair of numbers: the pair whose first member is the value of u for α, and whose second member is the value of w

for β. To any class Π of ordered pairs (α, β), there thus corresponds a class of ordered pairs of numbers, which may, or may not, be a function. For instance, to Π_1 there may, and to Π_2 there may not, correspond a function. Accordingly, w is said to be a function of u relative to Π_1 but not a function of u relative to Π_2. The question as to whether w is a function of u, *absolutely* rather than relative to a class Π, is incomplete and (except in the trivial case where the functor w is constant) unanswerable. Such is the question: If two pendulums have equal periods, but are out of phase by a quarter of a period, is the amplitude of the first pendulum a function of the amplitude of the second? The question admits relative answers only. For instance, relative to the class of pairs of observations of the two pendulums taken a quarter of a period apart from each other, the answer is Yes. Relative to the class of pairs of simultaneous observations, the answer is No. In physics normally the pairing of simultaneous acts is taken for granted.

The traditional literature, in the absence of a clear definition of fluent, confuses fluent variables with number variables, and thus applies the term *variable* indiscriminately to number variables, to fluents, and to fluent variables.[7]

NOTES

[1] Variables called parameters are also used in defining classes of functions; for example, one uses a and b as parameters in describing the class of all linear functions, ax + b, where x is the traditional, rather inadequate denotation of the identity function.

[2] In the literature, the term *functor* has been used in a variety of meanings.

[3] In defining the observed volume of the sun or even of the earth one must of course include what Bridgman calls paper-and-pencil operations among the acts of measuring.

[4] Cf. especially E. W. Adams, 'Comments on Menger's Variables, Constants, Fluents' following [1961, 5].

[5] Communications of observations in this sense are what some positivists call *protocol statements*.

[6] In contrast to *function variables*, the designation of a specific function would be a *function constant* in the sense of the logicians (See Section III of this paper). But since this term might lead to confusion with *constant function* one better dispenses with it (even though there is no word for *function symbols* analogous to 'numerals' for number symbols, which makes it easy to dispense with the term *constant* in regard to numbers).

[7] Using 'constant' the way logicians use this term (See Section III), and the adjective 'variable' for 'nonconstant', A. Wundheiler once told me about my distinction between what herein is called 'fluents' and 'fluent variables' "*I am greatly intrigued by your discovery that there are constant variables and variable variables.*" It was after this conversation that I decided to use the short term 'fluent' for what physicists and applied mathematicians refer to as 'variables' or 'variable quantities' and what until then I had called 'consistent classes of quantities' (See Section VI).

CHAPTER 11

WITTGENSTEIN ON FORMULAE AND VARIABLES

This chapter contains a discussion, announced in [1961, 1] p. 349, of the single reference to mathematical variables that I found in Wittgenstein's writings published by 1961. The chapter begins with a full description of the context in which the remark occurs.

In Wittgenstein's *Philosophical Investigations*, Section I, 143 ff. is devoted to the formation of sequences of integers, for example, of the odd numbers, and the application of rules such as $a_n = n^2 + n - 1$. Here, Wittgenstein is actually using the letter n as a number variable in the sense of Weierstrass, that is, a symbol that may be replaced by numerals, each replacement resulting in a numeral or a variable-free formula; e.g., the replacement of n by 4 yields $a_4 = 4^2 + 4 - 1$ (which equals 19). He returns to this topic in Sections I, 183 ff., where he distinguishes formulae that "determine a number y for a given x" and formulae that lack this property, giving $y = x^2$ and $y \neq x^2$ as respective examples.[1]

The *Remarks on the Foundations of Mathematics* begins with the same topic citing

(1) $y = x^2 + 1$

in its use as a formula of the first kind and mentioning

(2) $y > x^2 + 1, \quad y = x^2 \pm 1, \quad y = x^2 + z$

as examples of formulae of the second kind. From the point of view of logic it may be added that the possibility or impossibility of using these formulae in determining y for given x is based on (or expressed in) the propositions that for every x there exists one and only one number y such that (1) holds, and there exist unequal numbers, y_1 and y_2, satisfying (2).

In the symbolism of the elementary logical calculus,

(1′) $(\forall x)\{[(\exists y): y = x^2 + 1] \,\&\, (\forall y')[(y' = x^2 + 1) \rightarrow (y' = y)]\}$,

(2′) $(\forall x)(\exists y_1)(\exists y_2)[(y_1 \neq y_2) \,\&\, (y_1 > x^2 + 1) \,\&\, (y_2 > x^2 + 1)]$.

Wittgenstein's use of formulae of the first kind is closely related to the idea of *function*. The actual pairing to each x of the number that the formula "determines" for x is what is said to be *defining a function*. The result of such a definition, that is, the function defined, is a class of pairs of numbers of a certain kind.[2] In the case of (1), the function is

(1*) The class of all pairs (x, $x^2 + 1$) for any number x or, equivalently, the class of all pairs (x, y) such that $y = x^2 + 1$.

The number paired to x is called the *value* of the function for x.

Throughout the traditional literature, the class (1*) is referred to either by formula (1) as "the function $y = x^2 + 1$" or by its value for x as "the function $x^2 + 1$". Symbolic identification of a function (a class of pairs of numbers) and a value of that function (a single number) is of course objectionable.[3] It is also quite arbitrary that one chooses just the value for x rather than that for, say, y or a, and uses the formula $y = x^2 + 1$ rather than, say, $b = a^2 + 1$.[4]

Wittgenstein's third formula (2) can be used in pairing a number to each pair of numbers (x, z), e.g. to (4, 3) the number $4^2 + 3$, which equals 19. This use is likewise called *defining a function*. The result is

(2_z^*) The class of all pairs ((x, z), $x^2 + z$) for any number x and any number z.

In each pair, the first member is itself a pair of numbers; the second member is again a number, called a value of the function. Such classes are traditionally referred to as "*functions of two variables*". For these functions there does not exist a universal traditional notation either by formulae or by values.[5]

Wittgenstein's formulae (1) and (2) also play a great role in analytic geometry. There is a theory deserving the name of *arithmetical geometry* in which points of the plane are defined as pairs of numbers, and points of the 3-dimensional space as triples of numbers. Certain classes of pairs (including all those that are functions) are called *curves in the plane* – the class (1*) is by definition said to be a *parabola*. Certain classes of triples of numbers are called *surfaces in space*, for example, the class of all (x, y, z) such that $y = x^2 + z$ or, equivalently, of all triples (x, $x^2 + z$, z) for any x and any z.

Following Descartes' original treatment one also uses the same letters *x*, *y*, *z* in the analytic geometry of (physical or imagined) spatial objects. In that theory, *coordinates* are designated by these letters and are often referred to as, and confused with, variables[6]. Yet, in total contrast to number variables as used in algebra and arithmetical geometry, coordinates are *fluents*. A fluent pairs (as does a function) a number to each object of a certain kind. In the case of coordinates, the objects are points of a physical plane or space in which three points O, X, Y or four points O, X, Y, Z have been chosen in such a way that the segments OX, OY and OZ are pairwise perpendicular. The fluent x pairs to each point P the the distance $x(\mathrm{P})$ – briefly $x\mathrm{P}$ – of P from the line joining O and Y in the plane, and from the plane through O, Y, Z in space, the segment OX being in both cases the unit of distance. Analogously, one defines the fluents y in the plane, and y and z in space. In the plane,

(1**) the class of all points P such that $y\mathrm{P} = (x\mathrm{P})^2 + 1$

is a parabola. (This is the reason why in the arithmetical geometry of the plane the class (1*) is by definition called a parabola.) In space,

(2**) the class of all points P such that $y\mathrm{P} = (x\mathrm{P})^2 + z\mathrm{P}$

is a trough-like surface intersecting each line perpendicular to the plane through O, X, Z in exactly one point.

While using variables in the sense of Weierstrass in his remarks about sequences of integers and in his distinction of two kinds of formulae Wittgenstein never mentions the word 'variable'. The term occurs only once, at the very end of Section I, 1 of the *Remarks*. There, he first points out that one might ask a pupil. "Is $y = x^2$ a formula that determines y for a given x?"[7] in order to test whether he understands the use of the word *to determine*. Then Wittgenstein continues with the following puzzling remark: "Or it might be a mathematical problem to calculate (*berechnen*) whether there is only one variable on the right-hand side of a formula, as for example, in the case:

(3) $y = (x^2 + z)^2 - z(2x^2 + z)$."

There can be hardly any doubt that the "calculation" which Wittgenstein would have considered to be a solution of the problem is the following:

(4) $(\mathrm{x}^2 + \mathrm{z})^2 - \mathrm{z}(2\mathrm{x}^2 + \mathrm{z}) = (\mathrm{x}^4 + 2\mathrm{x}^2\mathrm{z} + \mathrm{z}^2) - (2\mathrm{zx}^2 + \mathrm{z}^2) = \mathrm{x}^4$

for any three numbers x, y, z, whence (3) implies and is implied by

(5) $y = x^4$.

On the right-hand side of (5), there is indeed only one letter.

But what had Wittgenstein in mind when saying that there is only one variable on the right-hand side of the formula (3)? And what did he mean by *variable* in this statement?

Certainly Wittgenstein did not mean variable in the sense of Weierstrass. For it is manifest by inspection (without any need for, or even any possibility of, "calculation") that the right side of (3) includes *two* variables in this sense: x and z. Both letters may be replaced by various numerals, for example, by 1 and 3, respectively, yielding $(1^2 + 3)^2 - 3 \cdot (2 \cdot 1^2 + 3)$. This statement is not affected by the fact that the result equals 1 in all cases where x is replaced by 1 regardless of what replacement for z is chosen; and is x^4 for any replacement of x no matter what z is. (Similarly, one cannot reasonably say that there is *no* variable in the sense of Weierstrass on the right side of the formula

$$y = z^2 - (z + 1)(z - 1),$$

even though each replacement for z yields 1.)

Nor is there any reason to believe that in his comment on formula (3) Wittgenstein had in mind variables of arithmetical geometry or fluents of the analytic geometry of our space,[8] even though (3) and in fact all his formulae are typical of these geometries[9].

What then *did* Wittgenstein mean by variable? It would be interesting if a consistent and comprehensive concept of variable could be defined according to which the right side of (3) included only one variable. For my part, I am inclined to doubt that this is possible. Even though Wittgenstein's strange comment on that formula (3) is contained in the part of his *Remarks on the Foundations* that the editors say is the only one which existed in typescript and "is certainly the most worked over of them all". I rather believe that when Wittgenstein wrote it his ideas on the difficult concept of variable had not crystallized.

NOTES

[1] What interests Wittgenstein is the *understanding* of such rules. He points out, correctly, the need for an explanation of the way formulae are to be used.

[2] It is a class such that no number is the first member of more than one pair belonging to the class.

[3] As a result of this misidentification, 'the function $x^2 + 1$' is traditionally said to assume the value $x^2 + 1$ for x, which is a totally unacceptable proposition since the function does not assume itself as a value (See Chapter 8).

[4] Both shortcomings can be corrected by introducing permanent symbols for the identity function (pairing to each x that same number x as its value for x) and the constant

functions (pairing to every x one and the same number) – I have proposed the letters *j* for the former and, for example, c_1 or *1* for the function pairing 1 to each number. Then $j^2 + c_1$ or $j^2 + 1$ are symbols for the function (1*) – for the function *itself*, which in the traditional literature has had no symbol (while the logarithmic and the cosine functions have had the symbols log and cos). At the same time one obtains variable-free and equation-free symbols for countless other functions rescuing them from what amounts to traditional anonymity. Of course, the choice of the permanent symbol *j* is also arbitrary, but only in the sense and to the extent that every name is arbitrary. Variables, on the other hand, are totally arbitrary symbols that may be replaced (and often must be replaced and interchanged) in any context and at any occasion.

5 The expression 'the function x^2 + z' is likely to be mistaken for a reference to

$(2^*_{y,z})$ the class of all pairs ((x, y, z), $x^2 + z$) for any x, any y, and any z,

which is totally different from (2^*_z), the first members of its pairs being triples of numbers. But (2^*_z) is of course the same as

(2^*_y) the class of all ((x, y), $x^2 + y$) for any x and any y.

Again, all this arbitrariness and alphabetic mysticism can be avoided as can the symbolic identification of functions and function values if one introduces two permanent symbols – I have proposed the letters *I* and *J* – for the functions pairing to each pair (x, y) the numbers x and y, respectively, or which is tantamount, to each (x, z) the numbers x and z, so that *I*(x, z) = x, *J*(x, z) = z and *I*(x, y) = x, *J*(x, y) = y. Then the class (2^*_y) or (2^*_z) is unambiguously and unmistakably the function $I^2 + J$.

6 In order to facilitate the distinction I have proposed (See Chapter 8) the use of Roman type for number variables and of italics for fluents and functions. This typography is used in the present paper except in references to the traditional literature and in quotations. In Wittgenstein's *Investigations* all single letters are in Roman type, in his *Remarks* they all are italicized.

7 In the bilingual edition of Wittgenstein's *Remarks* the translators render "*für ein gegebenes x*" (for a given *x*) by "for a given value of *x*" and several times repeat this insertion of the words 'value of', which have no foundation in the German original. Already Frege pointed out that one cannot speak of "a value of a variable". A fluent, on the other hand, can indeed be said to assume a value for each object to which it pairs a number. For example, the abscissa *x* assumes for the points O, X, Y, Z the values 0, 1, 0, 0, respectively.

8 It seems more likely that Wittgenstein was somehow thinking of *functions* without, however, clearly distinguishing j^4 and I^4.

9 In arithmetical geometry, (3) is an abbreviated designation of

(3*) the class of all triples (x, y, z) such that

$$y = (x^2 + z)^2 - z(2x^2 + z) \text{ or, equivalently, } y = x^4,$$

that is, the class of all triples (x, x^4, z) for any x and any z, or (a, a^4, b) for any a and b. (3*) is closely related to (but not identical with) the class of all triples (x, y, z^4) for any x and y or all (a, b, a^4) for any a and b. This class, in turn, is intimately related to the function I^4, that is, the class of all pairs ((x, y), x^4) or ((x, z), x^4) for any x, y or x, z.

In the coordinate geometry of spatial objects, (3) is an abbreviation of

(3**) the class of all points P such that

$$yP = ((xP)^2 + zP)^2 - xP \,.\, (2(xP)^2 + zP) \text{ or } yP = (xP)^4.$$

This is a trough-like surface lying along the line joining the points 0 and Z. (In plane geometry, $y = x^4$ is a parabola-like curve.)

PART IV

DIDACTICS OF MATHEMATICS

Readers unfamiliar with the integral calculus are advised to begin the reading of Part IV with Chapter 13.

INTRODUCTION

In every field, deepening of insight into the basic concepts eventually is bound to affect its presentation to beginners; and without contemplation and clarification of the fundamental ideas, didactics is threatened by stagnation.

In the field of mathematics, such clarifications and corresponding reforms of the didactics have been stymied on every level by the devotion to the term 'variable' and to the symbol 'x'. This attachment was dramatized for me in the mid 1950's by two experiences. A teacher of elementary mathematics asked me, "Why change the notation and write *j* instead of x?" – as though I were using *j* as a number variable and as though, however the clarified formulae might differ in appearance from their traditional counterparts, the changes were aimed at the notation! Then I expounded the differences between variables in logic and pure mathematics and variables in science and applied mathematics in (separate) conversations with a great logician and an eminent physicist. At first both men were startled by this distinction which was totally new to them; after some discussion, however, they came round to my point of view, even admitting the need for different terms; but each one insisted that without any question the term 'variable' ought to be retained in *his* field and be reserved for *it*.

The realization of the need for a clarification of the basic concepts and a reform of the didactics first came to me in the early 1930's in Vienna – but not in connection with the Circle. It was rather when teaching a course on calculus at the University that I was vexed by finding that the traditional treatment forced me to dodge some difficult questions. A few years later in America, as described in Chapter 12, some of those questions were raised by beginners whose initiation into calculus was my responsibility – in the mid 1940's in a particularly penetrating and constructive form by my student B. D. Fried (now professor of physics at the University of California in Los Angeles). In the 1950's, *Calculus. A Modern Approach* ([1952, 5], [1953, 3], [1955, 4]) appeared, the latest edition of the book utilizing some valuable comments by N. E. Steenrod, A. Wundheiler and others on the earlier writings. The reviews that came out were rather unusual in number and contents. But the great organizations for the furthering and the dissemination of mathematics and mathematical education were at that time so engrossed in propagandizing the reform (mainly of arithmetic and algebra) that has become known as 'the new mathematics' that they never arranged a discussion of the radically new approach to calculus. Teachers following that approach were naturally scarce and, moreover, discouraged by the attitude of their organizations.

Today the book is no longer available. But some of its ideas (especially the distinction between functions and function values, and the introduction of a symbol for the identity function) have entered a few more recent texts while the enthusiastic praises of 'the new math' and its effects begin to be mixed with critical voices. So the time may be ripe for reconsidering the effects of the counterpart of Occam's Razor on the teaching of calculus and, more generally, for studying the didactic implications of the resulting clarification of some basic mathematical ideas.

CHAPTER 12

A NEW APPROACH TO TEACHING INTERMEDIATE MATHEMATICS

A reprint of [1958, 4] with minor omissions and amendments. The writing of this paper was part of the work made possible by a grant from the Carnegie Corporation in New York for the development of a new approach to the teaching of mathematics.

Nothing is more distasteful to an active mathematician or scientist than discussions of symbolism and notation; and that dislike is perfectly understandable. After having overcome in his youth whatever difficulties the formal expression of ideas presents, the mathematician finds that certain ways of writing have become second nature and he regards any suggestion of a change, even if he recognizes its merits, as nothing but a trivial nuisance.

There are, however, situations in which a thorough discussion of such matters on the highest level is inevitable. They occur when, at turning points in the history of culture, it becomes imperative to make certain techniques and ideas of mathematics available to wider strata of the population. In the large groups to be initiated, many persons lack the ability to overcome the difficulties that the specialist overcame in his youth. Moreover, an immense collective benefit results if even persons with that ability are spared unnecessary complications.

Such a turning point affected arithmetic when, during the Renaissance, mercantilism and experimental science were born. In banks and laboratories, the letters introduced by the Greeks and Romans as numerals proved to be utterly inefficient, even though they had served arithmeticians for over 2000 years. Unfortunately, medieval mathematicians misinterpreted the specialists' manipulative facility as intrinsic simplicity of the ancient numerals and regarded the Hindu-Arabic ideas as a pure nuisnace. "Even in the 15th century," wrote G. Sarton, "there were still any number of learned doctors and professors who claimed that the Roman letters were much simpler than the Hindu numerals." Such prejudices confined the knowledge of arithmetic to a small elite and retarded its democratization as well as its progress. Eventually, however, as everyone knows, practical exigencies prevailed – incidentally, to the ultimate benefit of pure mathematics too.

The middle of the 20th century appears to be another such turning point. This is the time when scientific and technological progress has reached proportions necessitating the dissemination, on a large scale, of intermediate mathematics. A considerable part of the population should learn certain techniques of algebra, analytic geometry, and calculus, as well as some basic ideas of those theories. The attempts toward this aim, which follow traditional lines, are generally regarded as not sufficiently successful. In my opinion, the principal stumbling block is the fact that most of those great mathematical ideas and techniques are being presented in their 17th century form.

USES OF x

A principal feature of those antiquated formulations is the indiscriminate use of the letter x (as well as of the letter y) in diverse meanings and according to discrepant rules. What enhances the confusion are references to those diverse types of x and y by one and the same term, namely, *variables*.

Algebra

In algebra, beginners learn that in the formula $x + 1 = 1 + x$ they may replace x with numerals, thereby obtaining formulas such as $4 + 1 = 1 + 4$. But they find that this practice must not be applied to the statement that the function $x + 1$ is nonconstant, since replacement therein of the "variable" with 4 would lead to the false statement that the function $4 + 1$ is nonconstant. Beginners further learn that squaring the equation $y = x^4$ yields $y^2 = x^8$. But they find that the square of what often is referred to as the function $y = x^4$ is the function $y = x^8$. This contradiction is so blatant that many mathematicians altogether refrain from referring to the said functions as $y = x^4$ and $y = x^8$, and rather call them, briefly, the functions x^4 and x^8. As a result, however, x frequently has various meanings in one and the same statement as, for example, in the following sentence

(S): The function x assumes the value x for any number x.

This statement is being inculcated into beginners until many of them get used to it. Getting used to it means realizing that the function x does not assume itself as a value (even though it assumes the value x), that it is not any number (even though x stands for any number), and so on. In other words, accepting statements of the type of (S) means realizing that they must not be taken literally. Therefore, there are also many beginners who give up. They have heard so much about the perfect precision of the language of mathematics that, after failing to understand some mathematical statements when taken literally, they regard themselves as lacking any mathematical ability. This group includes persons of great intelligence.

Analytic geometry

In analytic geometry, the beginner learns that a certain parabola is the locus of all points (x,y) such that $y = x^2$ or (which is the same) of all points (a,b) such that $b = a^2$. For instance, this parabola includes the point $(3,9)$ but not the point $(9,3)$. Indeed, $9 = 3^2$ but $3 \neq 9^2$; in other words, if $a = 3$ and $b = 9$ then $b = a^2$, whereas if $a = 9$ and $b = 3$ then $b \neq a^2$. Of course, that parabola may also be said to be the locus of all points (b,a) such that $a = b^2$. This locus, too, includes $(3,9)$ but not $(9,3)$ since $b = 3$ and $a = 9$ imply $a = b^2$ whereas $b = 9$ and $a = 3$ do not. All this is in no way surprising. But ask the following question: May this same parabola also be described as the locus of all points (y,x) such that $x = y^2$? Of course, the answer is again affirmative. Yet this affirmative answer would utterly bewilder the beginner. He would be unable to reconcile it with the fact (likewise taught in analytic geometry) that the parabolas $y = x^2$ and $x = y^2$ are altogether different. The teacher's only hope is that the question mentioned will not be raised and, therefore, that the apparent contradiction will remain unnoticed. For within the classical frame of concepts it is impossible to explain that paradox, the explanation being that mathematicians traditionally use the same pair of letters x, y in discrepant meanings when talking about the parabola $y = x^2$ and about the locus of all (x,y) such that $y = x^2$. That in the latter case one may interchange x and y while in the former one must not is a mere symptom of conceptual differences which traditionally remain inarticulate.

Calculus

In calculus, the reciprocity of differentiation and integration – the very core of the theory – traditionally is expressed as follows:

$$\text{(1)} \qquad \frac{\mathrm{d}}{\mathrm{d}x}\int_a^x f(x)\mathrm{d}x = f(x)$$

for any continuous function $f(x)$ and any number a. Serious shortcomings of formula (1) become apparent in manipulating symbols, even though it has been frequently claimed that the classical symbolism (while perhaps obscuring some of the contents) certainly facilitates manipulative use. In its five occurrences in formula (1), the letter x follows altogether discrepant rules. It may, without any change of the meaning, be replaced with any other letter in its last two occurrences on the left side of formula (1) or in its other three occurrences. For instance, the formulas

$$\frac{d}{dx}\int_a^x f(t)dt = f(x)$$

and

$$\frac{d}{du}\int_a^u f(x)dx = f(u)$$

have exactly the same meaning as formula (1). In contrast, formulas resulting from other replacements, such as

$$\frac{d}{du}\int_a^x f(x)du = f(x)$$

and

$$\frac{d}{du}\int_a^u f(x)du = f(u)$$

are, in general, incorrect. The letter a in formula (1) may be replaced with a numeral as in

$$\frac{d}{dx}\int_1^x f(x)dx = f(x). \tag{2}$$

No such replacement of x in one or more of its occurrences yields a valid formula. For instance,

$$\frac{d}{d4}\int_1^4 f(4)d4 = f(4)$$

is utterly nonsensical, while

$$\frac{d}{dx}\int_1^4 f(x)dx = f(4)$$

is, in general, false. He who wishes to state that the two functions equated in formula (2) assume equal values for $x = 4$ may replace x with 4 only in its last occurrence. What he must write is:

$$\left(\frac{d}{dx}\int_1^x f(x)dx\right)_{x=4} = f(4). \tag{3}$$

In view of such examples, one may well question the traditional claim that the classical presentation is conducive to a purely mechanical handling of symbols.

Needless to say, these complications do not present the slightest difficulties to anyone who has mastered a traditional course in calculus. The reason for mentioning them is the present vital interest in *increasing* the number of people who master the ideas and manipulative techniques of calculus (though not necessarily in their 17th century form), reminiscent of the Renaissance interest in increasing the number of men able to perform multiplications and divisions (though not necessarily in Roman numerals). True, the time-honored formulation (1) of the Reciprocity Law, which goes back to Leibniz, has been successful during the three centuries that witnessed the activities of Euler, Gauss, and Poincaré. But the Greek numerals were successful during the eight centuries from Pythagoras to Archimedes to Diophantos.

QUESTIONS

In view of the quoted (and countless other) examples from intermediate mathematics, what is remarkable is how well teachers succeed in transmitting to many students a feeling for what is right in manipulating x and y and an instinct for which type of variable is present where. (Since the underlying distinctions do not attain the level of the articulate, feelings and instincts are all that can be transmitted.) That, notwithstanding all the teachers' efforts, many beginners, including talented students, give up, is not surprising. Clearly, various circumstances contribute to the unfortunate and dangerous situation in our current mathematical education. But the antiquated symbolic and conceptual frame in which mathematics is being presented certainly is in itself a sufficient reason.

When directing the mathematics instruction of large groups of Navy trainees during World War II, I collected questions that some beginners asked. I heard the same questions again from mature men in a metropolitan night school complaining that in their youth they had never received satisfactory answers – a fact on which they blamed the superficiality of their mathematical understanding. But within the conceptual and symbolic frame of traditional mathematics, most of these questions are simply unanswerable.

NEW PRESENTATION

It was in response to those questions that in the years from 1941 to 1956 I developed a new presentation of intermediate mathematics in numerous publications.[1] The essence of this presentation is an approach to mathematics that is based on common sense, which thereby furthers the understanding of the material, and which, moreover, results in truly mechanical manipulations.

1) Emphasis is laid on the possibility of talking about mathematical objects and their interrelations somewhat as one talks about people and their family relations, as in the sentence: The father of the paternal grandfather of a person is the paternal grandfather of the father of that person. The words 'a person' delimit the scope of the assertion and make clear that it is being proposed about all persons and not only, say, about all living men, or, on the other hand, about all mammals. If f, g, and $=$ were generally accepted symbols for the words 'the father of,' 'the paternal grandfather of,' and 'is the same as,' respectively, then one might write:

$$fg\mathrm{X} = gf\mathrm{X} \text{ for any person X.}$$

But the letter X in the formula is not self-explanatory and does not delimit its scope. This is why the formula, in order to render the sentence, must be amplified by the explanatory legend 'for any person X.' Similarly, the square of the fourth power of a real number equals the fourth power of the square of that number. The words 'a real number' delimit the scope of the assertion and make it clear that it is being claimed for all real numbers and not only, say, for all integers, or, on the other hand, for all in some way generalized numbers. Using universally accepted symbols and a letter in lieu of the words 'a real number', one may write

$$(\mathrm{x}^4)^2 = (\mathrm{x}^2)^4 \text{ for any real number x.}$$

Since the letter x is not self-explanatory, the mere formula would again leave uncertainty about the scope of the assertion; therefore it must be amplified by a legend. Even the very intent of a mere formula including a letter is in need of clarification. One and the same formula occurs, for instance, in the *assertions* that $x^2 - 9 = 0$ for x being 3 or -3, and in the *problem*: find x such that $x^2 - 9 = 0$. In case of an imperative legend, no assertion is intended, and the letter is referred to as the *unknown* of the problem, whereas the letter in a formula accompanied by a description of its scope is called a *variable* – in arithmetic, more specifically, a *numerical variable*. This is the only sense in which the latter term is used in the new presentation. If the writer of a formula that includes letters not designating specific mathematical objects fails to append a legend explaining the intent and the scope of the formula, then he forces his reader to do mere guessing – a procedure strictly shunned in the new approach.

2) The area (say, in square feet) of a square is the second power of the length (in feet) of the side of that square or, in a formula following these words:

(4) $a(Q) = (S(Q))^2$ for any square Q.

Here, Q serves as what might be called a square variable. In contrast, a and s designate definite mathematical objects of the type that Newton called *fluents*, namely, area and length in the realm of squares, each fluent resulting from the association of a number with any object of a certain kind. Traditionally, formula (4) is abbreviated to the formula $a = s^2$ connecting the two fluents themselves rather than their values for any square – a situation unfortunately obscured by referring to the fluents a and s as *variables*, and thereby adding yet another meaning to that highly equivocal term. Naturally, a and s must not, in formula (4), be replaced with letters designating any two other fluents (say, perimeter and diagonal), nor should they be interchanged. While $a = s^2$ is true, $s = a^2$ is false, just as, in the realm of numbers, $e < \pi$ is true and $\pi < e$ is false. Contrast $a = s^2$ with a statement about many numbers; for example, for any two positive numbers, a and s,

$$\text{if } \sqrt{\mathrm{a}} = \mathrm{s}, \text{ then } \mathrm{a} = \mathrm{s}^2.$$

Here, a and s do not designate fluents. Here, they serve as numerical variables and may, without any change of the meaning, be replaced (for example, with x and y) or even interchanged: for any two positive numbers a and s,

$$\text{if } \sqrt{\mathrm{s}} = \mathrm{a}, \text{ then } \mathrm{s} = \mathrm{a}^2.$$

In the clarified presentation, the conceptual difference between numerical variables and fluents is visibly reflected in a typographical distinction that the reader will note in paragraphs 1) to 5) of this section. Letters in roman type serve as numerical variables, while fluents and functions are designated by *italic* type. This device not only greatly facilitates intelligent reading of mathematics but forestalls a great deal of otherwise almost inevitable confusion. The class of all points (that is, pairs of numbers) (x, y) such that $\mathrm{y} = \mathrm{x}^2$ is the same as the class of the pairs (a,b) such that $\mathrm{b} = \mathrm{a}^2$ or of the pairs (y,x) such that $\mathrm{x} = \mathrm{y}^2$. On the other hand, the parabola $y = x^2$, that is, the class of all points P such that $y(\mathrm{P}) = x^2(\mathrm{P})$, is of course different from the parabola $x = y^2$. Here, x and y are fluents, the abscissa and the ordinate, whose values for the point P are $x(\mathrm{P})$ and $y(\mathrm{P})$. None of the paradoxes of the 17th century notation has to be explained, because in the new presentation none of them ever arises.

3) The (traditionally symbol-less) identity function, which for any number x assumes the value x, is a mathematical object of paramount importance, and it clearly deserves a permanent symbol. If j is used as its designation, then the obscure statement (S) is replaced by:

> j is the function such that, for any number x, the value $j(\mathrm{x})$ equals x.

Every statement made in the course of the new presentation may be – in fact, must be – taken literally. The function $j + 1$ is nonconstant, and $(j^4)^2 = j^8$. Here again it might be argued that since the lack of a symbol for the identity function has not impaired the success of analysis for the past 300 years, such a symbol must be superfluous. But, at about A.D. 500, Greek mathematicians could say that the lack of a symbol for zero had not impaired the success of their arithmetic for over 800 years. Yet the introduction of the Hindu cipher 0 made arithmetic even more successful and greatly furthered the development of algebra.

4) The distinction between numerical variables, fluents, and functions entails the distinction in calculus between the derivative of a function and the rate of change of one fluent with respect to another fluent – two terms traditionally considered as synonymous even though the derivative associates a function with a function, and the rate of change associates a fluent with two fluents. For instance, the derivative of the sine function is

the cosine function (the symbols for the sine and cosine functions are italicized, whereas x serves a numerical variable):

$$\mathbf{D}\, sin = cos \quad \text{or}$$

(5) $\mathbf{D}\, sin\, \mathrm{x} = cos\, \mathrm{x}$ for any x.

The rate of change of the distance travelled with respect to the time elapsed is velocity:

$$\mathbf{d}s/\mathbf{d}t = v.$$

For a harmonic oscillator,

(6) if $s = sin\ t$, then $\mathbf{d}s/\mathbf{d}t = cos\ t$.

Here s and t are specific fluents in contrast to the numerical variable x in formula (5). If s and t in formula (6) are misused as numerical variables, say, by replacing t with π, and s with 0, the result is an implication whose antecedent ($0 = sin\ \pi$) is valid, while its consequent $\mathbf{d}0/\mathbf{d}\mathrm{x} = cos\ \pi$ is utter nonsense. The derivative of a function is its rate of change with respect to the identity function: $\mathbf{D}f = \mathbf{d}f/\mathbf{d}j$, for any differentiable function f. The situation in integral calculus is analogous.

5) In the new presentation, symbols for operations and operators are introduced with great care, avoiding synonyms and equivocations, and in a way that is free of confusing ballast. For instance, the integral beginning at 1 of the function f might be denoted by $\mathbf{S}_1 f$, following the verbal pattern. This symbol bears a relation to the synonymous traditional symbols

$$\int_1^x f(x)\mathrm{d}x \text{ and } \int_1^x f(t)\mathrm{d}t$$

comparable to that of "1984" to "MCMLXXXIV" and "MDCCCC-LXXXIV." The introduction of any symbol is accompanied by articulate rules concerning its use. In particular, clear stipulations are made as to which part of a formula where that symbol appears is within its reach. One of them is the stipulation that within the reach of an operator symbol (all of which are printed in bold face, as $\mathbf{D}$ and $\mathbf{S}_1$) is only the immediately following function. On this basis it is clear that $\mathbf{D}\, sin\ 0$ is the value that the function $\mathbf{D}\, sin$ assumes for 0 and not the derivative of the (constant) function $sin\ 0$, which would be denoted by $\mathbf{D}\,(sin\ 0)$. Such rules, in conjunction with the use of a symbol for the identity function, make it possible actually to achieve what the classical treatment claims to achieve: to *manipulate formulas in a purely mechanical way*. In the traditional transition from formula (2) to formula (3), like letters in various occurrences are treated in altogether unlike ways. In contrast, the streamlined version of formula (2) – that is,

(2′) $\mathbf{D}\,\mathbf{S}_1 f = f$ for any continuous f

– implies $\mathbf{D}\,\mathbf{S}_1 f\mathrm{x} = f\mathrm{x}$ for any x; in particular,

(3′) $\mathbf{D}\,\mathbf{S}_1 f4 = f4$.

The transition from general statements to specific formulas proceeds by systematic substitutions and by replacements of variables with designations of specific objects. This technique results not only in a simplification of pure as well as applied analysis but in their complete standardization and automatization.

CONCLUSION

It goes without saying that, when initiating students into the great ideas of 17th century mathematics in any reformed presentation, one must not neglect to teach them to read the various classical notations, especially those going back to Leibniz, to Lagrange, and to Cauchy. (Our Renaissance ancestors, when disseminating the ideas of arithmetic in the reformed symbolism, taught their students also to read Roman numerals, which in fact are still being taught.) It has been my experience that this aim can be achieved without difficulty. Larger scale experiments would undoubtedly result in further improvements.

The main problem clearly lies in the instruction of *teachers*[2] who have been brought up to regard as nonexistent just those points that cause the beginners' crucial difficulties, and who themselves have never been provided with answers to their students' basic questions – the questions concerning the antiquated frame of intermediate mathematics – even though many teachers feel that those questions are justified. In other words, the problem is *to instruct those teachers to use symbols and basic concepts consistently and to transmit to their students the clarified techniques according to articulate rules*. Considering the remarkable, if partial, success of teachers along traditional lines, one may be confident that, equipped with adequate conceptual and symbolic tools, they will make intermediate mathematics available to such wider strata of the population as the present age demands.

NOTES

[1] [1953, 2], [1954, 6], [1955, 4], [1956, 2], [1957, 1] and later writings.

[2] B. F. Skinner points out (*Science*, **128** 1958, p. 176) that this problem exists in the case of every sweeping new method, mentioning as an example the approach to the Calculus here presented. He suggests the use of teaching machines to overcome the problem.

CHAPTER 13

WHY JOHNNY HATES MATH

A reprint of [1956, 4] with minor amendments,

1. THE PROBLEM

"Mathematics the most-hated Subject" was the title of an editorial that in the mid 1950's appeared in a large newspaper. "According to a national survey of high school seniors," says a report sponsored by the Carnegie Foundation, "twelve per cent of them had never taken any algebra or geometry; twenty-six per cent had quit studying mathematics after one year; and another thirty per cent had dropped the subject by the end of the second year." In view of the unprecedented need for scientists with an adequate basic training in mathematics "there is cause for great alarm," and "concern over the mathematical incompetence of the average – and even above-average – American has become almost a national preoccupation." All associations and foundations connected with either science or education have expressed grave warnings, and large public as well as private funds go to projects whose aims are to explain why mathematics is so unpopular and to alleviate the resulting crisis.

Discussion has brought forth a great variety of partial explanations. Experts have blamed the teachers and the students. They have criticized the universities, the teachers colleges, the secondary schools and the elementary schools. They have found fault with the choice of classical topics, their arrangement, and the absence of modern mathematics from the curriculum. They have emphasized that the current textbooks are obsolete, the classrooms overcrowded, and the teachers underpaid. They have, in other words, searched for the roots of the trouble practically everywhere – everywhere except in the procedures of mathematics itself.

Mathematics is looked upon with a mixture of awe and gratitude. And indeed everyone has reason to be grateful. Mathematics has been the decisive factor in understanding the universe and the most powerful tool in controlling nature. All this is beyond any doubt. What is on shakier ground, however, is the general belief that the tremendous achievements of mathematics are due to the clarity of its basic procedures and to the precision of its current language. The time has come when it must be frankly admitted that some mathematical processes are successful not because of those presumed qualities, but rather in spite of obscure foundations, ambiguous expressions, and lack of articulateness. None of these shortcomings – it goes without saying – are in any way bothersome to active mathematicians, who handle their material with a virtuosity that is based on intuition and experience. Mathematicians have the skill to cope with self-created difficulties. But Johnny, studying algebra in a secondary school or college, lacks that skill and is unable to understand part of what he hears from his teacher or reads in the textbook. and this is the reason why Johnny hates math.

2. THE FAULT IS NOT THE STUDENT'S

The mathematical symbols that the student encounters most frequently – on almost every line on every page of his texts – are letters, especially the letter x. Yet he is rarely taught clear and articulate directions for the use of letters. In the formula

(1) $x^2 - 1 = (x + 1)(x - 1)$

he may replace x, for instance, with 4 or with 7, since

$$4^2 - 1 = (4 + 1)(4 - 1) \text{ and } 7^2 - 1 = (7 + 1)(7 - 1).$$

But shame on Johnny if he replaces x with 4 or 7 in some of the other statements that he finds in his text, for example, in

(2) the function $x^2 - 1$ is nonconstant

or in

(3) $x^2 - 1$ is a quadratic polynomial.

For $4^2 - 1$ and $7^2 - 1$ are neither nonconstant functions nor quadratic polynomials. Johnny, however, resents being blamed for doing in (2) and (3) precisely what he had been expected to do in (1). Very understandably, he thinks that the game is unfair because he has not been told the rules. Indeed, where he may, and where he must not, replace letters by numerals is something the student has to find out himself. This is a difficult task.

Nor are the diverse uses of x in (1), (2), and (3) the only ones that the beginner encounters. In the elements of algebra he has to find the number x such that

(4) $x + 1 = 5.$

This task is called solving an equation. The student is expected to say, "The solution is $x = 4$."

In the elements of analytic geometry (usually incorporated in elementary algebra), a Cartesian frame of reference is chosen and a simple procedure taught according to which the line parallel to the second axis and 4 units to its right is called

(5) the line $x = 4$.

Thereafter, the student is expected to refer to $x = 4$ as the equation of the said straight line. This "equation" is worlds apart from the equation (4) or its solution, $x = 4$, and so is the letter x in (5) from the letter x in (4). Furthermore, both are different from the letters x in (1), (2), and (3).

It can be pointed out that physicists also use the same letter in different meanings. They write t for *time* as well as for *temperature*, for example. But they do so in different contexts; at any rate no student confuses the two meanings of t. However, the student is bound to confuse various mathematical meanings of x because these often occur in the same context and sometimes even in the same sentence. For instance, in his study of the line $x = 4$, the student may well have to solve equations such as (4) with the solution $x = 4$. Formulas such as (1) may well occur in the same context as statements (2) and (3), even though they are based on totally different uses of x. Mathematics suffers from x-itis.

The task is not made easier by teachers and texts referring to the same letter x by different technical terms. It is a *variable* in (1) and sometimes also in (2); while x in (2) and (3) is sometimes called a *function*. In other contexts x is called an *indeterminate* in (3). It is called an *unknown* in (4) and an *abscissa* in (5). Slightly more advanced parts of algebra and analytic geometry abound in references to x as *real part*, as *parameter*, and again as a *function* – but a function quite different from that in (2) and sometimes called the function x of x and y. The survivor of freshman mathematics finds that x in $\int_0^1 \cos x \, dx$ is called a *dummy variable* and, later, encounters references to the *operator* x. These distinctions are not very helpful since many students fail to grasp the meanings of these terms. Even though at least eight occur over and over and over again, some of them (especially indeterminate and parameter) are introduced without precise definitions, even the differences between them (especially between *variables*, *parameters* and *indeterminates*) are not clearly described, and articulate rules for the various ways to manipulate them (especially *variables*, *functions*, and *indeterminates*) are lacking.[1]

The student's task is made even harder if he is told[2] that the last letters of the alphabet are variables (standing for any one of many numbers), whereas the first letters of the alphabet are constants (each designating just one number). He cannot find any difference in meaning between the following two statements:

$$x + 1 = 1 + x \text{ for every number } x,$$

$$a + 1 = 1 + a \text{ for every number } a.$$

He cannot find any difference because there is none. The two statements have precisely the same meaning, "But how can that be?" he wonders, considering that the last letters of the alphabet are variables and the first letters are constants. He does not yet know that much worse inconsistencies are in store for him. After he has solved a few equations such as $x^2 = 4$ and $x^2 = 7$ he will be taught to solve the general equation

$$x^2 = c, \tag{6}$$

from which the specific equations result by replacing c with 4 or 7. The letter c in (6) thus stands for any one of many numbers. Yet c is referred to as a constant and not as a variable.

If Johnny is to understand mathematics he has to learn quite a few oddities of the mathematicians' traditional jargon. Figuratively speaking, white is occasionally referred to as "black."

3. NOR IS THE TEACHER TO BLAME

"Variable is perhaps the most mathematical of all notions," wrote Bertrand Russell, one of the great logicians of our time. "It certainly is also one of the most difficult to understand . . . In the present work [*The Principles of Mathematics*, 1903] a satisfactory theory as to its nature, in spite of much discussion, will hardly be found." The search for the solution of some of those difficulties is left to the high school and college teachers.

The last five decades have witnessed the rise of what is called modern mathematics. New branches of mathematical knowledge, undreamed of in 1903, have come into existence in a development so splendid that mathematicians have forgotten, or thought that they could ignore, the words in the *Principles*. Modern ideas make geometry much more lucid as well as much more interesting than it used to be. But just the difficulties Russell alluded to in 1903 have lingered on unabatedly. Mathematicians are eagles which have not freed, nor even tried to free, their legs from shackles that proved harmless in their upward flight. Johnny, however, does not fly. He is learning to walk and in this process, very understandably, he considers shackles on his legs a nuisance that he asks his teacher to remove. It appears, however, that, regardless of whether the teacher has seen the grandiose new vistas of modern mathematics, in some points he still cannot do better than Bertrand Russell did in 1903. Within the frame of the traditional mathematical symbolism he simply cannot answer some of the student's pertinent questions about variables and constants and about x and y.

Indeed, what shall the teacher say when the student (confusedly or clearly) asks in which cases x may, and in which cases x must not be, replaced by numerals? What shall he say when pressed for the difference between a variable and an indeterminate? How shall he reconcile the definition of a constant as a symbol that stands for just one number with the use of the word constant in reference to the letter c in the equation

$$(7) \qquad y = x + c$$

in analytic geometry? What shall he answer when asked whether or not c in (7) and c in, say, the equation

$$(8) \qquad x = c$$

are the same constant? And, for that matter, what is the answer to the question whether x in (7) and x in (8) are the same variable? More generally, when are two constants equal and when are two variables equal?

The teacher's position is anything but enviable since these questions are anything but stupid. Complacently silencing the inquiring student is not a way out of the crisis: Johnny will quit math after one semester. Saying "as you will learn later" now, and "as you have learned before" later, is neither honest nor efficient: Johnny will drop mathematics at the end of the second semester. Ridiculing those questions is of course the worst response: that makes Johnny hate math for the rest of his life.

But the teacher is not to blame if he cannot answer these and dozens of similar questions. He himself has never been told the answers.

4. WHERE THE TROUBLE LIES

The very reasons that exculpate the student and the teacher point to the true cause of the difficulties: the current symbolism and prevailing basic procedures of mathematics. These are creations of the seventeenth century, which, to quote again from Russell's *Principles*, "however fruitful of results, involved a diminution of logical precision and a loss in subtlety of distinction." This framework of mathematics has been left untouched for over a quarter of a millennium. Some parts of the foundation were frozen about 1700 and have remained so ever since. They still satisfy most mathematicians – for the simple reason that becoming a mathematician is practically tantamount to getting thoroughly familiar with, and used to, that antiquated framework and since, as Whitehead once so aptly put it, there is ". . . a passionate attachment of some leaders of thought to the cycle of ideas within which they received their own mental stimulus at an impressionable age". But that all is not well is clearly evinced by the difficulties that beset students and teachers of mathematics.

What requires a thorough analysis are the processes by which, in spite of all difficulties, so many beginners do grasp the subject or at least acquire an efficient manipulative skill. The situation may perhaps be described as follows.

Children learn with amazing ease some rather complicated arithmetical conventions, e.g., that $10 - 3 - 2$ equals $7 - 2$ (and not $10 - 1$), whereas $10 - 3 \times 2$ equals $10 - 6$ (and not 7×2). This and much more, if their teachers are successful, they learn in grade school. Not in explicit rules, though. Such rules would be entirely beyond a child's comprehension. Children learn the conventional use of minus and times signs by observing what the teacher writes and by guessing what he thinks while writing. And they begin to conform to the rules in tentative experiments and by imitating the teacher. This means that children learn arithmetical symbols just as, a few years earlier, they learned English words. They absorb conventions of arithmetic just as they absorbed the conventions of English grammar. The procedure is so thorough that both types of conventions become second nature to them.

In high school and during their first year in college, adolescents are expected to learn algebra and analytic geometry essentially in the same manner. The terms (variable, indeterminate, parameter, etc.) are repeated over and over again, some of them without precise definitions or even clear descriptions of the differences between them. This combined with the lack of consistent rules for the use of x and y simply does not leave the student any alternative method of learning. And some students do learn algebra. In Johnny's view, some grasp the stuff because they are geniuses, while others only pretend that they understand and make the teacher think they do so by their peculiar way of watching him do problems on the blackboard and then doing similar problems in tests. Johnny has a point there. Indeed, those who pass the course (including the geniuses) learn the use of those terms and symbols by observing what the teacher writes and by guessing what he thinks while writing. They begin to conform to the rules in tentative experiments and by imitating the teacher. Well, one might say, this should not be too hard for Johnny because it is precisely what he did as a child when learning to speak.

But the clairvoyance that makes children painlessly learn their native tongue is blurred in many adolescents studying a foreign language. Acquiring more than a smattering proves to be very, very difficult and necessitates formal training in vocabulary and grammar. While the adult is superior to the child in reasoning and in understanding explicitly formulated rules, he is inferior in guessing the implicit and in instinctively grasping the intent of words. In the normal process of maturing, the logical faculties sharpen, but the devinatory powers blunt. As far as mathematics is concerned, arithmetic is something of a native tongue, whereas algebra, analytic geometry and calculus today are foreign languages studied in adolescence. What Johnny really wants and (of course vaguely and inarticulately) asks for are, as it were, the vocabularies and grammars of these languages. But these the teacher cannot give him without, among other things, clarifying the differences (left unexplained in the current textbooks) between the basic concepts, refraining from occasional references to white as "black," and abandoning the indiscriminate use of x and y in totally diverse meanings. In other words, to satisfy Johnny's confusedly expressed but clearly justified demands, time-honoured mathematical practices must be changed.

But why, one might ask, are matters coming to a head just in the United States and in the 1950's?

The answer to the first question is that Johnny, who has a great deal of common sense and an incurable foible for calling white "white," has never enjoyed intellectual acrobatics as much as do his cousins abroad. Until recently no one cared. Now, all of a sudden, the nation desperately tries to sell him mathematics. On this buyer's market he dislikes the brand of mathematics offered to him and he balks. What may satisfy accomplished mathematicians is not good enough for Johnny. His naïveté, too often scoffed at, does not tolerate that "diminution of logical precision" mentioned by Russell. He is disgusted (to the point of dropping math) with what appears to him as a confusion beyond comprehension. Unconsciously he aims at regaining some of the "lost subtlety." Therefore, in the case of

Johnny vs. antiquated mathematical procedures, this writer wholeheartedly sides with the plaintiff.

That matters are coming to a head just in the 1950s is due to the postwar increase in the scale at which we disseminate (or at least attempt to disseminate) mathematico-scientific knowledge, thus, in a way, to the democratization of higher education.

These specific causes of the trouble, which are indeed gratifying, make it even more imperative to solve the grave problem at this time.

5. A PROPOSED SOLUTION

Simple questions about costs and return, which Johnny would rationally attack and readily answer on the street at any time, in mathematics classes often paralyze his brain. The idea of treating the problems in the context of algebra or calculus as he would treat them in any other connection does not even dawn upon him. For he regards mathematics as detached from, nay opposed to, common sense – that common sense of which he has more than his share – instead of realizing that mathematics is the epitome of common sense. For the reasons indicated he mistakes mathematics as an abracadabra and tries to guess the right incantation instead of approaching algebra rationally.

Under these circumstances, the presentation of mathematics must be absolutely and impeccably rational. Even the slightest inconsistency in practices or terms is intolerable. Any contradiction, no matter how trivial to the initiated, only serves to strengthen some beginner's misconceptions about magic. The solution of the grave problem created by this misunderstanding of mathematics is the expurgation of everything, even remotely resembling nonsense followed by an appeal to the students' common sense.

Various attempts are being made in the mid 1950's to revise the presentation of the classical system of mathematics. Only experiments on a large scale can decide merits and result in improvements of any particular approach. But it may perhaps be mentioned that students exposed to this writer's clarified treatment of the basic concepts and to his articulate rules for the use of the basic symbols eagerly absorb this vocabulary and this grammar of the mathematical language and respond to this new approach enthusiastically. The essence of this treatment is summarized in three points:

1. A clear distinction is maintained between numerical variables such as x in (1) and variable quantities such as the abscissa in analytic geometry,

as in (4). Naturally, the same symbols are not used for variables and variable quantities. For instance, one may use roman type for the numerical variable and italic for the abscissa as in

(1*) $\mathrm{x}^2 - 1 = (\mathrm{x} + 1)(\mathrm{x} - 1)$ for any number x

(4*) $x = 4$.

This distinction can easily be maintained on blackboards and on paper by writing

x, y, z

for variables and

$\mathscr{x}, \mathscr{y}, \mathscr{z}$ or X, Y, Z

for variable quantities or by underlining the symbols for the latter.

2. A clear distinction is maintained between the numerical variable x in (1*), on the one hand, and the identity function, as in (2), on the other. The identity function is defined by the process of associating with any number that very same number. The result of this process, the function assuming for any number x the value x, may therefore be considered as the class of all pairs (x, x) for any number x. While in this definition x is a variable, there is nothing variable about the identity function itself. Of course, the function is denoted by a symbol different from those used as numerical variables. I use the symbol *j* for the identity function.

3. A clear distinction is maintained between a numerical variable such as x in the statement

$$\frac{\mathrm{x}^2 - 4}{\mathrm{x} - 2} = \mathrm{x} + 2 \text{ for any number } \mathrm{x} \neq 2,$$

and an indeterminate. The variable is meant to be replaced by the designations of specific numbers (in fact, of any number ≠ 2), whereas an indeterminate is not. An indeterminate is rather of the nature of the imaginary i in

$$\frac{\mathrm{i}^2 - 4}{\mathrm{i} - 2} = \mathrm{i} + 2.$$

Here, i is not meant to be replaced by anything, whence no one would dream of qualifying the last formula by saying "providing i ≠ 2."

For numerous details of this new approach, such as the definitions of variable quantities, indeterminates, and parameters, the treatment of constants, etc., the reader must be referred to the author's books.[3]

In the long run, a teacher will be successful only if (in the spirit of Russell's humble admission of imperfection) he deepens his insight into mathematics to the point where, perhaps contrary to what he may have learned or believed, he realizes that the material he has to teach urgently needs clarification. No one should be surprised that in unfreezing the 250-year-old foundation, the introduction of two or three new symbols becomes indispensable. What is surprising, and actually a new triumph of Renaissance mathematics, is the fact that overhauling its colossal machinery calls for so few and so small adjustments to bring it up-to-date. It is as though in building a skyscraper instead of a seventeenth-century chateau all one needed were a few extra hammers.

The teacher must acquire a full understanding of the conceptual differences between

(a) variables,
(b) co-ordinates,
(c) the identity function,
(d) indeterminates,

(reflected in their dissimilar uses) and he must initiate the beginner into these distinctions. Since maintaining them is next to impossible within the framework of the current symbolism – that pan-x-ology in which all the diverse concepts (a)–(d) and others are denoted by x – *the teacher must denote unlike concepts by unlike symbols.*

This is a modest price for the resulting clarification. It is this clarification that enables the teacher to answer dozens of otherwise highly embarrassing questions. And most of today's unanswerable questions, which concern the equivocal use of basic terms and symbols, will not even arise.

6. IS A NEW APPROACH WORTH THE TEACHER'S WHILE?

The main obstacle to experiments along basically new lines is a reasoning, apparently prevailing among some research mathematicians which might be summarized as follows: "Since for the past two hundred years and to this day, all mathematicians and scientists have achieved complete mastery of mathematics with its time-honored procedures and in its traditional presentation, and since furthermore, the difficulties here discussed do not disturb any accomplished mathematician in the least, it follows that youngsters who cannot cope with these difficulties must be mathematically utterly incompetent. To revise procedures or symbols for their sake is not worth anyone's while since their study of mathematics cannot, under any circumstances, be profitable either to those mathematical morons themselves or to anyone else."

But this reasoning is a dangerous fallacy. For how can one rule out the possibility that some quite intelligent youngsters, deterred by incomprehensible basic material, have gone into other studies or occupations and thus never reach the mathematical level on which research men train talent? How can one even rule out the supposition that some logical minds are dissatisfied with mathematics as presented to them and abandon it at an early stage just because of their logical abilities? And how can a country with a need for trained mathematicians and scientists afford to go along with ideas that risk the loss of talent, perhaps even of especially valuable talent?

As a matter of fact, statistics seem to indicate that this loss of talent is very real. The national survey mentioned at the beginning of this paper shows that among the brighter high school seniors alone (that is, the top thirty per cent on a test of mental ability) more than forty per cent never get beyond the elementary phases of mathematics. And, passing from the statistics of masses to the individual, this writer will never forget the remark of an eminent legal authority, renowned for the acumen of his analytic mind, when he recalled "Mathematics is a field in which I wasn't any good in school. I never understood what x meant."

NOTES

[1] In [1956, 5], twelve different uses of x and y are listed. Cf. also Chapter 8 of this volume.

[2] According to O. Frink, this is the conventional usage. See O. Frink. *Mathematical Reviews* (June, 1955) pp. 575–576. See also G. Polya, *How to Solve It* (Princeton: Princeton University Press, 1945), p. 125.

[3] [1955, 4] and [1957, 1].

CHAPTER 14

ON THE FORMULATION OF CERTAIN QUESTIONS IN ARITHMETIC

By Herbert J. Curtis and Karl Menger

A reprint of [1956, 1] with minor amendments.

Everyone's answer to the question "$9 \times 4 = ?$" is "36." Yet "4×9," "$6 + 12 + 18$," "$100 - 64$," "6^2," and countless other answers would be equally correct. Why then does everyone say "36?"

Psychologically, the almost automatic response is, of course, due to the fact that as children the respondents memorized the multiplication tables. Similarly, because they have memorized the alphabet, they automatically answer "y" when asked which letter follows x in the alphabet. But there is more to answering "36" than being conditioned to this response. The answer "36" is actually different from all other possible answers. The difference arises from the fact that the symbols for integers fall into two classes: (1) *basic symbols*, in the Western world, the Hindu-Arabic numerals "1", "2," "3," and so forth; (2) *expressions*, such as "4×9" and "$6 + 12 + 18$" describing numbers by certain relations to certain other numbers in terms of basic numerals and symbols such as "$\times$" for multiplication, and "$+$" for addition.

The first point of this paper concerns the clarification of the nature of countless arithmetical questions including, besides $9 \times 4 = ?$, problems with preambles such as "find the product . . . ," "compute the sum . . .," "determine the quotient . . . ," *Each such question aims at evoking the basic symbol for a number described by an expression.*

It would seem to be highly desirable that students of arithmetic be acquainted with this situation at an early stage of their training. Students do not have to be very mature to grasp the difference between basic symbols and expressions. And only if the nature of such a problem is understood is the question "$9 \times 4 = ?$" above the level of questions such as "Which letter follows x in the alphabet?"

For rational numbers, there are two commonly used sets of basic numerals: decimal fraction symbols such as .5, .6875, .066 . . . 6 . . . or $.0\dot{6}$, and $.\dot{1}\dot{4}\dot{2}\dot{8}\dot{5}\dot{7}$; and common fraction symbols such as 1/2, 11/16, 2/30, and 1/7. In some cases the decimal fraction symbol is shorter, in others the common fraction symbol is shorter.

Each type of symbol conveys certain information about numbers better than does the other. For instance, a glance at the symbols $.\dot{6}\dot{2}\dot{1}$ and $.\dot{6}\dot{4}\dot{1}\dot{0}\dot{2}\dot{5}$ reveals that the first number is smaller than the second. It takes most people longer to deduce this fact from the common fraction symbols for these numbers, 23/37 and 25/39. On the other hand, the product of the two numbers mentioned is more easily found for common than for decimal fractions. As an extreme example, contrast the ease with which one learns to recognize the product of 7 and 1/7 as 1 with the difficulty in recognizing the product of 7 and $.\dot{1}\dot{4}\dot{2}\dot{8}\dot{5}\dot{7}$. That common fraction symbols are more suitable for multiplications and divisions is not surprising since they are defined in terms of multiplication alone, whereas decimal fraction symbols are defined in terms of multiplication and addition.

Which of the two systems one chooses in a given case will depend largely upon the nature and the purpose of the study. For instance, in printing comparative information, especially tables, decimal fraction symbols are preferred.

The second point of this paper, illustrated by the preceding examples, is to stress *the existence of various possible systems of basic numerals*. On the level of instruction where non-decimal positional symbols are introduced, dual and duodecimal numerals (using the bases two and twelve, respectively) supply further examples. If the aim of a question is to evoke the basic symbol for a number given by an expression, then it obviously presupposes a clear specification of a definite system of basic symbols.

Among the irrational numbers are some whose (necessarily non-periodic) decimal representations can be described by simple rules. An example is .101001000100001 . . . described by the rule that the

> 1st, 3rd, 6th, 10th, . . . , $\frac{1}{2}n(n+1)$st digits are 1 and all others are 0.

On the other hand, for the positive number whose square is 2, there is no simple rule describing the entire string of digits although one may compute as many digits as desired. True, a reader of the symbol 1.41412 . . . . imagines that a reference to the number in question is intended. But this is a mere guess, since countless other numbers begin with those same five decimals and since the symbol does not in any way indicate what digits follow (except possibly that an unending string is to be expected). The only complete and unambiguous description of the positive number whose square is 2 is by an expression such as $\sqrt{2}$, consisting of one or more symbols for relations or operations and of one of more basic numerals.

The expression 2^{32}, even for an integer, is short and clear while the basic symbol for that number is long and untransparent.

The third and perhaps most important point of this paper is to stress that, *for many numbers, expressions are not only the simplest symbols but the only available symbols that are complete and unambiguous.*

This insight seems to have an important bearing on questions aiming at the basic symbols for numbers. Current texts in arithmetic and algebra abound in problems such as: Compute $\sqrt{16} - \sqrt{9}$. Perform the indicated operations: $\sqrt[3]{8} + \sqrt[3]{27}$.

In these examples, expressions designating numbers for which simple basic numerals are available have been selected. But very similar expressions designate numbers for which simple basic numerals do not exist, such as $\sqrt[3]{7} + \sqrt[3]{26}$. It would, of course, be perfectly reasonable to ask: Find the first five digits of $\sqrt[3]{7} + \sqrt[3]{26}$. But it would not be reasonable to ask: Perform the operations indicated in $\sqrt[3]{7} + \sqrt[3]{26}$. What could one do but replace the complete and unambiguous expression by one of countless incomplete symbols none of which characterizes the number designated? On the other hand, the difference between the problems involving $\sqrt[3]{7}$ and $\sqrt[3]{8}$ does not concern the relations or operations. It lies in the fact that the expressions designate numbers one of which does, and one of which does not, possess a simple basic symbol.

The student of arithmetic and algebra should not only be trained in finding the basic numerals for some numbers described by expressions, but also in understanding that, for many numbers, expressions are the simplest, or even the only complete and unambiguous descriptions, and that in such cases replacement of the expression by a basic decimal numeral must be foregone.

We conclude with an example of what we believe is a reasonable and desirable formulation of problems of the type mentioned previously.

For each of the following numbers try to find the basic decimal numeral, or, if the latter is not available, a shorter expression (say, one that includes fewer root signs). Note that in some cases even such a "simplification" is impossible: a) $\sqrt{4.41} + \sqrt{4.84} + \sqrt{5.29}$; b) $3\sqrt{5}$; c) $\sqrt{3} - \sqrt{2}$; d) $\log_{10}\sqrt{10}$; e) $\sqrt{2 + \log_{10}100}$; g) $\sqrt{2}\cos 45°$; h) $1 + \cos 45°$.

Similarly, consider the sum, the difference, the product, and the quotient of the following pairs of numbers: i) $\sqrt{3} + \sqrt{2}$ and $\sqrt{3} - \sqrt{2}$; j) $\sqrt[3]{3} + \sqrt[3]{2}$ and $\sqrt[3]{3} - \sqrt[3]{2}$; k) $\sqrt[3]{3} + \sqrt[3]{2}$ and $\sqrt[3]{7} - \sqrt[3]{5}$.

NOTE

[1] We write .5 and $\frac{1}{2}$ instead of ".5" and "$\frac{1}{2}$" where the words "the symbol" make it clear that reference is made to the numeral and not to the number.

CHAPTER 15

ON THE DESIGN OF GROUPING PROBLEMS AND RELATED INTELLIGENCE TESTS

In this chapter, large parts of [1953, 1] are reprinted with indications of the places where sections of the original paper have been omitted.

Part I of the present paper contains a critical analysis of a particular test in Figure Grouping. In Part II, a general method is developed by which correct tests of this type can be constructed. Part III contains examples which considerably widen the scope of both the critical and the positive part.

PART I. THE LOGIC OF FIGURE GROUPING

1. *An Example*

An intelligence test administered during World War II to a large group of college students included a question which, as far as I remember, read as follows:

"Which of the five figures below has a property not possessed by any of the other four figures?"

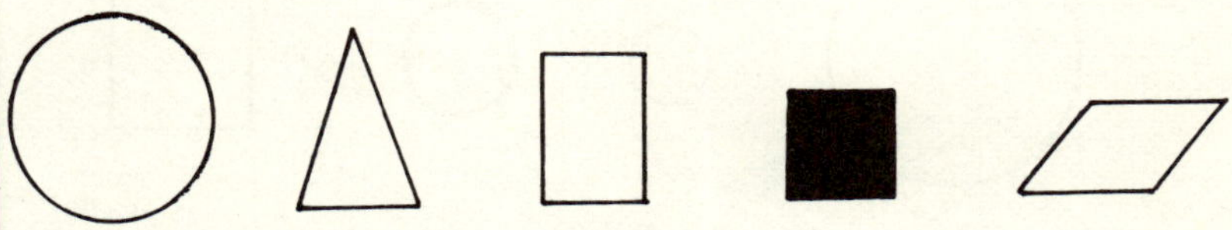

Fig. 1

Upon an inquiry, the testing agency revealed that the expected answer was 'The Square' – the square being the only one of the five figures which is black. Any other answer was held against the intelligence, more specifically, against the reasoning ability, of the tested person.

Wartime duties prevented me from pursuing the matter beyond a protest to the testing agency which had no effect. Recently, my concern was renewed when it came to my attention that questions like the one mentioned are still being used in intelligence tests.

In the example above, it is clear that each of the five figures has a 'distinctive' property, that is, a property not possessed by any of the other four figures. In fact, for each figure we shall list two properties neither of which can be claimed for any of the other four figures.

(1) The circle is the only one of the five figures which has (a) no corners, (b) infinitely many axes of symmetry.

(2) The triangle is the only one of the five figures which has (a) exactly three corners, (b) only one axis of symmetry (namely, a vertical axis of symmetry).

(3) The rectangle is the only one of the five figures which has (a) equal angles and unequal sides, (b) exactly two axes of symmetry (namely, a horizontal and a vertical axis of symmetry).

(4) The square is the only one of the five figures which has (a) equal sides and equal angles, (b) four axes of symmetry (namely, a vertical, a horizontal and two diagonal axes).

(5) The parallelogram is the only one of the five figures which has (a) four angles none of which is right, (b) no vertical symmetry.

Clearly, each of the figures has further distinctive properties. For instance, the blackness of the square has not been included in the above list. Moreover, the mere property of being a circle is distinctive for the first figure; the property of being a triangle, for the second; and so on.

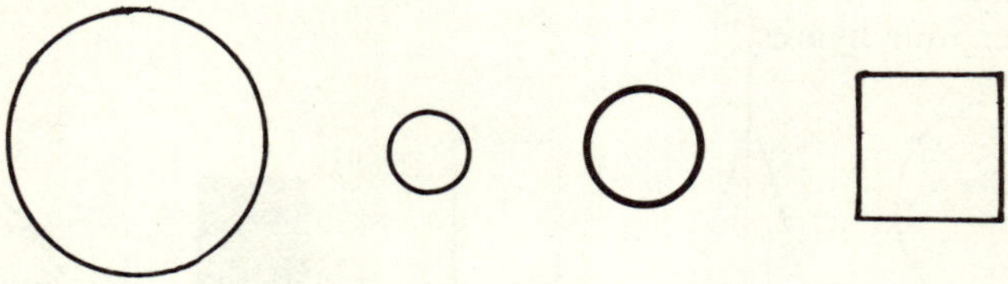

Fig. 2

The situation is by no means due to a shortcoming of the selection of these particular five figures. Consider, for instance, the group of three circles and one square in Fig. 2. The tested person might be expected, even more strongly than in the first example, to single out the square. But nothing is logically wrong with singling out the first circle as the only figure whose area exceeds a square centimeter; the second, as the only figure with an area less than one half of a sq. cm.; the third as the only circle with an area close to one sq. cm.

2. *The Logical Situation*

The logic of the matter can be described as follows: First consider two nonidentical objects. By Leibniz' principle of the identity of indiscernibles, the nonidentity of the two objects implies the existence of a property possessed by one of them, say the first, and not possessed by the other one. Clearly, the negation of this property is a property of the second object not possessed by the first.

Next consider a class of more than two objects no two of which are identical. By virtue of Leibniz' principle, for each pair of objects belonging to the class, there is a property possessed by the first but not possessed by the second. But the makers of tests are interested in properties of an object which are distinctive, that is, not possessed by any other object of the class. The question arises for a group of objects whether each has a distinctive property (as was the case in the examples mentioned above); or whether several have distinctive properties while others do not; or whether none has a distinctive property; or finally whether exactly one object has a distinctive property. That the last situation obtains is implicitly assumed in the test questions. While this assumption is often invalid, the last mentioned situation is, at any rate, the one desired by the makers of tests. We shall therefore call a class of objects a *test group* if exactly one of the objects of the class has a distinctive property.

If the realm of properties to be taken into consideration is limited and specified, then it is easy to show that each of the four above mentioned situations can actually occur. Consider, for instance, three categories: color, form of the contour, and number of sides; and in each category two properties: white-black, dotted-solid, square-triangular. Suppose that it is explicitly stipulated that only the above categories are to be taken into consideration. Then consider the following groups of three or four objects.

Example I. A *black* solid square, a white *dotted* square, and a white solid *triangle*. Each figure has a distinctive property – the one designated in *italic* print.

Example II. A *black* solid *triangle*, a white *dotted* square, and a white solid square. The first figure has two distinctive properties, the second has one, the third has none.

Example III. A black solid square, a black solid triangle, a white solid triangle, and a white solid square. None of the four objects has a distinctive property. In this example the contour of the four figures has not been varied. The example would remain valid if two contours were dotted and two left solid.

Example IV. A *black* dotted square, a white solid triangle, a white solid square, and a white dotted triangle. Only the first object has a distinctive property, namely, blackness.

Example IV is the only one that admits a unique answer to the question "Which object has a property not possessed by the other two?" It represents the only test group. But even in this example one has to limit and specify the properties to be taken into consideration in order to be able to say that the first object is the only one with a distinctive property.

The first object may cease to have this distinction

a) if properties other than those specified (such as position relative to the margin of the page or order of the objects) are taken into consideration (for instance, the second object might be characterized by the property of being immediately to the left of the vertical median line of the page);

b) if different specified properties (such as color and shape) are combined into one (for instance, the fourth object might be characterized as the only white square[1] in Figure 2).

To my knowledge, however, in intelligence tests the realm of properries to be taken into consideration is not usually limited and specified.

As an empirical extension of Leibniz' logical principle, the following law might be formulated: In every finite class of objects no two of which are identical, each object has a distinctive property.

PART II. A METHOD OF CONSTRUCTING CORRECT GROUPING TESTS

3. *The Construction of 5-Element Test Groups*

If the realm of properties to be taken into consideration is limited and specified, then the construction of five objects O_1, O_2, O_3, O_4, O_5 in which exactly one object has a distinctive property is a simple combinatorial problem.

First consider five categories each containing two properties (such as white-black, dotted-solid) or as we shall say, briefly, *binary categories*. We shall indicate each category by a Roman numeral. In each category, we denote one property by $+$, the other by $-$. If O_1 is the only object with a distinctive property, namely, the property $+$ of category I, then O_2, O_3,

O_4, and O_5 must have the property – of category I. Besides, we shall let each of the objects O_2, O_3, O_4, O_5 share with O_1 the property – of one category to keep them from having distinctive properties. This leads to the following scheme:

	I	II	III	IV	V
O_1:	+	−	−	−	−
O_2:	−	−	+	+	+
O_3:	−	+	−	+	+
O_4:	−	+	+	−	+
O_5:	−	+	+	+	−

Of course, the objects, the categories, and the properties may be permuted.

In a more economical way, a test group of five elements can be constructed by means of three binary categories. Again, let O_1 be the only object with a distinctive property; namely, property + of category I. The other objects shall, in addition to property – of category I, have the following properties:

$$O_2: + +; \quad O_3: + -; \quad O_4: - +; \quad O_5: - -.$$

What properties O_1 has besides property + of pair I is immaterial.

Suppose, for instance,

I is type of contour (dotted +, solid −),
II is the color (white +, black −),
III is the shape (square +, triangular −).

In the above class, O_1 is the only object with a distinctive property (dottedness). (Fig. 3.)

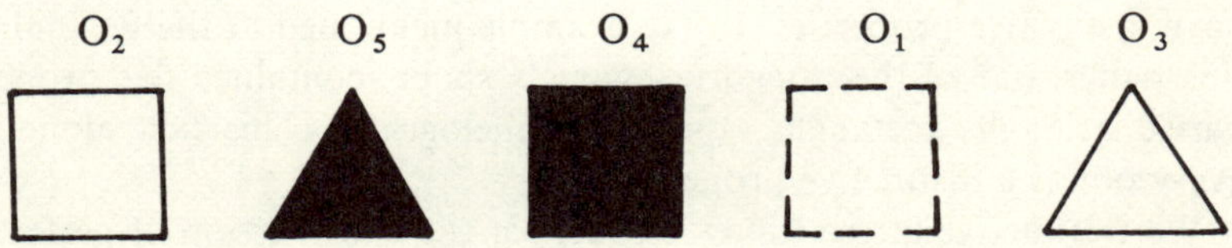

Fig. 3

Next consider a *ternary category*, that is, a category containing three properties such as white-gray-black. If the three properties are distributed among five objects, then at least one object will have a distinctive property.

For instance, the five objects may have the colors white, white, gray, gray, black, in which case the last object has the distinctive property of blackness. If the colors were chosen according to the scheme white, gray, black, black black, then two objects would have distinctive properties: exactly one object would be white, and exactly one would be gray. A test class cannot contain two such objects. Moreover, it is clear that if two ternary categories are selected, then either at least two objects have distinctive properties (which is forbidden), or at least one object has two (or more) distinctive properties.

We shall call a test group *perfect* if the only object possessing a distinctive property has only one distinctive property.

The combination of a ternary category (white-shaded-black) and a binary category (square-triangular) leads to the most economical construction of a perfect test group of five. The example in Fig. 4 includes three squares of all colors and two triangles of different colors. The (black) square with the color not matched by a triangle is the only object with a distinctive property.

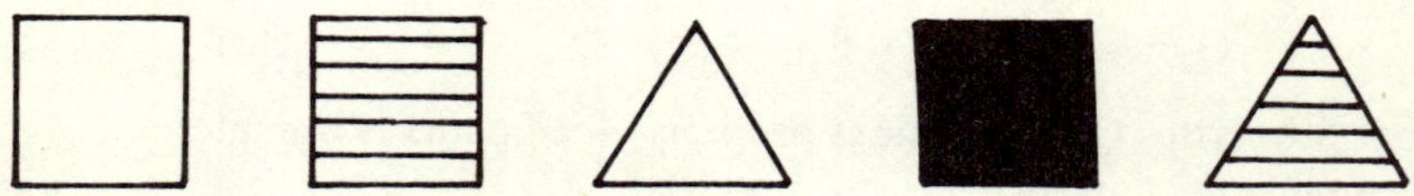

Fig. 4

A perfect 5-element test group cannot be constructed with categories of more than three properties. For if four properties are distributed among five objects, then only one property occurs twice and at least three objects have distinctive properties. In the example mentioned at the beginning of this paper, one of the categories, namely shape, contained five properties (circle, triangle, rectangle, square, parallelogram). This fact alone gave every object a distinctive property.

We thus arrive at the following rule for the construction of perfect test group of five objects with specified properties to be taken into consideration: Only binary categories (such as white-black, solid-dotted) can be used with the possible exception of one ternary category (such as white-gray-black). If a ternary category is used, then it supplies the distinctive property.

4. *The Construction of Perfect Test Groups of More Than Five Elements*

Since the variety of perfect test groups of five is small, it may be of interest to see how larger perfect test groups can be constructed. Sextuples can be obtained by means of four binary or one ternary and two binary or two ternary categories. Below, each category is indicated by a Roman numeral; the properties of binary and ternary categories are denoted by +, − and +, 0, −, respectively.

	I	II	III	IV		I	II	III		I	II
O_1:	+	+	+	−		+	+	−		+	−
O_2:	−	+	+	+		−	+	+		−	+
O_3:	−	+	−	+		−	−	+		−	0
O_4:	−	−	+	+		0	+	+		0	0
O_5:	−	−	−	+		0	−	+		0	−
O_6:	−	−	−	−		0	−	−		0	+

An example of a perfect 10-element test group follows. It is based on a quaternary and a ternary category: A, B, C, D meaning 6, 5, 4, 3 sides, and a, b, c meaning black, striped, white.

O_8 O_7 O_2 O_9 O_4 O_6 O_1 O_{10} O_3 O_5

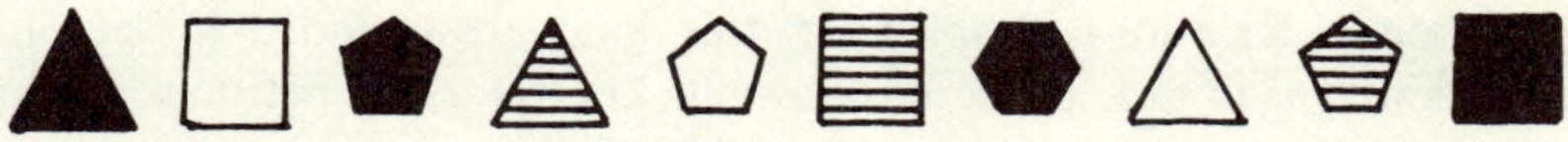

Da Cc Ba Db Bc Cb Aa Dc Bb Ca

Fig. 5

O_1 is the only figure with a distinctive property (having six sides).[2]

PART III. EXTENSIONS

5. *Applications to Other Types of Tests*

It is clear that the ideas developed in both preceding parts apply to numerous types of tests other than Figure Grouping.

A recent test in Letter Grouping includes the four groups

AABC ACAD ACFH AACG

"Three of the groups are alike in some way . . . Mark the one that is different." The testee is supposed to mark ACFH, because it is the only group one which does not have two A's. But clearly, AABC is the only group including only the first three letters of the alphabet, the only one including the letter B, and also the only one in which any two consecutive letters are neighbors in the alphabet; ACAD is the only group in which the letters are not in alphabetic order, and also the only one including the letter D; finally AACG is the only group including two consecutive letters which determine an interval of three letters of the alphabet between them (C . . . G), and also the only one including the letter G.

A unique solution can be expected only if the properties to be considered are limited and specified, and a test class is constructed by the method explained in Part II. Otherwise there is the danger of a plurality of reasonable solutions due to hidden relations which can easily be overlooked in formulating the test.

Suppose a child is asked to underline the word not belonging in the group

apple, pear, strawberry, cloud, plum.

The boy underlining strawberry, because it is the only one of the five objects to which he usually has to look down, may be the potential genius in the group of tested children.

The principles developed in the preceding parts apply also to tests in Eduction of Relations. Problems of this kind are included in reasoning tests as well as in tests concerning specific knowledge. A fictitious example will illustrate the point.

van Gogh: van Beethoven = Raphael: X.

Suppose the following four choices are given for X:

a) Galileo; b) Mozart; c) Chaucer; d) Milton.

In favor of a) it might be said that Beethoven was of the same (Flemish-Dutch) extraction as van Gogh, and that only Galileo was of the same (Italian) extraction as Raphael. In favor of c) it might be said that Beethoven lived before van Gogh, and that only Chaucer lived before Raphael. b) Mozart is the only composer. By choosing Mozart one would, however, emphasize analogous properties rather than educe identical relations since it is artificial to speak of a painter-to-composer relation. One might support even the choice d) by pointing out that only Milton had an infirmity (blindness) comparable to Beethoven's deafness, although here again, healthy-to-handicapped is an artificial relation.[3]

6. *A Comparison with Number Series Tests*

A number series test consists of a row of numbers which "follow one another according to some rule." The testee is expected to "find the rule and fill in the blanks to fit the rule." For instance, in 2, 4, 6, 8, – he is expected to notice that the given series consists of the first four even numbers in their natural order and to write 10, the fifth even number, in the blank.

Most mathematicians are opposed to tests of this type for the following reason: If a series of k numbers, $n_1, n_2, \ldots, n_k$ is given, one can, for any number N, find a rule according to which one should write N in the blank. Hence, whatever number is written in the blank, it can be said to fit some rule.

Yet two important points may be made in support of number series tests. The ability to derive, by induction, a simple rule in a short time is valuable in many domains of extra-scientific activity as well as in science and, to a limited extent, even in mathematics. The motivation of an unexpected continuation of a given number series, as well as the actual operations leading to the unexpected number, are in most cases time consuming. For instance, to obtain 30 as the continuation of 2, 4, 8, 16 certainly takes longer than to obtain 32, and to obtain 29 would take still more time.

If, at the beginning of the test, the testees are told that they will be credited with the number of solutions attained within the allotted time, then probably few unexpected answers will be submitted. It is doubtful whether even students of mathematics would turn in many unexpected answers together with the motivating law, both obtained within the allotted time.

In contrast, in the grouping test illustrated in Figure 1, it does not take longer to notice that the first figure is the only round one than that the fourth is the only black one. In grouping tests the time element is much less important than it is in the number series problems.

The author believes that ambiguities of the type criticized in this paper ought to be carefully avoided in designing tests. If in the example in Figure 1 the testee is expected to mark the square, then he simply is expected to guess what the designer of the test happened to have in mind. With the testee's intelligence, in general, and with his reasoning ability, in particular, the expected answer has nothing to do.

Even if there should be a high correlation between the expected answer in an ambiguous test and high achievement in a valid intelligence test, the author would not admit that an answer different from the one which happens to be expected, supplies any relevant information against the intelligence of the testee.

He would even go so far as to suggest that, on the contrary, unexpected groupings may well be a sign of superior intelligence. In order to make an answer based on a deliberate unexpected grouping distinguishable from a random answer, the testee would, of course, have to indicate the motivation or the principle of classification which has guided his choice. Unfortunately this requirement rules out children, and makes the grading and the interpretation of the test difficult.

For a small number of more mature persons an ambiguous grouping test may be an excellent criterion of intelligence. But it would have to be phrased somewhat like this: "Find a group of four (or, if you can, several groups of four) among the following five items which have a common property not shared by the fifth. In each case state the property." Then the items would follow: words, figures, groups of letters, as the case may be.

NOTES

[1] As Dr. J. K. Senior pointed out, language habits have to be taken into consideration in formulating grouping tests. If in Example IV, squares and triangles are replaced by horses and cows, the German language has, indeed, a single word for "white horse" (Schimmel).

[2] In the original paper, Part II concludes with three perfect 12-element test groups based on a quaternary category in conjunction with a quaternary, a ternary and a binary, and two binary categories, respectively.

[3] In the original paper a discussion of the very interesting Spearman tests follows.

CHAPTER 16

THE GEOMETRY RELEVANT TO MODERN EDUCATION

This chapter contains the *Introduction* to [1971, 2] – a paper that was first presented at an international conference on the teaching of geometry in Carbondale, Illinois, 1970.

For some time now, the shortcomings of Euclid's Elements have been recognized in precollege teaching; and the age that identified the study of geometry with reading Euclid's book is definitely past. But it has also become clear that a presentation of the whole of Euclidean geometry with modern rigor would require more time than secondary schools can apportion to this task. As a result of this situation, many educators feel that they are forced to adopt one of two courses: either to spend too much time on the teaching of geometry or to present the subject without meeting the requirements of modern rigor – a task in which they are supported by textbooks some of which actually have much greater shortcomings than Euclid's Elements. Two ways out of this dilemma have been proposed.

The first is to give up the teaching of geometry altogether – a step welcomed by those students who doubt that geometry is, to use a term in vogue, relevant. But young people who are not taught geometry miss the great experience of seeing a deductive theory developed before their eyes – a theory, moreover, that is accompanied by mental pictures and drawings on paper which motivate its assumptions and illustrate its conclusions. The study of rings and fields in algebra cannot compensate for this loss.

The second proposed way out of the dilemma, initiated by G. D. Birkhoff, is the deduction of geometry from postulates involving arithmetical ideas, in particular the profound concept of real number. This amounts to a synthetic presentation of something akin to analytic geometry. A related method is the embedding of geometry into linear algebra and its presentation as a theory of vector spaces over the real numbers. But although these ideas can be presented in the form of deductive systems they deprive the student of the experience of developing such a theory from assumptions about really simple and purely geometric concepts such as points, lines, and incidence. And just such a treatment is of particular importance because of certain applications.

The applicability of geometry to physics, chemistry, and biology, to engineering and economics and, of course, to other branches of mathematics has been stressed, as has the beauty of geometry, by all those who oppose the discontinuance of the teaching of geometry. It is true, that as far as some of these applications are concerned, a postulational theory of vector spaces is preferable to one in terms of points, lines, and incidence. What I wish to emphasize in this paper, however, are applications of a totally different kind, namely applications to reasoning in general – thus, as it were, applications of the *spirit* of synthetic (synthetic *synthetic* rather than synthetic *analytic*!) geometry to a wide spectrum of topics ranging from the most abstract to the most practical – from philosophy to politics. There is no better immunization against the pseudo-inferences and sham proofs that abound in extra-mathematical reasoning and in daily life than to be exposed in one's youth to a truly airtight deductive theory, however modest in scope. And, as the ancient Greeks discovered, there is no better material for a rigorous treatment than facts about space. *The study of a modicum of synthetic geometry is an experience no one should miss in the course of a modern secondary education: it has far-reaching social relevance.*

I propose a third way out of the dilemma. It is based on the conviction that geometry should be included in the curriculum, that it ought to be taught with perfect rigor, and that it should be presented within a reasonable span of time. The way to achieve these aims is *to abandon the idea of teaching Euclidean geometry in its entirety and to present only a part or an aspect of Euclidean geometry – but that part or aspect with absolute rigor – as well as some simple related theories which, from the points of view of various students, are "relevant".*

As the material to be presented I have selected the following chapters:

I. The elements of affine plane geometry.

II. The theory of a self-dual fragment of the affine plane[1] with applications to kinematics[2].

III. Certain principles of noneuclidean geometry.

If the quality of the students permits it and more time is available, then these chapters should be supplemented by expositions, strictly separated from the deductive theories, of stimulating intuitive material on linear and planar patterns and polyhedra, on sphere, torus, and Möbius strip, and on some interesting simple curves as well as the most general curve concept[3].

This choice of material should make it clear that no attempt is being made to convert the course in geometry into a course in logic. Nor is it my purpose to make the students either over- or under-estimate deductive reasoning, but rather simply to let them see how this mental process works. I actually believe that such a course will, when competently taught, implant in students more deductive methodology than they could get from a first course in logic (on the algebra of classes and propositions). In addition, the course offers some (to the students) unexpected and exciting revelations about space.

The course outlined in this paper is designed for everyone receiving a modern secondary education. Those who plan to specialize in mathematics, science or engineering will need more geometry – additional chapters and additional methods. But the course contains elements that they, like everyone else, should know and understand and yet rarely learn. Hence, if the course were confined to those who do *not* plan to specialize in science or engineering, then some of these young people might well develop a better insight into certain aspects of geometry and general deductive methodology than would some future scientists and engineers and perhaps some future teachers of mathematics, who then might perpetuate the underrating of the social relevance of the study of geometry. Moreover, it is quite likely that, when an inspiring teacher teaches such a course some non-scientifically minded students might change their minds.

Before outlining the material I wish to discuss from educational, psychological, and sociological points of view what this approach achieves.

(1) The student learns that the deductive geometric system must start from unproven assumptions in terms of undefined concepts and that, within that theory, these initial (so-called *primitive*) elements *remain* unproven and undefined. He can easily be made to realize that this is a fundamental feature of any deductive theory.

(2) He learns that the original models of the primitive elements of the geometric theory are objects and facts of the physical world, which the undefined concepts and unproven assumptions are designed to reflect. Perennial occupation with these objects has created in man something called geometric intuition, which is a second, if less tangible, motivation for the development of deductive theories about space.

(3) He is shown that, within the totality of the propositions of a theory, the postulates are not distinguished as such; that they may be replaced by other assumptions from which they and their consequences can be proved; and that one may even choose different primitive concepts, in terms of which those originally chosen can be defined. Of course, in such an alternative development of the theory, *its* primitive elements remain unproven and undefined.

(4) He learns that definitions in terms of undefined words can, when skillfully chosen, be of great practical importance as abbreviations that

simplify theorems and proofs; but that, in principle, definitions can be dispensed with since the definiendum can always be replaced by the definiens.

(5) He learns what proving really means: combining and transforming the assumptions according to the rules of logic. The particular choice of the material makes this nature of the proofs fully transparent since not only are the assumptions free of references to numbers but their logical combinations and transformations do not utilize the theory or, except for pairs, triples, and quadruples, even the concept of sets. For example, a line is not defined as a set of points. What rigor of geometric proofs means can be illustrated in the not infrequent situation where a student wants to use in a proof some plain and simple idea or some evident proposition but is told by the teacher: "That idea may be intuitively clear to you, and that proposition can indeed be verified in drawings on paper. But even so, since they have not been included among the basic concepts and the postulates of the theory, you must not use them unless you either derive them from that basis (by a definition or a proof) or formulate them explicitly as additions to that foundation." The fact that experience and intuition, while essential in inspiring and motivating the formulation of the foundations, are excluded from the strictly deductive process is very important for the student's understanding of deductive theories in general.

(6) He learns not only that, given a set of assumptions, certain propositions can be proved but also a complementary fact of paramount importance: that certain other propositions *cannot* be proved from those assumptions. He sees examples of systems (including planes containing altogether only a finite number of points) that satisfy certain postulates but not certain other propositions – a fact that *proves* that the latter cannot be proved from the former. The significance of this insight for the intellectual education and even the practical life of the student cannot be overestimated.

(7) He learns that the self-dual fragment of the affine plane is a sufficient basis for pre-limit calculus[4]. In that fragment, one can determine the area polygons of a step line, and the slope step line of a polygon, and establish their inverse character, which is at the core of calculus.

(8) He finds that one and the same theory may be capable of a variety of interpretations – an illustration of what Ernst Mach called the *Denkökonomie* of mathematics. By means of a simple dictionary, the geometry of the self-dual fragment of the affine plane is translated into Galileo's kinematics of a two-dimensional world (that is, a one-dimensional space in time).

(9) The study of the principles of non-Euclidean geometry brings the student another revelation of fundamental importance: that one may vary the assumptions of a theory substantially and develop an incompatible sister theory, either one self-consistent if the other one is. Any mystery is dispelled by displaying to the students non-Euclidean geometries in situations with which they are perfectly familiar; one such geometry on a sheet of paper, and another one on the streets of a modern metropolis governed by what I have called the *taxicab geometry*[5] whose circles are squares (See Chapter 18).

(10) The student will become familiar with *various applications of geometry to nature*, in fact, with the possibility of applying even mutually incompatible theories. The teacher can at least report, if not explain, to him that just as the self-dual fragment of the affine plane can be applied to Galilean kinematics, a certain non-Euclidean plane (more specifically, a fragment of the fragment just mentioned) can be applied to the kinematics of Einstein's special theory of relativity.[6]

(11) It is perhaps not too much to hope that the preceding ten points, when competently presented by able teachers, may convince a good many students of the relevance of geometry. Without counting intrinsic interest and beauty of geometry at all they may realize its relevance in two directions: in direct applications to science and in the formation of a *Weltbild* that far transcends geometry and science. But one should also try to reach anti-intellectuals and diehard fighters for practicality, who might even say that they don't need relativity theory and that certainly the study of planes that contain only finite numbers of points is outright ridiculous. "Indeed", the teacher should admit, "the discovery of such planes at the turn of the century began as a quite esoteric study. But 25 years later, statisticians found that the tool they needed in order to devise the most efficient tests to compare certain competing substances as to their relative merits were just those planes with only finite numbers of points. And today the results of the geometry of such planes are used by the United States Department of Agriculture in testing fertilizers." This is a use of ivory tower mathematics that even advocates of superpracticality should find earthy enough.

NOTES

[1] Cf. [1949, 4] and [1970, 1] Chapter 5.
[2] Cf. [1970, 1] Section 5.4.
[3] Cf. [1932, 1] and [1970, 1] Part 4.

[4] Cf. [1955, 4] Chapters 1 and 2.
[5] Cf. [1952, 4].
[6] Cf. [1970, 1] Section 5.13.

PART V

PHILOSOPHICAL RAMIFICATIONS OF SOME GEOMETRIC IDEAS

CHAPTER 17

ON DEFINITIONS, ESPECIALLY OF DIMENSION

Part I of this chapter is a translation of the pages in Chapter 2 of my book *Dimensionstheorie* [1928, 9] that deal with definitions in general and defining dimension in particular – passages often quoted by philosophers. In notes (added in bold-face brackets) I bring some of my views on the subject up to date. Part II is a reprint of sections of [1943, 1], a paper from an interesting series that L. R. Ford arranged in the *American Mathematical Monthly* when he was editor of that journal. Each paper of the series had in the title a question beginning with the words *What is* . . . ? Ford asked me to answer the question for *dimension*. (As to the genesis of my ideas, See Chapter 21.) Part III summarizes some basic methodological ideas propounded in Chapter 1 of [1928, 9].

I

The most important characteristic of a spatial object is its dimension. Curves, surfaces and solids are the fundamentally distinct main categories of spatial forms encountered in daily life; and dimension is the equally basic principle of classification in every kind of geometry.

The geometric terms 'curve', 'surface' and 'solid', taken from the everyday language, are among the words with which everyone uttering them associates certain ideas (*Vorstellungen*). That a person associates an idea with a given word means, generally speaking, that there are certain entities he unhesitatingly designates with the word; and certain others of which he unhesitatingly says that they are not to be so designated.[1] For in the course of his life, especially in childhood, he has learned the use of the word by being told: this is designated by the word; that is not. In between floats, so to speak, a third class of entities about which he is uncertain whether he should or should not so designate them. Passing from an individual to the group of all persons using such a word we see again: There are certain entities for which everyone, and others for which no-one uses the word. Besides, there is a floating intermediate class of entities.[2]

The word 'curve', for example, is used by everyone for an arc, a circle and a lemniscate; and by no-one for a cube or the surface of a sphere. For when beginning to use the word a person learns that 'curve' is employed for wire-, thread- and streak-like forms; 'surface' for plate-, sheet- and skin-like objects; and 'solid' for blocks. . . . A classification of *all* spatial objects according to their dimension does not exist in common language. The question of the dimension of certain objects of modern geometry would be left unanswered by some people and elicit conflicting answers from others. Nor are most people capable of stating precisely what they mean by a curve, a surface, a solid; that is, they cannot explain these words in terms of other words. Nonetheless these words serve a useful purpose in everyday life because there the complicated objects of the floating class are rarely (if ever) observed; and in the daily intercommunications between people about specific cases a completely precise general definition is unnecessary.

Before the development of dimension theory, the situation was not much different in geometry. The fundamental significance of dimension as a principle of classification was clearly recognized by some eminent mathematicians of all periods. Yet for the vast majority of spatial objects dimension could not be defined. Mathematics did not possess an all-comprising dimension concept.

If a word with which an idea is associated in daily life is to be made precise by a scientific definition, then there is no reason why the latter should be at variance with the everyday use; that is, why things generally denoted by the word should be excluded from the concept or why things generally excluded from it should be included. Thus one may stipulate as a *formal postulate* concerning the scientific definition of a word borrowed from everyday language *that the definition make that usage precise, and complete it where it is incomplete without contradicting the common usage.*

Such a definition and its completion can, however, be formulated in various ways – notably not only as to form but, in view of a floating class, also as to content. Floating entities may be included in or excluded from the concept to be defined without contradicting the common usage. The establishment of a definition thus involves arbitrariness. This arbitrariness can be justified only by the fertility of the definition. For while a word in everyday life contributes to the intercommunication of people a scientific definition is to serve as the starting point of deductions.[3] Definitions are dogmas; only inferences from them are cognitions. One thus may stipulate as a *material postulate* concerning a scientific definition *that it be a source of cognition by yielding an extensive, aesthetically satisfactory theory.*[4] The fulfilment of this material postulate represents the only possible justification of the introduction of the concept by its definition. If the concept is designated by a word borrowed from the vernacular, then the fulfillment of the *formal* postulate, while of secondary importance for the concept itself, justifies its *designation*,[5]

What has been said is particularly valid for dimension. And even the consideration of special objects suggests that there is an abundance of beautiful facts and interesting problems to be associated with a precisely defined (general) concept. What is the dimension of the union of two n-dimensional sets? What is the structure of an n-dimensional set, in particular the distribution therein of the points in which the set has different dimensional character? [for example, of the points in which a solid has the character of a surface or a curve? (See p. 216[6])] Do sets of various dimensions offer various possibilities of decomposition? What are the relations between the concepts of dimension and of connectivity? How does the dimension of an object react to transformations of the object? What are the dimension properties of the subsets of a Euclidean space? We shall develop a dimension concept which will be justified by the deduction of an extensive theory answering all these and many other questions. In conclusion it will be shown that the objects ordinarily called n-dimensional are also n-dimensional in the sense of our theory. While the immense fertility of this concept justifies this definition the fact shown in conclusion will justify the designation of the concept, namely as dimension.

In comparing a solid, a surface and a curve we notice structural differences. The study of these differences suggests that it is impossible to convert by certain transformations such as bending, for example, a surface into a curve or into a solid. At the same time it is plausible that this non-convertibility of objects of different dimension is merely a consequence of certain structural properties common to all objects having the same dimension. For if one designates an object as a surface, this act is by no means preceded by attempts to transform it. Dimension is rather a geometric characteristic of an entity as it is.

Formerly, even in the narrow domain of objects to which a dimension could be associated at all this characteristic was defined as their transformability (satisfying certain conditions) into certain other objects, especially into objects somehow built up of n-dimensional cubes or simplexes. Hence even for the special objects for which dimension was defined its geometric character was not fathomed. At the same time the idea of transformation as the starting point for the definition of dimension entailed, from the outset, a considerable restriction of the domain of objects to which a dimension could be associated. For transformations of the kind that cannot convert cubes of different dimension into each other transform cubes only into very special objects. Thus for the sake of both intuitiveness and generality of the dimension concept one must concentrate on intrinsic structural characteristics of objects of different dimension.

II. WHAT IS DIMENSION?

Strictly speaking, all material objects are 3-dimensional. Yet, only such objects as a metal sphere, a wooden block, or a rock are considered to be typical representatives of 3-dimensional entities (solids). A piece of sheet-iron, paper, and a membrane approach what we mean when we speak of 2-dimensional objects (surfaces). Wire, threads, and streaks of chalk represent our idea of 1-dimensional entities (lines).

What is the difference between objects of different dimensions? Originally, mathematicians believed it to be a difference in quantity, in the sense that a surface contains more points than a line and less points than a solid. Primarily the words 'more', 'less', and 'equally many' are restricted to finite sets while surfaces, as well as lines and solids, contain infinitely many points; but Georg Cantor extended their use to all sets. According to him, two sets – finite or infinite – contain equally many elements if there exists a one-to-one correspondence between their elements. Cantor found that two infinite totalities do not necessarily contain equally many elements. For instance, among geometrical objects a line segment contains more points than the set of all points on a straight line with integer distance from a certain point. On the other hand, a line segment, a square, and a cube do contain equally many points. Since these objects are of different dimensions, it follows that dimension is not a quantitative property.

Later, geometers thought that the difference between a 1-dimensional and a higher-dimensional object lay in the fact that the former, but not the latter, can be traversed by a continuously moving point. Indeed, lines on a paper or a blackboard are drawn, that is, traversed by the point of a pencil or chalk. But Peano discovered that a continuously moving point can traverse a square surface or a solid cube though nobody would call these objects 1-dimensional. On the other hand, 1-dimensional objects were found which cannot be traversed by a continuously moving point. The fact that an object is the path of a point, while interesting in itself, thus has no bearing on the question of the dimension of the object.

When one-to-one, as well as continuous, mappings had proved to be inadequate bases for the definition of dimension, mathematicians attempted to characterize dimension of a totality T as the least number of real numbers required to describe the elements of T topologically, that is, in a one-to-one and bi-continuous way. Each point of our ordinary space can be characterized by three, but not less than three, real numbers, for example, by its Cartesian or spherical coordinates; each point of a simple surface by two, but not less than two, real numbers – the points of a sphere by longitude and latitude; each point of a simple line by one number. Thus, according to the last definition, our space is 3-dimensional, simple surfaces are 2-dimensional, simple lines are 1-dimensional. Similarly, each color sensation of a normal eye can be topologically characterized by three, but not less than three, real numbers, namely the quantities of three standard colors whose mixture produces an identical sensation. Hence, the totality of color sensations of a normal eye is 3-dimensional while the corresponding totalities for a partially or totally color blind eye are but 2- and 1-dimensional, respectively. In the same way, a totality of all mixtures of four ingredients which cannot be obtained by mixing less than four of them is called fourdimensional. In fact, in this direction lies our only elementary analytical approach to the fourth dimension and the higher dimensional spaces.

Unfortunately however, these definitions apply only to very simple spatial objects, namely to those which can be obtained by means of a simple transformation from a straight segment, a square, or a cube (or sphere). Such entities are called arcs, discs, and topological spheres. In our space and in the plane, arcs and discs form only a very small part of the lines and surfaces studied in modern geometry. Even if objects that are unions of a finite number of arcs and discs are admitted the domain is still very restricted. For instance, the line mentioned above that cannot be traversed by a continuously moving point does not belong to this domain since it is not a sum of a finite number of arcs. In fact, it is the sum of infinitely many arcs. But all sets that are sums of infinitely many arcs cannot possibly be called 1-dimensional since the square and the cube are sums of infinitely many straight segments.

To formulate the intuitive difference between lines, surfaces, and solids we can devise a simple experiment the outcome of which depends upon the dimension of the object to which it is applied. We cut out from the object a piece surrounding a given point. If the object is a wooden block we need a saw to accomplish this, and the cutting is along surfaces. If the object is a membrane a pair of scissors suffices, and the cuts are along curves. If we deal with wire we may use a pair of pliers and have to pinch the object in dispersed points. Finally, in a dispersed object such as a heap of sand no tool is required to perform our experiment, since nothing needs to be dissected. This characterization of dimension leads from n-dimensional to (n-1)-dimensional objects. It ends with dispersed sets, naturally called 0-dimensional, and, beyond these, with "nothing," in set theory called the "empty set." It is, therefore, convenient to consider the empty set as -1-dimensional.

To make this idea precise only two simple auxiliary concepts are needed: neighborhood and boundary. In our 3-dimensional space we call a set N a *neighborhood* if each point of N is the center of a sphere (though perhaps a very small sphere) all of whose points belong to N. The interior of a cube is a neighborhood, whereas a cube with its faces is not; for even the smallest sphere about a point of a face contains points not belonging to the cube. Nor is a plane a neighborhood in our space; for each sphere about each point of a plane contains points not belonging to the plane. The *boundary* of a neighborhood N is the set of all points that do not belong to N but are centers of arbitrarily small spheres containing some points of N. For the interior of the cube the boundary obviously consists just of the six faces.

In terms of these concepts the result of our recursive dimension experiment can be explained as follows: A set S of points of the space is *at most* n-*dimensional* if each point of S lies in arbitrarily small neighborhoods whose boundaries have at most (n-1)-dimensional intersections with S. The set S is n-*dimensional* if it is at most n-dimensional but not at most (n-1)-dimensional[6]. The empty set, called -1-dimensional, is the starting point of the recursive definition.

At the end of the paper [1943, 1], I ask, "Have we answered the question *What is dimension*?" and I say, "In one sense, we have. We have explained which sets are 1-dimensional, which are 2-dimensional, and so on. In fact, with each subset of our space (and even of much more general spaces), we have associated an integer – the dimension of the set. This result is also expressed by saying that dimension has been defined as a set function. But there are many other set functions." We may, for example, associate with each subset of the space the (finite or infinite) number of pieces (in some sense) of which it consists or a measure of some kind (if this is possible). In this connection, the problem as to what is dimension assumes an aspect different from the question (answered above) as to which sets are n-dimensional. "It may be interpreted in the following sense: *By which properties is dimension characterized among the countless set functions*? So far this question has only been answered for the plane (See [1929, 1], [1930, 1]) that is, among the functions defined for the subsets of the plane." In the realm of these functions, I proved dimension to be characterized by the following five properties.[7]

1. The value of the function for a subset of a set S never exceeds its value for S.
2. A set S cannot be split into denumerably many closed subsets for each of which the function assumes a smaller value than for S.
3. For any two sets that are homeomorphic (that is, can be topologically mapped onto one another) the function assumes the same value.
4. Each set S can be topologically transformed into a subset of a closed set for which the function assumes the same value as for S.
5. For some square, some straight line segment, some set consisting of a single point and the empty set the function assumes the value 2, 1, 0 and −1, respectively.

III

Chapter 1 of my book on dimension theory begins with a brief sketch of the historic development of the universe of all geometric objects – an often neglected aspect of the history of geometry.[8] During the *first* stage, in antiquity, this totality was quite small; and it remained restricted as long as exclusively synthetic methods prevailed. In the *second* stage, it was enormously enlarged when Descartes and Fermat described points by coordinates – the point P in a plane by numbers denoted by $x(P)$ and $y(P)$ or rather, briefly, by x and y – and geometric objects by arithmetical relations between the coordinates of their points – for example, a certain parabola by the relation $y(P) = (x(P))^2$ or, briefly, $y = x^2$. This method (See Chapter 8, Sections 3 and 6) has become known as *analytic*. Except for a few isolated, more complicated curves and surfaces, only the objects described by the very simplest (namely linear and quadratic) relations had been known in antiquity. The others constituted a new geometric world – the Milky Way compared to our planetary system. And a century later, even local ("differential-geometric") aspects of all these objects were considered by applying the differential calculus. The *third* stage was launched by G. Cantor with another immense extension of the world of geometry through his creation of the theory of point sets. He introduced objects of a variety undreamed of even at the second stage – the whole universe of billions of galaxies compared to our Milky Way. Only the simplest of these objects had been amenable to the methods of the second period.

Though at the outset mainly applied to functions and measure, point set theory first entered the domain of intuitive geometry with the study of connectivity and of arcs; then with dimension theory and the theory of curves; and soon thereafter, with the study of general metrics. Since all this work is plainly concerned with the world of shapes and forms I have always preferred the designation *set-theoretical geometry* to *point set theory*. The former indicates that these studies are more closely related to geometry than to the theory of sets. I wished to stress this fact because set theory has been beset by doubts ever since the discovery of antinomies connected with such notions as the set of all sets, while the theories of dimension and connectedness have nothing whatever to do with those notions or even with ultrasets (See Chapter 6). Actually, the designation *set-theoretical geometry* has become quite common during the past fifty years.

But what I attempted in Chapter 1 of my book was much more than a merely terminological rapprochement of the new ideas with the traditional concepts. Fréchet and others had introduced various kinds of general spaces and I recognized (as did Urysohn, See Chapter 21), some of these spaces as adequate media for the study of dimension. But whereas analytic geometers and even Cantor operated in Cartesian spaces in which points are numbers or pairs of numbers or n-tuples of numbers, any general space was simply introduced as a set of undefined elements (a procedure that raised in many traditional geometers the doubts about sets mentioned above)[9] – of course as a set in which, depending upon the particular type of general space to be introduced, a particular structure was assumed to be given. Introducing, for example, what is called a topological space one postulates following Hausdorff that there be given a system of subsets, called neighborhoods, satisfying assumptions modelled after the properties of the open sets in Euclidean spaces. But this is not the way by which one introduces the prototype of all spaces – the 1-dimensional Cartesian space, that is, the line of real numbers. Classical introductions of the line start with the rational numbers or pairs of rational numbers, called rational intervals; then the points of the line (that is, real numbers) are defined in terms of the elements of either of these denumerable sets: as cuts in the rational numbers or, for geometric purposes preferably, as certain nested sequences of rational intervals; and finally the line is introduced as the set of these well-defined points. In the first quarter of this century, Huntington and A. N. Whitehead, followed by Nicod, developed Euclidean geometry in terms of undefined spheres or other solids introducing points as nested sequences of these elements. What I attempted in Chapter 1 of [1928, 9] was to establish a rapprochement of set-theoretical with traditional geometry by developing all the topological spaces considered in my book (and throughout most of the new geometry) in analogy to the classical introduction of the line: starting with a denumerable set of neighborhoods or lumps generalizing the rational intervals; then defining points as certain nested sequences of such lumps; and finally defining the space as the set of these well-defined points. I only succeeded, however, in thus introducing a certain type of space (called compact). But in 1930, my student A. Wald, later famous as a statistician, solved the problem completely by demonstrating that all topological spaces considered in my book can be introduced in the way just described and thus in analogy to the classical introduction of the line of real numbers.[10]

NOTES

[1] [*Addition in the Translation:*] Moreover, when uttering a word such as 'curve' most persons have mental images; that is, they imagine either clear pictures of specific examples or have more hazy and confused visions of many curves combined. – At the suggestion of the phenomenologist F. Kaufmann, I sent a reprint of my paper [1925, 2] on the curve concept and the theory of curves to E. Husserl. In January 1926, in a note of thanks Husserl called the paper "a beautiful sign of the so necessary turn that also mathematics has begun to take toward an ultimate fondamentation which goes back to the primordial sources of its concepts in intuition (*Anschanung*) – in other words, toward a phenomenological foundation".

[2] [*Addition in the Translation:*] More precisely, this floating class is made up of five subclasses some of which may in some cases of course be empty: a class of objects belonging to everyone's floating class; three classes of objects belonging to some (but not all) persons' floating class and in which furthermore there are a) only objects that the other people include; b) only objects that the others exclude; c) objects that some include and other exclude; finally a class of objects included by some and excluded by others but not belonging to any person's floating class.

[3] [*Addition in the Translation:*] There are also scientific definitions whose sole (and more superficial) purpose is to obviate the need for the frequent repetition of some long phrases. Such a definition introduces a single term instead of a particular combination of many words,

[4] [*Addition in the Translation:*] As a corollary of the facts described above, in some cases one may find several definitions of a concept satisfying both postulates – a situation that may result in a proliferation of definitions.

[5] [*Addition in the Translation:*] That the fulfillment of the formal postulate is only of secondary importance is illustrated by Cantor's use of the terms *closed* and *open* in geometry. Following him one calls, for example, a cube including its six faces and a sphere including its surface closed, while these solids completely stripped of their surfaces are called open. The underlying general concepts are of the utmost importance even though the terms taken from the vernacular are at variance with the common use. This common usage is rather typified by the title of de Musset's comedy *Il faut qu'une porte soit ouverte ou fermée*. In contrast, the empty set is both closed and open, as is the whole space; while a cube from which only one face has been deleted is neither closed nor open, as are in fact most sets. Hence many beginners find themselves, if only for a moment, baffled by Cantor's terms; but presently everyone comes to appreciate their creation by definitions regardless of the terms employed to designate them.

[6] **[**In our 3-dimensional space, there are only sets of the dimensions 3, 2, 1, 0 and −1. Examples of sets with dimensions exceeding 3 exist in higher-dimensional Euclidean and more general spaces.**]**. That S is not at most (n − 1)-dimensional means that S contains at least one point at which S is at least n-dimensional, that is, a point which does not lie in arbitrarily small neighborhoods whose boundaries have at most (n − 2)-dimensional intersections with S; the boundaries of all sufficiently small neighborhoods of such a point have at least (n − 1)-dimensional intersections with S.

[7] The utter simplicity of the five characteristic properties of dimension (which may serve as the postulates of an axiomatic theory of dimension in the plane) is certainly remarkable. They all of course correspond to theorems of the dimension theory developed by Urysohn and myself from our definition (See Chapter 21): 1 and 3 reflect two of the most elementary statements of the theory; 2, the so-called sum theorem; 4, a deep result of Hurewicz; and 5 is a simple standardization comparable to the choice of a unit in a metric theory.

[8] Yet it has been known since Pasch and Hilbert that even synthetic Euclidean geometry must start from a set of undefined elements. Actually, though unnecessarily (See Chapter 19), Hilbert starts with three sets of undefined elements: points, lines and planes. Traditional geometers, however, are more familiar with the interrelations between these elements than with the relations between open sets reflected in the postulates of a topological space.

[9] Cf. also [1949, 2].

[10] Cf. Wald, Ergbnisse e. mathem, Kolloquiums, **3**, 6–11 and my report in [1940, 1], III, Section 4.

CHAPTER 18

SQUARE CIRCLES (THE TAXICAB GEOMETRY)

Square circles or round squares have haunted many and diverse philosophical writers as the archetype of the impossible and the absurd; they were assigned a place near – or rather below – golden mountains, unicorns and mermaids. They have been discussed by Thomists, existentialists and linguistic philosophers as well as by Herbart, Bergson, Russell and many, many others.

By a circle, philosophers cannot very well mean anything different from what geometers mean; and a geometer means by a circle or, more specifically, by the circle in a plane Π with the center 0 and a radius of r units the class of all points in Π that are at the distance of r units from 0. But philosophers probably have been unaware of the following facts: 1) that the shape of that circle essentially depends on the method by which distances in the plane Π are measured; 2) that, as Hermann Minkowski indicated at the beginning of this century, there is a consistent and perfectly reasonable definition of distance according to which a circle is indeed a square; and 3) that, as I pointed out in lectures and writings decades ago[1], many people actually encounter square circles in their lives. So does, for example, anyone who uses a taxicab in a modern city built in square blocks. Chicago is such a city.

Indeed, for a ride from an intersection 0 to a point P that is 4 blocks east and 3 blocks north of 0 the cab driver will, according to his meter, charge a fare for 7 blocks (See the Figure). To little avail will the passenger protest that in a right triangle having sides of 3 and 4 length units the distance between the end points of the hypotenuse is 5 (and not 7) units. In vain will he plead that this fact was already known to the ancient Egyptians and that it is a special case of what is probably the most famous of geometric propositions – the theorem of Pythagoras. *He must pay the fare for 7 blocks.*

And where are the other points that he could reach at the same cost, that is, the other points at a taxicab distance of 7 blocks from 0? Or, to put it in a different way, where are the other points of the circle about 0 of radius 7 in the taxicab plane? They are (See the Figure) at the intersections

3 blocks east and 4 blocks north,
2 blocks east and 5 blocks north,
1 block east and 6 blocks north,
7 blocks due north,
1 block west and 6 blocks north,
2 blocks west and 5 blocks north,
and so on . . .

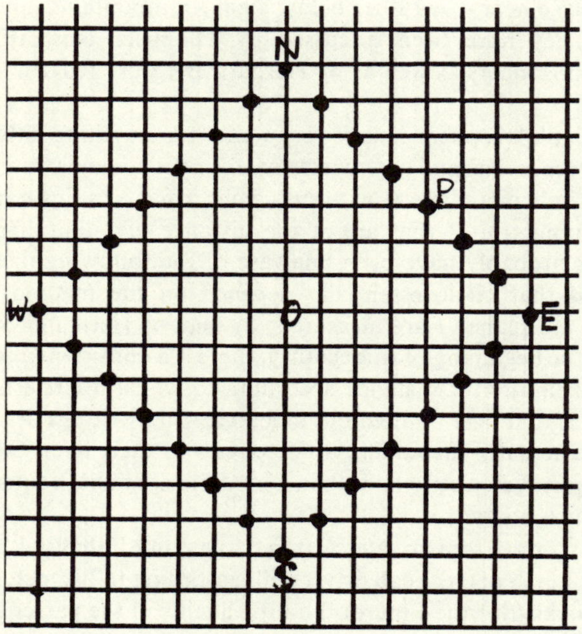

in all, 28 points lying on a square in diamond position to the streets. Its corners are 7 blocks straight north, east, south, and west of 0. The measuring of distances in the taxicab geometry thus leads to square circles – circles even of a certain practical significance, namely for the passengers' purses.[2]

"But the distance travelled along broken paths," a philosopher might object, "is not what we mean by distance. We measure distances as the crow flies." Measuring distances as the crow flies is indeed a useful convention in most, but not in all situations. A cab moving from 0 to P between city blocks obviously cannot proceed along the diagonal which those have in mind who say that the distance between 0 and P is 5 blocks.

Nor, incidentally, can even the crow! For in order to profit from a diagonal shortcut in the air, the crow must first somehow fly above the houses and finally down again. But such a course exceeds the length of 5 blocks; and if the houses are very tall it may even add up to 7 blocks or more[3].

Clearly these facts are of only limited importance for philosophers who, for their purposes, may replace the square circle by some other quasi-geometric *bète noire*. But they illustrate a new state of affairs. In the past, mathematicians have pleaded inability to associate meaning with certain metaphysical ideas. In dealing with square circles they, conversely, associate meaning with (and even give a practical interpretation to) the philosopher's paramount example of absurdity.

NOTES

[1] Cf. especially [1952, 4].

[2] The taxicab geometry is also (Cf. [1971, 1]) "an approach to the idea of non-Euclidean geometry, though of a variety quite different from Bolyai's and Lobachevsky's ... Students are quite interested in the taxicab geometry – in fact, so much so that they are even eager to study the still stranger and more complicated taxicab geometry in a city with one-way streets." (About the latter, cf. H. C. Curtis, *Am. Math. Monthly* 60 (1953) 416.)

[3] In this latter case, the shortest flight of the crow would be along the path of the cab.

CHAPTER 19

THE ALGEBRA OF GEOMETRY

Projective geometry has always been called the geometry of joining and intersecting to contrast it with Euclidean, affine and other geometries which besides consider congruency, perpendicularity and parallelism. When preparing a course on projective geometry at the University of Vienna in 1927/28 I therefore looked for an axiomatic foundation that would reflect this principal feature of the theory, in other words, for a development based on assumptions about joining and intersecting. But nowhere in the immense literature could I find what I was looking for. So I formulated such a foundation myself.[1]

Some aspects of this development may interest students of methodology and philosophers of science.

(1) In order to obtain a truly complete foundation I soon realized that I had to make such utterly simple assumptions as that the join of the geometric objects A and B is the same as the join of B and A – trifles that geometers usually did not even deign to mention, let alone to elevate to the role of postulates. Yet I discovered that such trivia were not only necessary but, jointly, sufficient for the development of projective geometry.

(2) For all their simplicity, these ideas brought about an extensive synthesis of various branches of mathematics – from the outset including geometries of all dimensions, the theory of finite sets, Boolean algebra, aspects of measure theory and probability theory à la Keynes. This list was later vastly expanded by Garrett Birkhoff and others.

(3) The development by-passed the traditional foundation on various categories of entities or undefined concepts: points, lines, planes, and diverse kinds of hyperplanes in higher-dimensional geometries. Instead, as especially J. von Neumann emphasized[2], my theory was based on the assumption of a single kind of geometric object or a single undefined term, 'entity'. These entities were assumed to be connected by two, as it were, algebraic operations, whence I later spoke of an *algebra of geometry*. Its main features are commutativity and associativity of the operations (commutativity of joining has been mentioned above); their idempotency, that is, the

fact that each entity, A, joined or intersected with itself yields A; the existence of a vacuous entity, V, which joined to any entity does not change the latter, and of a universe, U, which does not affect entities by intersection; and what I called an absorption law[3] that makes it possible to define a part relation for which transitivity and all the other properties attributed to a part relation in geometry follow from the assumptions mentioned. By definition, an entity A is a part of B if A and B have the join B and the intersection A. At the same time and independently, Fritz Klein in Germany developed a similar theory, more complete with regard to algebra but neglecting geometry. Klein also devised a felicitous designation for classes of entities with two such operations: he called them *Verbände*, while I had no suggestive name for them. In the American literature, G. Birkhoff a few years later most successfully introduced the term *lattice*.

(4) Commutativity and associativity are also properties of the addition and the multiplication of numbers while idempotency and the absorption law are not. But what struck von Neumann most was, as he told me (though he did not to my knowledge emphasize this point in his writings), the *nondistributivity* of the algebra of geometry: I did not assume that join and intersection in general satisfy for three geometric entities the analogue of the so-called distributive law concerning addition and multiplication:

$$(a + b) \cdot c = (a \cdot c) + (b \cdot c) \text{ for any three numbers a, b, c.}$$

For example, if A and B are points and C is a line having neither A nor B as a part, then the (geometric) join of A and B is a line whose intersection with C is a point, whereas the intersections of A and C as well as of B and C are vacuous, and so is the join of these two intersections (See the Figure). If however, distributive laws are adjoined to the other assumptions of the algebra of geometry then, while geometry proper is excluded, the result is a foundation of the Boolean algebra of classes which, therefore, is a special case of the general algebra of geometry.

Besides the laws of commutativity, associativity, idempotency and absorption and the existence of V and U, I did, however, assume an analogue of subtraction or complementation; namely, if A is a part of D, the existence of an entity having with A the join D and the intersection V. There may of course exist many such complements of A to D. For example, if D is the line labelled 'join (A, B)' in the Figure, then B as well as any third point that is a part of D satisfies the assumption just mentioned. As G. Bergmann proved in 1928, the assumption that in all cases only one

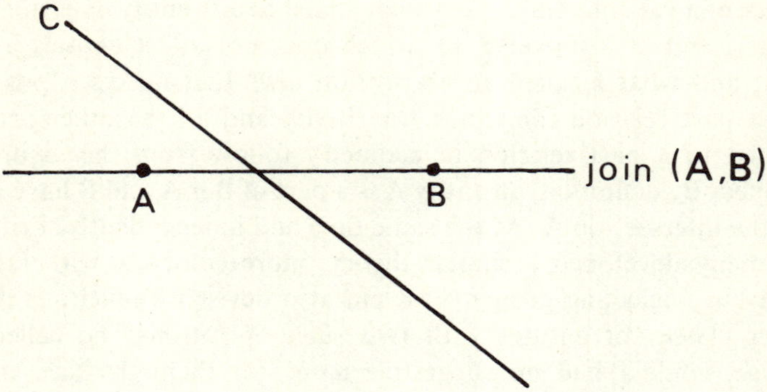

such complement exists is equivalent to the validity of distributive laws and characterizes Boolean algebra[4].

(5) More interesting perhaps to historians of geometry is my introduction of points. A point is an entity having only the vacuous entity and itself as parts. "This introduction of points," I wrote[5], "obviously makes precise Euclid's initial words in his *Elements: a point is that which has no part*." In the algebra of geometry, Euclid's sentence is an abbreviation of the kind that we use in arithmetic when sometimes saying, "a prime number is an integer that has no factor" meaning of course an integer having only 1 and itself as factors. For every integer has 1 and itself as factors, and every geometric entity has V and itself as parts. But the essential fact about the definition of point in the context of the algebra of geometry is its integration in the deductive development of the theory whereas in Euclid's system, for the lack of a definition of parts or assumptions about them, the celebrated initial sentence of the *Elements* is doomed to isolation as a mere intuitive remark detached from and inaccessible to Euclid's logical reasoning.

(6) For the development of some aspects of the theory, I finally postulated the existence of a *dimension function* pairing (according to certain conditions) an integer to each entity as its dimension. The vacuous entity V is given the dimension -1, just as is the empty set in set-theoretical geometry (See p. 212); and points are called 0-dimensional. Further conditions include the important assumption that for any two entities, A and B,

$$(*) \qquad \dim A + \dim B = \dim \operatorname{join}(A, B) + \dim \operatorname{intersection}(A, B).$$

For example, if both A and B are 1-dimensional (that is, lines), then their join is 3-, 2-, or 1-dimensional according as their intersection has the dimension -1, 0 or 1. Formula (*) is also valid for measurable sets, A and B, if dimension is replaced by measure, as well as for events if dimension means probability. In a terminology later developed in the United States, the classes of entities that I introduced in 1927 are called *normed lattices* (dimension playing the role of a norm). In 1928, G. Bergmann[6] succeeded, through penetrating studies, in *proving* under very general assumptions the existence of a dimension function (which I had to *postulate*). He thus defined lines, planes, etc. in terms of parts (and thus ultimately in terms of joining and intersecting), just as I had defined points.[7]

Formula (*), as has been mentioned, is valid in many fields – a fact that accounts for the synthesizing power of the theory, but it does not hold in the dimension theory of point sets discussed in Chapters 17 and 21. There, for example, the union of two curves (continua of dimension 1) is 1-dimensional regardless of the dimension of their intersection.

The probing methodologist is bound to ask how one can reconcile conflicting results of two theories of dimension having in common important features – among others, that points, lines and planes have the respective dimensions 0, 1, and 2. The answer is two-fold:

a) The dimension theory of point sets deals with all subsets of the space whereas the algebra of geometry only considers linear entities – in a 3-dimensional universe only points, (straight) lines, and planes. Each linear entity corresponds to a point set whereas a cross, a lemniscate and a paraboloid are point sets that do not correspond to linear entities.

b) The join of two linear entities does not always correspond to the union of the two corresponding point sets; in general, it corresponds to a much larger set. For example, if the entities L and M are (1-dimensional) lines whose intersection is a point, then their join is the (2-dimensional) plane spanned by L and M, whereas the union of the two corresponding point sets is a (1-dimensional) cross-shaped point set which does not correspond to any entity considered in the algebra of geometry. The join of two points, A and B, is a line which in general has many parts, namely, points besides A and B, whereas the union of two sets A and B each consisting of a single point is a pair of points which has no subsets consisting of single points other than A and B. In the algebra of geometry all parts of a line L (besides V and L) are points, whereas in set-theoretical geometry the subsets of a line L include all finite and infinite sets of points of L. But the intersection of two linear entities does correspond to the intersection of the corresponding point sets – a fact that accounts for some of the similarities of the two theories.

NOTES

[1] [1928, 4], the fourth of the *Bemerkungen zu Grundlagenfragen.*

[2] In his paper "Continuous Geometry" (*Proc. Nat. Acad. Sci. U.S.A.* 1936), von Neumann writes: "The suggestion of replacing . . . distinct classes of "undefined entities" [points, lines, planes] by a unique class which consists of all linear subspaces of a given space was first made by Menger."

[3] The postulate states that two entities, A and B, have the join B if and only if they have the intersection A.

[4] "Zur Axiomatik der Elementargeometrie", *Monatshefte Math.und Phys.* 36, (1929).

[5] loc. cit. [1], p. 323.

[6] loc. cit.[4].

[7] In the following years, in connection with courses at the University of Vienna, I further developed the theory in collaboration with my students F. Alt and O. Schreiber (Cf.[1936, 2]).

CHAPTER 20

GEOMETRY AND POSITIVISM
A PROBABILISTIC MICROGEOMETRY

This chapter contains a reprint of the first half (entitled "Positivistic Geometry") of the paper read at the Mach Symposium in 1966 (See Chapter 9).

Euclid's *Elements* are an idealized description of the space in which we live and experiment. But Euclid's assumptions transcend the realm of the observable in two directions; in the large and in the small.

With regard to our space *in the large*, Euclid assumed the possibility of extending straight lines without ever coming to an end and without ever returning to the starting point. The former assumption is made in the *Elements* explcitly; the latter, tacitly. A further hypothesis about space in the large, somewhat more complicated and less plausible, is the famous Fifth Postulate, about parallel lines, which is the foundation of the theory of similar figures and which accounts for the theorem about the sum of the angles in a triangle. While that axiom can be neither verified nor falsified directly (i.e., by rods or even light rays) some of its consequences do lend themselves to tests by observation. One can, for instance, measure the angles in a triangle. So, when Bolyai and Lobachevsky challenged Euclid's postulate by using a different assumption as the basis of a non-Euclidean geometry, Gauss actually did measure those angles, undisturbed by Kant's doctrine of the *a priori* necessity of Euclid's system. Obviously, Gauss considered geometry as the science of our physical space; and, in the spirit of positivism, all he claimed as the result of his measurements was a tentative confirmation of Euclid's theory: a confirmation within the limits of errors of observation and confined to terrestrial triangles, which are macroscopic but not megascopic figures.

Riemann developed Bolyai's, Lobachevsky's, and his own non-Euclidean geometrics into a general theory of space, which is differential-geometric in character. But the word *differential* should not mislead us into the belief that he modified Euclid's geometry in the small. Sufficiently small triangles in Riemann's geometry differ from Euclidean triangles as little as we please. Significant deviations from Euclidean geometry occur only in the large. This is why it took more than a century until these modifications of Euclid's geometry were applied to nature, namely, in connection with the general theory of relativity, and why, somewhat paradoxically, the applications of this differential geometry concern astronomical distances and cosmology.

Euclid's assumptions about space *in the small* may be said to be conspicuous by their absence. Because of the laws about similar figures, his theory simply amounts to an interpolation of the macrocosmic facts into microcosmic domains; and, for all practical purposes, the same can be said about the non-Euclidean and Riemannian geometries. Perhaps because of their implicitness, these assumptions of Euclid have never been challenged as seriously as were his explicit postulates about space in the large.

It is in this respect that a positivistic approach calls for revisions, in fact, for changes that are much more profound than those that have led to Riemannian geometries. Looking at space in the small from a positivistic point of view we must question the very framework of Euclid's system and his most basic concepts.

The chapter on the continuum in Mach's *Prinzipien der Wärmelehre* contains the following passage: "All that *appears* to be a continuum might very well consist of *discrete* elements, provided they were sufficiently small compared with our smallest practically applicable units or sufficiently numerous." (Parenthetically, it may be pointed out that the quoted passage favorably compares with less fortunate assertions of Mach in his polemic against atomism. Some of those assertions border on a dogmatism conflicting with Mach's own general philosophical ideas and subsequently have been disproved by experiments. But it should also be emphasized that Heisenberg's scheme of confining atomic theory to the connection of observable quantities restored Mach's program; and that today's view of atomic physics as a method which, from present observations, derives probabilistic predictions about future observations is the very epitome of Mach's ideology. Yet, perhaps because of Mach's dogmatic fight against atomism, his programmatic views are rarely quoted in quantum mechanics – certainly not as frequently as in relativity theory.)

As far as geometry is concerned, the passage quoted from the *Wärmelehre* actually says more than is justified in a strictly positivistic geometry. With regard to the ultimate structure of space (whatever that may mean), a positivistic geometer should out-Mach Mach and refrain from making any statements whatsoever. But the concepts of classical geometry that he must challenge include the notions of point, line, and congruence as well as the post-Euclidean concept of numerical distances.

Since points and lines do not exist in the macrocosm one might at first think that these very concepts represent the idealizations of macrocosmic objects such as dots or cross hairs and rods or light rays that microgeometry requires. Actually, however, points and lines cannot be ascertained in the microcosm any more than in the macrocosm. How can one locate, identify, and recognize a point and determine straight segments (if in the small there is anything deserving this name!) or ascertain that a point and a line are incident and that three points are collinear?

It has often been said that, from an empirical point of view, the application of irrational numbers to space is illusory since no observation can distinguish irrational from rational coordinates. But neither is a rational number or even an integer ascertainable as the abscissa of a point on a line. Whereas in counting discrete elements, say apples in a basket, one may arrive at the number 12 and then is sure that the contents are not 11.97 or 12.02 apples, the statement that the abscissa of a point on a line is 12 cannot be more strongly supported than the claim that the abscissa be 12 plus some sufficiently small but otherwise arbitrarily chosen number, which may be rational or *irrational*. From the positivistic point of view the difficulty in describing continua does not lie in the use of special kinds of numbers. It lies in the identification of the individual elements, in particular, in their description by numbers of any kind. This difficulty precludes of course a positivistic application of the limit concept. What we actually observe or immediately derive from observations are difference quotients and product sums, which can be treated by numerical methods. Whereas, in mathematical analysis, difference quotients and product sums are considered as approximations to derivatives and integrals, from a positivistic point of view, most statements involving derivatives and integrals are unverifiable approximations to assertions about difference quotients and product sums – but very convenient approximations, because calculus has a simple algebraic structure and is well developed.

No one was more keenly aware of the problems connected with the notion of the physical continuum than Poincaré, who emphasized the difficulties on many occasions, e.g., in *La science et l'hypothèse*, *La valeur de la science*, and *Des fondements de la géométrie*. In all these books he characterized the physical continuum by the formulae

(1) $A = B, \qquad B = C, \qquad A \neq C,$

symbolizing the fact that an element of a physical continuum may be indistinguishable from two others that can be distinguished from one another. The *mathematical* continuum, according to Poincaré, is a construct whose main purpose it is to overcome the formulae (1), which he called contradictory or at least repugnant. He finally returned to the problem in *Dernières Pensées*, where he proposed the following definition of a physical continuum. It is a system S of elements, any two of which can be joined by a chain of elements of S such that each element of the chain is indistinguishable from the next element.

While, after Poincaré, these ideas of his have not in my opinion been sufficiently developed, two steps have been taken in the direction of a positivistic geometry.

The first step consisted in inverting the traditional procedure of taking *point* as the primitive concept and of subsequently defining spatial objects as sets or unions of points. Whitehead and Nicol started with *lumps* or *pieces* and defined points as certain sets or intersections of lumps. Pursuing a course suggested by dimension theory (See Chapter 17, III) in the early 1930's Wald defined points of topological spaces as certain denumerable sequences of lumps. R. L. Moore applied similar ideas to the points in the plane. Somewhat related theories, which, however, were essentially oriented toward algebra rather than geometry, were developed by Stone, Wallman, and Milgram, who introduced points as (indenumerable) sets of elements of a Boolean ring, a distributive lattice, and a general partially ordered set, respectively. A brief summary of all these theories from a geometric point of view and a bibliography are contained in my lecture "Topology without Points" (See [1940, 1] III). But as that title indicates, these theories were confined to the topological aspects of geometry and did not include the ideas of congruency, distance and the like.

The second step was the development of a statistical or probabilistic metric geometry, which I outlined in the early 1940's.[1] Whereas traditionally one associates, with any two points p and q of a space, a number $d(p,q)$, called the distance between p and q, the probabilistic theory is based on the assumption that with any two elements a distribution function F_{pq} be associated. Its value for x, that is, the number $F_{pq}(x)$, may be interpreted as the probability that the distance between p and q be $\leqslant x$. From the fact that distances are certainly non-negative it follows that, if p and q are any two elements, then $F_{pq}(x) = 0$ for all $x < 0$. With increasing x, the values $F_{pq}(x)$ monotonically increase from 0 to 1. If $F_{pq}(x') = 0$ for some $x' > 0$, then the distance between p and q certainly exceeds x'. If $F_{pq}(x'') = 1$ for some x'', then the distance between p and q certainly does not exceed x''. If $F_{pq}(x) < 1$ for all x, then, with increasing x, the numbers

$F_{pq}(x)$ asymptotically approach 1. A numerical distance $d(p,q)$ between p and q, as assumed in the classical theory, may be said to exist in the special case where $F_{pq}(x) = 0$ if $x < d(p,q)$, and $F_{pq}(x) = 1$ if $x > d(p,q)$. Normal distributions, on the other hand, are ruled out by the fact that $F_{pq}(x) = 0$ for at least all negative x.

The symmetry of the distance is expressed by the assumption that, for each pair of elements, $F_{pq} = F_{qp}$; that is to say, $F_{pq}(x) = F_{qp}(x)$ for all x. Further assumptions are needed in order to reflect the basic geometric fact that for any three elements none of the three distances exceeds the sum of the other two. They can be expressed in inequalities connecting the distributions F_{pr} with the distributions F_{pq} and F_{qr}. Extensive studies of such triangle inequalities (one of which was proposed by Wald[2]) and of related subjects have been carried out during the past few years by Schweizer and Sklar[3], by Šerstnev[4] in Russia, and by others. A report by Schweizer on this new branch of geometry is in progress.[5]

In contrast to many postulates in classical geometry, some assumptions about the elements of a probabilistic space are such as to be valid if $p, q, r, \ldots$ are thought of as points or as lumps or as certain other entities. If p and q are lumps, then one may interpret $F_{pq}(x)$ as the probability that the average distance between p and q be $\leqslant x$. In this case, of course, one has not only $F_{pq}(0) = 0$ for any two distinct lumps but even $F_{pp}(0) = 0$. Another way would be to interpret $F_{pq}(0)$ as the probability that p and q overlap and then to proceed to the definition of distances by a chain method that will be discussed in a subsequent paragraph.

Especially remarkable is a probabilistic metric space studied by Schweizer and Sklar[6], which consists of random variables whose values are points in an n-dimensional Euclidean space E_n. The elements in this geometry may be thought of as cloud-like objects in E_n, which may not only overlap but completely interpenetrate. In the small, the behavior of lines is quite different from that of lines in a Euclidean space. In the large, the geometry asymptotically approaches that of Euclid.

Instead of elaborating on mathematical details of these interesting theories I here wish to discuss further steps suggested by the positivistic point of view. It certainly is in the spirit of Mach that these considerations comprise psychophysical as well as physical spaces, although from the excellent *Handbook of Mathematical Psychology* edited by Luce, Bush, and Galanter it appears that psychophysicists have not yet taken cognizance of, or at least not found use for, probabilistic geometry as a tool.

Unfortunately, the claim that the distance between two elements has (or probably has) a definite positive value does not in general seem to have a verifiable interpretation. If a unit has been chosen which is, say, approximately 100 times the size of the smallest discernible object, then there is no verifiable statistical correlate to the concept $F_{pq}(0.0001)$. But the more qualitative assertion that two elements p and q probably have *some* positive distance can be related to observations. If p and q are points, then $F_{pq}(0) = 0$ may be interpreted as meaning that p and q are always found to be *distinct*.

Extending this idea one may look upon the inequality $0 < F_{pq}(0) < 1$ as meaning that there is a positive probability, namely $F_{pq}(0)$, that p and q cannot be distinguished, and the positive probability $1 - F_{pq}(0)$ that they can. Using some (more or less arbitrary) criterion one may speak of *probably distinguishable* and *probably indistinguishable* elements, say, according as $F_{pq}(0) \leqslant \frac{1}{2}$ or $> \frac{1}{2}$. More specifically, for any number i in the interval $[0, 1]$, two elements p and q may be said to be *at least i-indistinguishable* if $F_{pq}(0) \geq i$. Statistically speaking, p and q are at least i-indistinguishable if they cannot be distinguished in at least the i-th part of the observations directed to these elements.

If $F_{pq}(0) = 1$, then p and q are certainly indistinguishable as to their location in space. The case arises of course if p is a point and the symbol q is another designation of p. Interpenetrating elements, such as the cloud-like objects mentioned before, may well have the same location in space and yet be distinguishable in some extrageometric way. Further important examples arise in psychophysics. If an observer, O*, at different times touches the same two spots, p and q, on the skin of a blindfolded person, B*, with the two points of a compass, then it may well be that B* experiences sometimes two sensations and sometimes only one. The relative frequency of the latter cases is the empirical correlate of $F_{pq}(0)$. For two spots that are very close to one another or very far apart, that number will, in general, of course be 1 or 0, respectively.

This probabilistic approach supplies us with the necessary complementation of Poincaré's ideas. The situation he symbolized by the formulae (1) may, along these lines, be described by

$$(2) \qquad F_{pq}(0) > 0, \quad F_{qr}(0) > 0, \quad F_{pr}(0) = 0,$$

formulae that can hardly be called repugnant, let alone contradictory. All they express is the fact that a certain relation (which is neither the identity nor equality!) is not transitive. The relation in question is, to repeat it, the indistinguishability of two elements in at least some of the observations directed to those elements.

This relation is clearly defined in the experiment described above in which the reactions of the blindfolded B* are superimposed on the geometry of the observer O* – a geometry which, though itself strictly speaking only probabilistic, is considerably more subtle than that of B*. The observer may draw a coordinate system on B*'s skin in ink in order to describe the location of the spots $p, q, r, \ldots$ and, at any rate, it is O* who decides whether the number of spots simultaneously touched on B*'s skin is 2 or 1. (Occasionally B* reacts to a single stimulus with two sensations.) For B* it would be, to say the least, much harder to construct a cutaneous geometry merely out of his own sensations. This problem leads back to the question of identifying elements. What for B* would be the meaning of the relation of indistinguishability considering that in cases of indistinguishable sensations he has only one?

In cases where a relation of indistinguishability is clearly defined it can be made the basis of a probabilistic metric that is quite different from the one outlined before. For any two elements, p and q, one may determine the smallest integer n such that there exist n elements $p_1, p_2, \ldots, p_n$, which form a chain connecting p and q similar to the chains considered by Poincaré but with a probabilistic complementation of Poincaré's idea; that is to say, a chain such that $F_{p_k p_{k+1}}(0) > 0$ for $k = 0, 1, \ldots, n$ if we set $p = p_0$ and $q = p_{n+1}$. This integer n may be called the *probabilistic distance* between p and q. If $F_{pq}(0) > 0$, then one may say that p and q have the distance 0. While a space with a probabilistic metric (i.e., with distribution functions F_{pq}) does not in general contain any finite chain joining two points, a space that deserves to be called a continuum does. G. Cantor defined connected sets by the property that any two points of such a set, for any number $d > 0$, can be joined by a chain of points in which any two consecutive elements are less than d apart.

The preceding probabilistic definition of distance can be greatly refined. If p and q are two elements, then for each number i in the interval [0, 1] one may try to determine the distance between p and q by the probability standard i or, briefly an *i-distance* $d_i(p,q)$ as the smallest integer n such that p and q can be joined by a chain of points satisfying the condition $F_{p_k p_{k+1}}(0) > i$ for $k = 0, 1, \ldots, n$. Clearly, by increasing the probability standard one lengthens these chains; that is to say, if $i' > i$, then $d_i'(p,q) \geq d_i(p,q)$. The distance by the highest probability standard, i.e., the number $d_1(p,q)$ is the smallest number of points in a chain joining p and q such that consecutive elements are always indistinguishable – if there is such a chain; for it may well be impossible to uphold this highest standard of probability.

The *i*-distances so defined are integers that do not necessarily satisfy the triangle inequality. If $d_i(p,q) = 2$ and $d_i(p,r) = 3$, then $d_i(p,r)$ maybe as much as 6. This actually is the case if the shortest chain joining *p* and *r* consists of the two points between *p* and *q*, the point *q* itself, and the three points between *q* and *r*. The numbers $d_i(p,q) + 1$, $d_i(q,r) + 1$, and $d(p,r) + 1$, however, clearly do satisfy the triangle inequality.

Another example of distances defined by chains is contained in Schrödinger's colorimetry[6]. Schrödinger sets the distance between two colors, C and C', equal to 1 if C and C' are *barely distinguishable*, and equal to the integer n if there exists a sequence of elements, $C_0 = C, C_1, C_2, \ldots, C_{n-1}, C_n = C'$ such that any two consecutive elements C_{i-1} and C_i are barely distinguishable while there does not exist any shorter chain of this kind. According to Schrödinger's assumptions about the color cone, each element C (represented by a vector in a 3-dimensional space) is the center of an ellipsoid whose interior points cannot be distinguished from C while the points on the surface of the ellipsoid are barely distinguishable from C. But such well-defined neighborhoods do not seem to exist in the color cone any more than on the human skin. One must first define distance 1 in a probabilistic way.

Perhaps the most difficult problem in microgeometry is, as I have already emphasized, the identification of individual elements of the space. In 1951, I suggested that, besides studying well-defined sets, it might be necessary to develop a theory in which the element-set-relation is replaced by *the probability of an element belonging to a set*. In a Paris note [1951, 2], I called such an object, in contrast to an ordinary or rigid set, *ensemble flou* (= hazy set). In a slightly different terminology, this idea was recently expressed by Bellman, Kalaba and Zadeh[8] under the name of *fuzzy set*. (These authors speak of the degree rather than the probability of an element belonging to a set.) Since, however, the concept of probabilistic sets still requires the identification of elements, I now believe that in geometry it ultimately will have to be combined with that of lumps, which can be more easily identified and distinguished than points. Moreover, lumps admit an intermediate stage between indistinguishability and apartness, namely that of overlapping. It is, of course, irrelevant whether the primitive (i.e. undefined) concepts of a theory are referred to as *points* and *probably indistinguishable*, or as *lumps* and *probably overlapping*. All that matters are the assumptions made about the primitive concepts. But the assumptions made in the two cases would not be altogether identical and I believe that the ultimate solution of problems of microgeometry may well lie in *a probabilistic theory of hazy lumps*. The essential feature of this theory would be that lumps would not be point sets; nor would they reflect circumscribed figures such as ellipsoids. They would rather be in mutual probabilistic relations of overlapping and apartness, from which a metric would have to be developed.

Since, as I have already intimated, a positivistic geometry cannot start with anything ultimate, it must start in the middle, which means that it must lend itself to extensions in two directions – to more and to less refined theories.

(1) There should be a formalism for the refinement of the geometry corresponding to the potential refinements of instruments and methods of observation, typified by the transitions from observations with the naked eye to the use of magnifying glasses, of microscopes, of electromicroscopes, . . . A refinement of a geometry resolves some of its indeterminacies in the small. The elements P, Q, . . . of the geometry – points or lumps – are replaced by finite sets of elements, $p, p', p'', \ldots; q, q', q'' \ldots; \ldots$. These sets need not be disjoint. How the distance between P and Q should be related to the distances between $p, p', p'', \ldots$ and $q, q', q'', \ldots$ has not yet been fully clarified.

(2) There should be a formalism of superimposing a less refined on a finer geometry corresponding to the blindfolded person's superimposing his subjective geometry of skin sensations on the observer's geometry of that surface (i.e., the observer's idea of points, distances, betweenness, and the like). Sets of points or hazy sets of points of the observer are combined into single sensation points of the blindfolded man. These sets are not in general equivalence classes since they need not be disjoint and, as has been remarked, indistinguishability on various probabilistic levels is not a transitive relation.

As far as the physical space is concerned, the outlook for the application of a positivistic geometry to atomic and particle physics is not at present very bright. For those physicists who express interest in the idea wish – understandably! – that mathematicians develop a large number of different geometries from which they (the physicists) can choose the one most suitable to their purposes. But it seems to be difficult enough to develop one geometry of such an entirely new type. The statements of macrogeometry should result from assumptions about space in the small through averaging processes somewhat like macrocosmic gas phenomena result from assumptions about molecules in kinetic gas theory. Therefore, one of the guiding principles in formulating assumptions about space in the small should be comparable to the correspondence principle in Bohr's atomic theory and guarantee that, in the large, the geometry asymptotically approach that of Riemann. Although such a development has hardly begun one may perhaps venture a certain hope that one day proper assumptions about space (and time!) in the small may geometrize microphysics somewhat like Riemann's geometry in the large has geometrized gravitation in the general theory of relativity.[9]

NOTES

[1] [1942, 1], [1951, 1], [1951, 4].

[2] A. Wald, "On a Statistical Generalization of Metric Spaces", *Proc. Nat. Acad. Sci. U.S.A.*, **29** (1943), 196–197.

[3] B. Schweizer and A. Sklar, "Espaces métriques aléatoires", *Comptes Rendus Acad. Sci. Paris*, **247** (1958), 2092–2094; and "Statistical Metric Spaces", *Pacif. J. Math.* **10** (1960), 313–334.

[4] A. N. Šerstnev, "On the Concept of a Stochastic Normalized Space", *Dokl. Akad. Nauk SSSR* **149** (1963), 380–283.

[5] For a brief summary, see B. Schweizer, "Probabilistic Metric Spaces – The First 25 Years", *The New York Statistician* **19** (1967), 3–6.

[6] B. Schweizer and A. Sklar, "Statistical Metric Spaces Arising from Sets of Random Variables in Euclidean *n*-Space", *Teoria Veroyatnostey* **7** (1962), 456–465.

[7] E. Schrödinger, *Ann. Phys.* **63** (1920), 397–456.

[8] R. Bellman, R. Kalaba, and L. Zadeh, "Abstraction and Pattern Classification". *J. Math. Anal. Appl.* **13** (1966), 1–7.

[9] Cf. [1949, 2].

PART VI

L. E. J. BROUWER

CHAPTER 21

MY MEMORIES OF L. E. J. BROUWER

1

In March of 1921 – I had just completed my first semester at the university, studying physics – H. Hahn joined the mathematics faculty. His first release was the announcement of a seminar on the curve concept (*Neueres über den Kurvenbegriff*), one two-hour meeting weekly during the spring semester. I hesitated, but finally mustered up my courage to audit the first session. There, without introduction, Hahn formulated the problem of making precise the idea of curve that everyone has but that no-one had been able to articulate. Various attempts had been made. G. Cantor thought of defining curves as one-to-one images of a segment; C. Jordan then defined them as its images by continuous mappings. But Cantor himself proved the first, and Peano showed the second, of these definitions to be unacceptable since both include squares and even cubes, which no one calls curves. Thereafter arcs were studied – topological images of a segment; that is, obtainable from it by mappings that are both one-to-one and continuous. Among the arcs one does not find squares or cubes, but neither does one find ellipses or lemniscates. Hence more complicated objects, called graphs, were built up by joining arcs in certain ways; but even this concept was too narrow. Other attempts were unsatisfactory for other reasons. So, Hahn said, the seminar would culminate with the realization that the problem was still unsolved and that a satisfactory definition of the curve concept still was unknown.[1] In conclusion of the first session, Hahn presented Cantor's most elementary concepts of point set theory.

I left in a daze; for I clearly saw that curves were different from surfaces and solids. Should I not then be able to make precise *how* they differed and to *define* curves? After a week of complete engrossment in the question I presented Hahn a solution in terms of the simplest set-theoretical concepts (See Ch. 17) before the beginning of the second session of his seminar – immediately adding the ideas of the ramification order of curve points and of the recursive continuation of the definition to higher dimensions. Hahn agreed that mine was a promising attack on the problem. So I began feverishly to elaborate these ideas – before long, feverishly not only in the figurative sense of the word; for through overwork[1a] combined with post-war undernourishment and studies in ice-cold libraries I contracted the grave respiratory disease then rampant in Vienna. This was to keep me away from the university for more than three semesters – most of the time in a sanitarium in the mountains of Styria.

In my studies there, for the first time I ran into the name of L. E. J. Brouwer as the man who had proved that no topological (that is, continuous one-to-one) mappings exist between cubes of different dimension. I fully realized the fundamental importance of this result; it seemed to me, however, that its primary significance was for the theory of topological mappings rather than for the idea of dimension. Suppose that early in this century someone had constructed a topological mapping between cubes of different dimension. The idea of dimension would have emerged unscathed just as it survived Cantor's revelation of one-to-one mappings between such cubes. What such a discovery would have seriously damaged is the role of topological mappings in the field of geometry, just as Cantor's revelation practically removed one-to-one mappings from that field. Of course, the latter mappings remained the basis of a theory of cardinality and the former might well have lived on as the central concept of an extra-geometrical topology. Geometers, however, would have continued to study dimension as the most important intrinsic property of geometric objects while continuing to look for a class of mappings narrower than that of topology or perhaps overlapping it, but provably free of mappings between cubes of different dimensions. Thus *whereas Brouwer's result cannot be said to have saved dimension* (which never was in need of redemption) *it was fundamental in making topology a part of geometry*. As I later found out, this point of view (which I have maintained ever since) is different from that of most mathematicians. But certainly nothing short of gross misunderstanding can regard it as belittling Brouwer's or Cantor's work.

For my part, I have always studied geometric objects *intrinsically* regardless of how they react to transformations or mappings and, correspondingly, have always preferred (as more intuitive) intrinsic definitions of classes of objects to the (more or less arbitrary) choices of a paradigm and of methods by which the other members of the class are obtainable from it. Such definitions have seemed to me even less intuitive when they include (besides transformations) certain ways of building up objects from transformed prototypes. But this is exactly what is done in defining curves as graphs – certain unions of topological images of a segment. This criticism is of course without detriment to the enormous importance of the graph concept in its own right. And it was in a similar situation that in developing my ideas on dimension I found Brouwer's name next: in his definition of n-dimensional manifolds, pseudomanifolds, complexes, etc. as certain unions of topological images of an n-dimensional cube – concepts which for all their great importance in their own right, offered narrow and to my mind unintuitive solutions of the dimension problem. Thus the very first paper on dimension that between 1921 and 1922 I submitted to the Viennese mathematical journal, *Monatshefte*[2] began with the following paragraph:

In contrast to Brouwer (who defines dimensionality as a property of some point sets by bringing into play the possibility of topologically mapping those sets into certain others), in what follows an attempt is made to define dimensionality as a property of every subset of a finite-dimensional space,[3] independent of the space and on the mere basis of the most elementary concepts of the theory of point sets. The aim is to include among the sets defined as n-dimensional as many as possible of those ordinarily called n-dimensional and as few as possible which ordinarily are denied this quality. We shall in this way arrive at a recursive definition of dimensionality and then show that the dimension is for every set invariant under topological mappings.

The manuscript (together with a minor correction that I submitted from my sickbed in February of 1922) ended with my definitive definition of n-dimensional sets by recursion starting with the empty set (called -1-dimensional) and applicable to all subsets of Euclidean spaces (See [1921–1923, 1] No. 2). In a footnote, I wrote: "After completing the present paper I found a similar idea in Poincaré (*Dernières Pensées*), if only imprecisely formulated due to the lack of the tool of modern point set theory." I then quote Poincaré's recursive definition, "which, however, today is untenable. . . ." (According to it, a double cone is 1-dimensional being decomposable by the deletion of a single point, namely the vertex.) But in Poincaré I found authoritative support for the general, as it were, programmatic ideas that I had formed about intrinsic definitions.

In my mountain retreat, I elaborated the theory[4] and when I returned to Vienna in April of 1923 I was already busy writing the paper in the form in which it appeared in the *Monatshefte* at the end of that year. It was at that time that I first learned about a short, but important paper published in 1913 in *Crelles Journal* in which Brouwer corrected the obvious error in Poincaré's recursive definition mentioned above thereby formulating the first tenable definition of n-dimensional sets – a definition very close to, though not equivalent to, the one on which I had started building a systematic theory. Brouwer justified his definition by proving, if in a far from transparent way, the n-dimensionality (in the sense of his definition) of the n-dimensional Euclidean space without, however, giving any hint at the potentialities of the idea propounded. Being the chief exponent of the method of defining dimension by mappings and transformations Brouwer of course in no way identified with Poincaré's critique of the unintuitiveness of these definitions. So, in 1922, it was in Poincaré that I found the only support for my programmatic ideas about dimension, while Brouwer's paper, for all its importance, appeared to me to be written as though the author was acceding to a fancy of Poincaré. Published outside of the *Mathematische Annalen*, where he had brought out all his other geometric papers, the article in *Crelles Journal* remained virtually unknown. It was not even mentioned in Kerékjártó's book on topology published in 1923, which discussed practically all of Brouwer's geometric work[5]. Nor had Brouwer himself ever returned to the 1913 paper.

While reading the proof sheets of my article for the *Monatshefte* I learned about a talk delivered in September 1923 at the annual meeting of the German Mathematicians' Association in Marburg by a young Russian, Paul Urysohn, communicating results on curves and dimension that he had published in the *Paris Comptes Rendus* in September 1922. As soon as I could get hold of the issues I eagerly read Urysohn's two splendid notes there – in time to add a reference to his work at the end of my paper [1921–1923, 2]. The equivalence of our basic concepts was obvious though I felt that the form of my definitions was somewhat simpler.

As I continued working on the second part of my paper[6] I became acquainted with another phase of Brouwer's work. Since 1920, in connection with the rise of the theory of relativity, Vienna had resounded with talk about Mach and positivism, the role of hypotheses and the verification of scientific propositions. When in 1923 I read about Kronecker and heard about Brouwer's insistence on the (at least virtual) exhibition of examples as proofs of existential propositions, their views appeared to me to be the counterpart in mathematics of Mach's positivism in science. I became even more interested when learning about Brouwer's attempt to develop an intuitionistic set theory. His basic definition of set, however, was incomprehensible to me; and none of the mathematicians whom I asked about it (including Hahn and Schreier) had the faintest idea of what Brouwer meant. Moreover, upon hearing about constructive proofs I wanted to know exactly what was meant by construction. It occurred to me that if I could sometime absent myself from Vienna for a few weeks I might visit

Amsterdam and try to clarify my ideas on intuitionism, though it was my dream to work in Paris.

Early in 1924, I sent out reprints of my paper. Brouwer, members of the Warsaw school and Fréchet in Paris were the first to reciprocate with reprints of their own work. Brouwer sent me a 1923 note from the *Amsterdam Proceedings*[7] – essentially a republication of the 1913 paper from *Crelles Journal.* Being still a student I was quite elated to receive many reprints and wrote letters of thanks. In September, at the annual meeting of mathematicians in Innsbruck, I was appalled to learn from Emmy Noether the tragic news of Urysohn's drowning that she had received from P. S. Alexandroff, a witness of the accident.

Later in 1924, I received from Brouwer a note just published in the *Amsterdam Proceedings* in which he proved the equivalence of what he called "the natural and the Menger–Urysohn dimension." A correspondence developed and since I wanted to publish my article on curve theory (*Grundzüge einer Theorie der Kurven*) in the *Mathematische Annalen*, where Brouwer was on the editorial board, I submitted the manuscript to him. The great interest that he expressed in my results prompted me to write him about my plan of a visit to Amsterdam in order to clarify my thoughts on intuitionism. We agreed on the spring of 1925, when he also expected Alexandroff and several other guests.

So in March 1925 I left for Amsterdam, full of enthusiasm and hope.

2

When I met Brouwer in 1925 he looked older than his 44 years. His figure was lank and youthfully agile; but his hollow-cheeked face, faintly resembling Julius Caesar's, was extremely nervous with many lines that perpetually moved changing his expression from one moment to the next. His features seemed incessantly to reflect intense inner reactions to what he was hearing and seeing, just as outward intensity in speech and movement and action was the hallmark of his personality.

When in Holland, Brouwer lived in a simple suburban house or in a cottage behind it, where he usually walked about in light sandals or barefoot. But his life was rather restless. He spent many short and long vacations abroad – in those years most of them in Locarno. He loved, as he once told me, riding in fast trains. At home, like all Dutch people of that time, he rode a bicycle. He was a strict vegetarian.

Brouwer spoke an almost accent-free German only rarely using an unidiomatic term or, in great excitement, an ungrammatic form.[8] His English and French were also very good, especially the latter. Whichever the language, Brouwer talked very fast and with great precision, though often in long and involved sentences. This he also did in his lectures, where he went into minute details. But he delivered them with an impressive emphasis.

My first meetings with Brouwer did not quite live up to my expectations. They were always in the company of several other mathematicians. The conversation centered on organizational matters such as the publication of papers, the distribution of reprints, lectures at future meetings in Germany, Alexandroff's requests for official letters from Brouwer to Moscow, and Brouwer's precarious relations with the French officials of the *Union Mathématique Internationale*. Brouwer also talked a great deal about other mathematicians' reactions to his topological and intuitionistic work. He gave the impression of being oversensitive and somewhat testy. In particular, I remember him complaining about Hausdorff, who once at a mathematical meeting appeared to him to suggest that the proofs of his topological theorems were obscure and would have to be completely redone. Alexandroff, always complaisance personified in his dealings with Brouwer, protested that this could not be Hausdorff's meaning; but Brouwer persisted. (I myself had at that time not yet had any contact with Hausdorff.)

After a few weeks, Brouwer asked whether I would become his assistant, mainly to help him read some manuscripts that he had to referee. (He received all geometric papers submitted to the *Mathematische Annalen* and some to be published in the *Amsterdam Proceedings*.) Also, I would be a *docent* (lecturer) at the University of Amsterdam and could start a course in the fall of 1925. I accepted with pleasure, proposing the theory of curves and dimension theory as the topic of my first course.

Thereafter, although from the beginning I had been aware of attempts to undermine our relationship, I frequently saw Brouwer alone and came to know more about his habits and tastes. I never found him depressed, but I rarely saw him laugh or display signs of a sense of humor. He usually manifested a high-strung personality – least tense when listening to or talking about music. He loved classical music and frequently attended concerts, especially of Beethoven's works, about which he spoke beautifully. In painting he seemed to favor the style of the Renaissance and was fond of portraits of that period. He looked with obvious distaste at the expressionistic graphic with social themes by the Dutch and Flemish artists of the 1920's (which I greatly admired) – even at Masereel's marvellous novels in woodcuts. They lacked the kind of beauty he seemed to expect from art. He also looked for comeliness in handwriting and developed in his middle years an aesthetic style of writing. But even more interesting were his ways of crossing out passages in manuscripts. In order to delete an entire paragraph he would draw many diagonal lines in both directions forming networks of perfect regularity; some had wider meshes, others were quite narrow grills. Single words or sentences he blocked out with solid black lines or rectangles. Some corrected manuscripts of his were quite picturesque (in a remotely Mondrianish way). The handwriting was also of interest to him in letters that he received; but there, his chief concern was the consistency of style of writing in a person's consecutive letters[9].

D. van Dantzig, who was a student in my first courses at the University of Amsterdam, once lent me one of Brouwer's first publications, a booklet *Leven, Kunst en Mystiek* (Life, Art and Mysticism, 1905), in the mid 1920's little known even in Holland. But it did not impress me as an important book. All I can recall from my (it is true, rather casual) reading is a general mystical trend, antisocialistic passages, and views about women that reminded me of those developed, likewise in the first years of this century, by the Austrian misogynist O. Weininger and widely read in the Vienna of my early youth. In particular, I remember a passage from Brouwer's booklet about the debasement of male activities (including smoking) when or because women adopt them. Van Dantzig mentioned that Brouwer did not like to talk about the booklet.

In his political views, Brouwer was decidedly antisocialist, far to the right and of a militant Teutonic chauvinism. He told me that in Brussels he insisted even in French stores on being waited upon in Flemish, and from Milan he once wrote me that its inhabitants are no less Teutonic than, for example, Munich's. But never did Brouwer during the years 1925–27, when I saw much of him, express or show any hostility to Jews, although he had at least one close friend who was violently antisemitic.[10] On the contrary, Brouwer furthered the careers of some young Jewish mathematicians. Only Einstein was an object of his strong animosity, but mainly because of the physicist's articulate opposition to German nationalism.

Brouwer's hatred in those years was concentrated on the French; and these feelings, which greatly bothered me, also tainted his mathematical judgment. He spoke in a highly uncomplimentary way about Poincaré as well as Lebesgue. This aversion against anything French was kept boiling by his endless and very unpleasant correspondence with the *Union*

Mathématique, where soon after World War I he had begun to defend the German cause.

Brouwer's creative activity in geometry had ended years earlier; but he kept up with the latest progress of point set theory and topology following it with intense interest, though not contributing to its explosive development. Once in a conversation during that period he seemed to me to take a mental note of some problem that I brought up, as though he wanted to think about it later – I have forgotten the details – but he never came back to the question.

In the late spring and the fall of 1925, I reported to him my latest results almost every week and immensely enjoyed these occasions. He never formulated problems to me; but he listened with an eagerness that I found very stimulating. Through the perpetually changing expression of his face he had the rare ability to be, as it were, even silent with intensity. He also showed concern for my then still unstable health and expressed warm sympathy when I suffered a bereavement. He warned me against overworking.[11] "A mathematicians' mind is like a fruit tree," he once said to me. "Not all blossoms turn into fruit; and if one year they all do, then the tree bears no fruit the following year." I revered him with a deep affection during those months.

Mathematical creation Brouwer regarded as a decidedly youthful activity. "Even Gauss," he once remarked, "spent his later years on boring numerical computations." I observed that according to what I had heard, Gauss enjoyed numerical computations from childhood on, all his life. "Mathematicians' old age is sad," Brouwer answered somberly, "Gauss' old age was sad, too." It was the only time that I ever discovered in him something like gloom. Only much later did I realize that these remarks were the key to understanding Brouwer's personality, especially several traits of his that I was soon to discover and which (together with some of his political views) I would find incompatible with the idealized picture that in youthful enthusiasm I had formed of him.

3

Brouwer was involved in many time-consuming controversies and quarrels – scientific, academic and political. All his attacks were phrased with the formality of legal briefs, some of them strewn with whereases. "Dossier" was one of the most frequently used words in his vocabulary. I regretted his spending energy on petty formalities and still less cared for his habit of motivating each of his aggressive actions (about which he obviously enjoyed talking) with some moralistic principle: justice, equity, historical accuracy or the sanctity of a document.

At the beginning of 1926, a new element entered our relations. Brouwer began to insist that the papers on dimension theory submitted to him contain mutual references. This, he said, would enable him to exert a meticulous watch over Urysohn's rights, which he called a precious trust committed to him. I had of course no objection; in fact, I admired his righteous protective attitude toward the deceased, though in combination with his penchant for legalism it wasted a great deal of time and energy of his as well as of mine.[12] Nor did I raise objections when in most arbitrations between authors he suggested that both insert references to his own 1913 paper. On the contrary, I found it gratifying that Brouwer tried more and more intensely to connect himself with Urysohn's and my ideas and to minimize the obvious differences between his definition and ours.[13] I did not yet suspect to what extremes he was prepared to go in achieving this ambition.

But soon Brouwer's preoccupation with references became an obsession. The opportunities to report new results to him, which had been such beautiful experiences, became ever more rare because he at once turned the conversation to footnotes in papers recently submitted. This development also disturbed W. Hurewicz, a brilliant Polish mathematician, a few years my junior, who had studied at the University of Vienna and written his first papers on dimension-theoretical problems that I had formulated to him before going to Amsterdam. In the spring of 1926, he followed me there and was soon employed by Brouwer.

On several occasions, I urgently asked Brouwer not to bother with those matters so much – certainly not for my sake. "They are not worth causing you annoyance, let alone excitement," I said. Once – I believe he was incensed about references or the absence of references to his or my or Hurewicz' work in a Russian paper submitted to him – I remember telling and writing him, "I concern myself with continuing my work; and this is a thousand times more important to me than superficial formalities."

In mid-1926, a long report on dimension theory ([1926, 4]) that I had written for the German Mathematicians' Association appeared, which made dimension theory widely known.[14] Brouwer, who had carefully read the manuscript, praised it highly; and when I dedicated the offprints to him he seemed very pleased. During that summer, he and Hurewicz were gone. I was alone, working incessantly on curve theory, dimension and metric geometry, and I obtained more geometric results than in any comparable period of my life.[15] In September I went to the annual meeting in Düsseldorf and presented a summary of all theorems on dimension known at the time including those that I had found in the summer, ([1927, 2]).

In between, I also pondered again over intuitionism, for the sake of which I had come to Amsterdam. In the spring of 1925, Brouwer had delivered weekly lectures about the intuitionistic theory of order which I had attended while reading or rereading all of his intuitionistic writings published up to that time. But I had not clarified my thoughts on the nature of construction. Various possibilities for defining it occurred to me and the idea of logical tolerance (See Part I) began to take shape in my mind. As to set theory, it was only after reading notes that R. Weitzenböck had taken in a course given by Brouwer in the early 1920's, that one day in the summer of 1926 it suddenly dawned on me that Brouwer's basic concepts actually were classical concepts in an unusual and complicated terminology. In particular, I recognized Brouwer's mysterious "sets" as essentially identical with what are traditionally called analytic sets while his "species" are ultimately nothing but ordinary sets (See p. 85) – of course, all concepts treated without the use of the law of the excluded middle. There was a two-fold reason why no-one had noticed this correspondence. Set-theoreticians, in particular Hausdorff (a specialist on analytic sets) and Fraenkel, told me that they failed to understand Brouwer's phraseology,[16] while others, in particular Weyl, who understood some and seriously tried to understand all of the formulations, because of their life-long aversion to set theory did not seem to be familiar with analytic sets. As I myself pointed out in presenting the intuitionistic–formalistic translations, the dictionary is of no significance for the technical development of the intuitionistic theory. I felt, however, that the translation would demystify its content (which it did, as Hausdorff, Heyting and others later wrote me). I therefore hoped that everyone interested in clarity and in a wider dissemination of the ideas would be happy with the result as I myself was. But to my disappointment, Brouwer seemed to dislike it; and to this day intuitionists, especially Heyting, in writing about set theory do not even mention the dictionary. (See p. 86, Note [0].)

In the fall of 1926, when Brouwer and Hurewicz returned, the difficulties intensified. In discussing the meetings we had had with Brouwer – jointly or separately – Hurewicz and I jokingly spoke of footnote-itis. But soon the matter became much too serious for jokes. Brouwer seemed no longer satisfied with being associated with the new dimension *concept* and more and more clearly aimed at being referred to as the founder of dimension *theory*. And Brouwer's talking himself into ever greater excitement had a still worse effect: as we felt with increasing vexation, we gradually became infected ourselves – first I, then even Hurewicz, one of the most lovable personalities I have ever met. I began to pay too much attention myself to those "superficial formalities" from which, not long before, I had tried to divert Brouwer; and on some occasions when he happened not to start discussing them, orally or in writing, then I did. Moreover, at about this time two grave events occurred. For weeks I had found alterations and insertions in Brouwer's characteristic handwriting in my manuscripts and proofsheets. When on one of these occasions I tried, of course with utmost restraint, to expostulate with Brouwer, he terminated the conversation after a short while; and the next day, a friend of his looked me up to tell me that Brouwer was infuriated by my remonstrations and that he declared that he would hurt me with all the harm that he could do to me if I should not go along with his suggestions. I did not repeat my objections or change my tone, but my feelings toward Brouwer became ambivalent. About the same time, by coincidence, it came to my attention that even in Urysohn's posthumously published work, references to Brouwer were being added to and changed without the alterations being in any way designated as such.[17]

4

Though deeply disillusioned, I of course still retained (and in fact have retained to this day) also positive feelings for Brouwer. But I wanted to leave Amsterdam and to work far from the atmosphere that had developed in the initially so pleasant surroundings. At the end of 1926, just when the idea of getting away began to obsess me I received a letter from Hahn asking whether I would be interested in returning to fill the chair for geometry at the University of Vienna.[18] During a visit to my hometown in the spring of 1927 it was decided that I should start lecturing in the fall.

My first course in Vienna was on dimension theory; I also decided to summarize all that was known about this subject at the time in a book – a book that would also include methodological ideas; a book that I hoped would make inspiring reading; and a book free of footnotes. Using manuscripts that I had already completed and following the arrangement of my 1926 report at the Düsseldorf meeting I finished the work in the first half of 1928. It appeared under the title *Dimensionstheorie* in September of that year.

There are indeed no footnotes in the book. At the end of each section a short postscript in small print describes the historical development of the material just presented.

In the postscript following the introduction of the dimension concept, the greatest shortcoming is my failure to mention Bolzano. Many mathematical studies by the great philosopher were not yet known at that time and have been published only more recently.[19] I should, however, have mentioned at least the penetrating remark in his *Paradoxes of Infinity* (See [1]). But oddly enough, if I had read that remark (which I do not remember having done until much later) I somehow did not connect it at that time with my own ideas although the latter were more closely anticipated by Bolzano than by anyone else.

As to Brouwer, I should have more strongly emphasized (as I did above, p. 239) that in correcting Poincaré's definition he formulated the first tenable general dimension concept. I was, however, correct in saying that Brouwer's definition was close to (but not equivalent to) the one introduced by Urysohn and myself (See p. 245). In their excellent book, *Dimension Theory* (1941), Hurewicz and Wallman referred to the difference between our concept and Brouwer's as "important improvements". That in my book I stressed differences rather than similarities was a reaction to my unsettling experiences mentioned at the end of Section 3, which made me see some traits and actions of the Brouwer whom I had idolized in a quite new light. The image of an arbiter of supreme punctiliousness that he had systematically created in me, the immense fastidiousness with which he had scrutinized cross-references in the papers submitted to him, his frequent emphasis on a watch over Urysohn's trust – all this and more became problematic when I saw the different standard that he himself applied to the work of the deceased. Yet I refrained in my book from even hinting at the irregularities in the way the *Mémoire* had been edited.

5

In 1928, Brouwer's long-standing conflict with the *Union Mathématique Internationale* came to a head.

When sponsoring international mathematical congresses at Strassbourg (1920) and Toronto (1924) the *Union* had restricted its invitations to countries belonging to the *Conseil International des Recherches*. Despite the demand of the western and northern democracies that all political discrimination be ended, the *Union* interdicted the dispatch of invitations to Germany (which had not yet joined the *Conseil*) even when scheduling the next congress to be held in Bologna, in September, 1928. The Italian organizers of the congress, however, in keeping with the conciliatory spirit of the mid-1920's, invited mathematical societies as well as individual mathematicians from all countries including Germany and Austria.

In March, Brouwer came to Vienna to deliver two lectures that I had helped arrange. I tried to make this visit a memorable experience for him; but during almost his entire stay he spoke violently against the congress because the interdictions mentioned above had not yet been retracted. The debate climaxed at a party in Hahn's home. The host and the physicist H. Thirring tried to calm Brouwer. "Deep historical antagonisms are not terminated by retractions; they are gradually *forgotten*," Hahn said. "This is how the conflict between Catholics and Protestants faded away and so will the hostility between France and Germany." Brouwer however persisted.

During the summer of 1928, David Hilbert, the supreme representative of German mathematics, made it known that he had accepted an invitation to deliver one of the conferences at Bologna. About two weeks before the Congress I received from Brouwer a circular letter quoting past anti-German utterances of the *Conseil* and ending with the suggestion that "every mathematician may consider for himself how participation in the scheduled congress is possible without an insult[20] to the memory of Gauss and Riemann, the cultural character of mathematics, and the independence of the human spirit."

I wrote Brouwer that I disagreed with his letter – that I disagreed with his position he of course already knew – but that, out of loyalty to him, I had decided not to attend the meeting that he so violently attacked, even though Hahn urged me to go. Overwork and strain connected with the publication of my book, which was just coming out, had made me very tired.

Hilbert, outraged by Brouwer's circular letter at once reorganized the editorial board of the *Mathematische Annalen* dismissing Brouwer despite his violent protests and despite a flood of letters that he sent to mathematicians and publishers. As common friends wrote me from Amsterdam, Brouwer was so completely beside himself that they feared for his mental stability.

6

It was at this time that Brouwer became incensed over my book, *Dimensionstheorie.* As Hurewicz wrote me, a point that particularly angered Brouwer was my remark that if dimension were not topologically invariant this fact would be more disadvantageous to topological mappings than to dimension. But I had emphasized this very point of view from the begining (See p. 238f), in particular also in my 1926 Report; and after carefully reading that manuscript Brouwer had seemed delighted to accept the dedication of the reprints. Now, Hurewicz wrote, Brouwer heatedly complained that the said remark belittled his achievements – a misunderstanding (See p. 238) and an inconsistency for which only his overexcitement can account.

In this frame of mind, Brouwer published in the *Amsterdam Proceedings* (1929) a note "Historiography of Dimension Theory" which began with the declaration that he had founded dimension theory in his 1913 paper. "There followed Urysohn in 1922, who . . . found [two] theorems and in connection with them clarified numerous dimension-theoretical questions, and Menger and Alexandroff who since 1924 and 1925, respectively, have further elaborated and deepened the theory."

In my book, I had represented Urysohn as (with me) one of the two founders of dimension theory and characterized his work as "great" and "fundamental". When I read the quoted passage in Brouwer's note I felt the view that Urysohn's work was a continuation of Brouwer's might be supported by the references to Brouwer's 1913 paper in the *Mémoire* (and indeed Brouwer himself quoted one of these allegedly Urysohnian references). At least a brief clarification seemed necessary. So in my reply ("*Antwort auf eine Note von Brouwer*", [1930, 8]) I first answered in detail all the allegations in Brouwer's note following his quoted initial declaration – invalidating them point by point with a few minor exceptions for which in my reply I expressed regret – partly oversights, partly overreaction to the disillusionment described at the end of Section 4. As to the *Mémoire*, all that I included was a short remark to the effect that "the references to Brouwer in Urysohn's *Mémoire* were posthumously added to and changed without any of these additions and changes having been designated as such." I expected that this would close the controversy for good since I did not think that Brouwer would care to challenge my quoted remark in print; and indeed, as far as I know, neither of the two editors of the *Mémoire* ever did. Nor did I, for my part, intend ever to return to the matter in a journal or in a book – as indeed I have not for almost half a century – unless and until it should be brought up again in print, as it was, wantonly, in Brouwer's *Collected Works* in 1976.[21]

For in thinking about Brouwer, I try to forget the dark chapter of our controversy and its ramifications. I rather like to recall other, brighter memories: of his mathematical achievements; of his interesting personality; of the way he spoke about mathematics and music; of the good he did for some young mathematicians, among them, during the first year of our personal relations, myself; and, above all, of the beautiful experience of watching him listen to reports of new discoveries.

NOTES

[1] Much later when learning about the following three facts I was surprised that Hahn had never mentioned them in any of his lectures that I attended nor – perhaps in order not to discourage me – in our conversations. Hausdorff in his famous book *Mengenlehre* (Set Theory, 1914) expressed doubt that, considering the various uses of the term *curve*, one could formulate a satisfactory definition at all. Janiszewski, the founder of the Polish school of geometers, pointed out to the International Congress in Cambridge (1912) that nothing positive could be said about the distinction between curves and surfaces, and stressed the negative character of his remarks at the beginning and the end of his paper. The great Austrian philosopher-mathematician Bolzano in his *Paradoxes of Infinity* (1848) made some remarks about curves and dimension that came closer to the definitive solution of the problem than Poincaré's more than fifty years later (See Section 4, below). That Hahn failed to mention this fact to me is particularly odd since he was about to publish a new edition of Bolzano's booklet.

[1a] Also in connection with [1921–23,3].

[2] The full title of the journal was at that time *Monatshefte für Mathematik und Physik* (later, the words *und Physik* were dropped). As to the paper submitted between 1921 and 1922, See [1921–23,1] Nos. 2 and 2a. Before writing that paper, during a new attack of my illness, I remembered a conversation I had had with my father a few years before his death concerning the services of scientific academies for the public. He mentioned that among other things academies acted as trustees of people's ideas by accepting sealed envelopes and then later testifying to their contents and the date of their deposition. Uncertain as to when, if ever, I would be able to elaborate my ideas I decided in the fall of 1921 to avail myself of this opportunity and deposited with the Viennese Academy of Sciences a letter with some of my thoughts on curves. This letter included the definitive definition of curves and stated five rather important theorems. One of them – that a curve cannot consist exclusively of end points – sounds ridiculously simple because for graphs it is trivial; but to me it was particularly significant since its proof led me to a method ("the modification of neighborhoods near their frontiers") which I recognized as an important tool and which during the next couple of years and even later indeed enabled me to solve some of the most difficult problems of dimension theory. In concluding I sketched in a few words a recursive definition of higher dimensions. (Poincaré's *Derniéres Pensees* was at that time not yet known to me.)

[3] "Space" here of course means Cartesian or Euclidean space.

[4] From Styria, I sent to the *Monatshefte* another paper in November, 1922 (See [1921–1923, 1] No. 3). Only after my return to Vienna in April, 1923 did I get it back with the suggestion that I clarify one highly technical point in one of the proofs using my method of the modification of neighborhoods.

[5] Among others, Kerékjártó mentioned the extremely important fixed point theorem; the generalization of Jordan's famous theorem (that a closed curve divides the plane into two parts) to surfaces in higherdimensional spaces; Brouwer's studies of mappings and transformations, especially the concept of the degree of a mapping; and the discovery of continua with amazing properties, among them a continuum that is the common boundary of three (or even more) domains in the plane. The logical critique of the (often misleading) common geometric intuition which prompted the construction of these domains led to the development of a refined intuition that can be logically substantiated. It was Kerékjártó's book that first brought these truly great achievements of Brouwer to my attention in my student days.

[6] [1921–1923, 2] and [1924, 3] combined constituted my doctoral thesis at the University of Vienna in the spring of 1924.

[7] This is the abbreviation herein used for *Proceedings of the Royal Academy of Science in Amsterdam.*

[8] Speaking about an ill-received lecture he had given in Göttingen a year or two before, Brouwer (still very angry) reported that during a subsequent visit Felix Klein had commented on the low level of the discussion following that talk with the words "*Wie ich höre, hat sich die Diskussion auf einem sehr niedrigen Niveau bewogen*". Attributed to Klein, *bewogen* instead of *bewegt* sounded especially droll, somewhat as though G. H. Hardy were quoted as saying, "A poor discussion, to judge from the report that I readed." (Brouwer told this same story – *bewogen* and all – twice.)

[9] When a foreign mathematician who formerly had written in longhand sent Brouwer a printed letter – a style he always used when under strain – Brouwer broke up the correspondence informing the writer that he could not continue an exchange with a man of two personalities.

[10] Once after his friend in a conversation with me had extolled someone as an Aryan lumen and condemned another man as a corrosive Jew, I pointed out to him that he happened to be mistaken about the racial backgrounds of *both* men, and that the situation was the exact opposite of what he assumed; to which, after a short silence, all he replied was, "The trouble is that Jews are not black" – a remark with such a wealth of implications that it has lived on in my memory. Then he changed the topic of conversation.

[11] Brouwer's very moderate assignments to me were essentially confined to translations of some older writings of his on intuitionism from Dutch into German.

[12] For example, he asked me to produce what amounted practically to affidavits from Hahn and Schreier attesting that I had been developing my ideas on dimension since 1921.

[13] Both in one of his first letters to me and later in his 1924 note, Brouwer referred to one of the differences between his 1913 definition and the one given by Urysohn and myself as a *Schreibfehler* (slip of the pen). Without at that time closely looking into the matter I repeated that term in [1924, 3]. Actually, however, Brouwer had not made a

slip at all, but had formulated a perfectly *valid* definition – though not quite the one on which Urysohn and I built our theories. (See, for example, Hurewicz and Wallman, *Dimension Theory*, p. 4).

[14] Urysohn's great work on dimension theory the first half of which was just then being posthumously published in the *Fundamenta Mathematicae* was not in its title connected to dimension theory. It was entitled *Mémoire sur les Multiplicités Cantoriennes* after a concept (related to connectivity as well as dimension) that was of great importance in Urysohn's particular presentation of the theory of dimension.

[15] Early in 1926 I had proved that any curve, no matter what the dimension of the space in which it is located – be it even Hilbert's infinite-dimensional space – can be topologically embedded in (that is, mapped on a subset of) our 3-dimensional Euclidean space ([1926, 2]). In fact I constructed a particular curve in our space such that every curve can be topologically embedded into a subset of this "universal curve". In the summer of 1926, I discovered a theorem on curves ([1926, 1]) to which (or rather to the special case dealing with graphs) D. König was to devote an added chapter of his well-known book on graphs – more recently, it has found use in electrical engineering; proved two basic theorems on dimensional structure, ([1926, 3]); and completed the first two in a series of Investigations into General Metric in which I introduced and studied convexity and geodesics in metric spaces and characterized the subsets of Euclidean spaces among them, ([1928, 2]).

[16] Neither did Alexandroff who had written an interesting paper on analytic sets seem to associate Brouwer's "sets" with them.

[17] Urysohn's *Mémoire* (See [14]) came out in volumes 7 and 8 of the *Fundamenta* (1925/26). More recently, Alexandroff (*Nieuw Archief voor Wiskunde*, 1969) has described the close cooperation between him and Brouwer, in bringing out Urysohn's posthumous papers – Brouwer particularly doing the proof-reading of the *Mémoire*. In the *Fundamenta*, however, no editor was mentioned which at that time suggested that no editing whatever had taken place – an impression corroborated by the fact that each of three relatively minor appendices (including supplementary notes, a list of new terms, and a table of contents) is marked "*redigée par P. Alexandroff*". But in 1926 I became aware that actually the *Mémoire* had been edited in both volumes. In volume 7, a footnote referring to Brouwer's 1913 paper at a crucial point had an essentially different wording than it had in preprints; and in volume 8, besides several references to Brouwer's 1913 paper, an entire such paragraph seemed unlikely to have been authored by Urysohn, being at variance with a preceding statement in the *Mémoire* by that punctilious writer. When, in a 1929 paper attacking me, Brouwer quoted one of these allegedly Urysohnian references to his 1913 paper I asked the editors of the *Fundamenta* 1) to explain the discrepancies mentioned above, 2) to quote the original wording of the footnote in volume 7, and 3) to describe the sources from which the *Mémoire* in volume 8 was printed. These three questions were answered as follows **[**my comments in heavy brackets**]**:

1) Soon after Urysohn's death, the editors of the *Fundamenta* received a letter from L. E. J. Brouwer in which, on the basis of an authorization from Urysohn's heirs, he demanded changes in the *Mémoire*, some of them in the section of volume 7 consisting of the sixteen pages 33–48. But as these pages were already printed Brouwer had a new typesetting of that entire section carried out at his own expense. Offprints of the *Mémoire* in volume 7, however, had already been sent out. **[**What I had read was one of these preprints.**]**

2) The footnote in volume 7 [following the author's claim to elucidate for the first time the question as to what dimension is] stated in Urysohn's manuscript that after completing his *Mémoire* the author learned about the 1913 paper "in which Brouwer intends to solve that question. Now Brouwer's proof," Urysohn continued, "contains an error that appears to me to be irreparable and which weakens all his results (*une erreur qui me semble incorrigible et qui infirme tous ses résultats*). For the details, I refer to a supplement to be found at the end of the first part of this *Mémoire*." Following instructions received from Urysohn in the summer of 1924, the reference to the error in Brouwer's proof and to the supplement was replaced by the statement (on p. 37) "I hope to return to the relations between these methods." Also in this version, just as in the original manuscript, Urysohn stated that in the 1913 paper "Brouwer intends to solve the question (*se propose de résoudre la question*)." [This is the final version received from Urysohn in the summer of 1924 as found in the preprints. In the recomposed section in *Fundamenta*, volume 7, the footnote is on pp. 37/38 and also refers to Menger's (that is, my own) 1923 publication; and it states that in the 1913 paper Brouwer "solves the question (*résout la question*)."]

3) The manuscript of the part of the *Mémoire* that was to appear in volume 8 was sent, upon Brouwer's demand, to him; and in the name of Urysohn's family he requested that they should receive it for reverent preservation (*zwecks pietätvoller Aufbewahrung*). The text in volume 8 was printed from a copy of the manuscript by Brouwer. [In the Russian version of the *Mémoire* in Urysohn's *Works* (edited by Alexandroff Moscow and Leningrad, 1951), which is otherwise practically identical with the French *Fundamenta* text, numerous references to Brouwer are missing. In particular, four *Fundamenta* references to Brouwer's 1913 paper – two in the text and two in footnotes (including the paragraph of intrinsically dubious authenticity as well as the one quoted as Urysohn's in Brouwer's paper attacking me) – are missing in the Russian version.]

[The supplement dealing with Brouwer's 1913 paper mentioned in Urysohn's original manuscript is not among those edited (*redigées*) by Alexandroff at the end of the *Mémoire* or in Urysohn's collected works. Historians of geometry will regret that it has not been published. Its publication (of course with a proper editorial note) would undoubtedly still be appropriate, since the author's untimely death did not allow him to return to the matter as he had promised. Even in the unlikely case that the supplement should include an inaccuracy it certainly would be worth reading since errors by mathematicians of Urysohn's caliber are usually of importance and always of interest.]

[18] As an economy measure, the chair had been left vacant after the departure of the geometer K. Reidemeister, a member of the earliest Circle of Schlick.

[19] A detailed study of Bolzano's ideas about dimension will be published by Dr. Dale M. Johnson.

[20] Brouwer used the somewhat incongruous term *Verhöhnung*, which means *deriding*.

[21] Immediately after my reply appeared in the *Monatshefte* Brouwer demanded (and at once received) from the editors the permission to publish a refutation of my short remark about the references to him in Urysohn's *Mémoire*. But he never availed himself of this opportunity and in fact never answered the remark in print at all. In conversations with several mathematicians, however, especially during the summer of 1932, Brouwer heatedly complained that I had made what he called a grave charge without substantiating it. (Actually, my remark was a simple statement of fact.) So in a response to Brouwer's

oral complaints, in the fall of 1932 I documented the remark in a short note, ([1932, 3]) "*Über die Hinweise auf Brouwer in Urysohn's Mémoire*" (On the References to Brouwer in Urysohn's *Mémoire*), by stating the three questions and answers mentioned in [17]; and I announced that note (and its availability on request) in the *Monatshefte*, [1932, 4]. But though I might have published there the note in its entirety I confined myself to bringing it out as a private print in keeping with my decision not to return to the matter in a book or journal unless it should be raised again in print.

PART VII

ECONOMICS. META-ECONOMICS

CHAPTER 22

THE ROLE OF UNCERTAINTY IN ECONOMICS

The following essay was essentially completed by 1923. In 1927, I presented the paper to the Viennese Economic Society; in 1934, it was published in the Viennese *Zeitschrift für Nationalökonomie* under the title 'Das Unsicherheitsmoment in der Wertlehre', [1934, 8 and 8a].

The reaction to my talk in 1927 was mixed. L. von Mises was one of the few Viennese economists who knew of Daniel Bernoulli's suggestion that a man appreciates an addition to his wealth, W, inversely proportional to W (hence evaluates W logarithmically); and Mises expressed particular interest in my proof (See Section 5) that just for the purpose for which Bernoulli had conceived it his splendid idea was inadequate, although for 200 years it had been accepted as a solution in all books on probability (including Poincaré's *Calcul de Probabilité*, 1896). On the other hand, Hans Mayer, at that time in charge of the *Zeitschrift*, explicitly advised me against publishing the talk, so that I submitted the paper only years later when O. Morgenstern, by then on the editorial board, urgently asked for the manuscript in which he had expressed great interest ever since my talk.

The essay proposed to economists and psychologists a program: to describe, after questioning people and observing their choice, how various individuals evaluate the chance of adding an amount D to their wealth W with the probability p; to summarize these results in a general function $F(p, D, W, \ldots)$, where the dots indicate parameters reflecting circumstances and character traits of the evaluators; to average those parameters and to interpret the individual deviations from these averages. Some economists have construed my simple remarks as having the same aim as Bernoulli's idea of logarithmic evaluation. But apart from the fact that both studies try to explain the so-called St. Petersburg paradox, their general tendencies are quite different. As has been pointed out on p. 6, Bernoulli introduced a value-theoretical economic consideration into a problem of the theory of probability, whereas I rather tried to introduce the study of subjective probability into economic value theory.

The most interesting paper 'Truth and Probability' by F. P. Ramsey, published in the collection *The Foundations of Mathematics* that R. B. Braithwaite edited in 1931, was unknown to me at the time when [1934, 8 and 8a] came out. Neither does it appear to have been known to von Neumann and Morgenstern, when in 1944 they presented an axiomatic treatment of utility in their *Theory of Games and Economic Behavior*. (They elaborated this treatment in an appendix to the second edition of that great book.) On the other hand, in their paper 'John von Neumann's Work on the Theory of Games and Mathematical Economics' (*Bull. Amer. Math. Soc.*, **64**, no. 3, pt. 2, 1958), H. W. Kuhn and A. W. Tucker write: "Professor Morgenstern reports that the article of Karl Menger *Das Unischerheitsmoment in der Wertlehre* played a primary role in persuading von Neumann to undertake a formal treatment of utility." My article certainly broached the question as to what sure addition to his wealth W a man equals in value to a gain, D, with probability p.

But the von Neumann-Morgenstern axioms as well as Ramsey's were based on the traditional concept of mathematical expectation and on the assumption that a chance which offers a higher mathematical expectation is always preferred to one for which the mathematical expectation is smaller. My study was not. It took into account observations such as the fact that in casinos the mathematical expectation 36/37 is preferred to 1. Indeed at roulette tables, countless people pay 1 dollar for the chance of winning 36 dollars with the probability 1/37 or, in other games, 2 dollars with the probability 18/37. The pleasure taken in risk by some individuals as well as the fear of risk in others are factors strongly influencing their economic behavior, and not only in gambling situations. Within the framework of the von Neumann-Morgenstern theory, however, as the authors admit, the specific utility of gambling cannot be formulated free of contradiction.

There followed a surge of interest in the role of uncertainty in economics. Important papers in *Econometrica* by K. Arrow, J. Marschak, G. Tintner and others were devoted to this topic in the 1950's. From a quotation in his paper, *Econ. Rev.* (*Tokyo*) **1** (1959), 167–173 I was pleased to see that my essay had also come to the attention of P. Samuelson, who later in *Internat. Econom. Revue* (1960), pp. 30–37, propounded a theorem very similar to the main result of Section 6.

There are three points of contact with the Vienna Circle: R. Carnap, with whom I discussed my talk in 1927, expressed a lasting interest, as reflected in his exposition of some of the ideas years later in his book *The Logical Foundations of Probability*, 1950; what I regard as an explanation of the Petersburg paradox in Section 2 as well as some of the methodological remarks in Section 10 are written, I think, in the spirit of Mach and Schlick; and a passage of the latter's *Problems of Ethics*, 1930, is quoted in Note[16].

The Viennese essay of 1934 was translated into English in [1967, 1]. For the present volume, I combined [1934, 8] and [1934, 8a], shortened the paper and rearranged a few paragraphs of the last sections, while Dr. Phyllis Kittel revised large parts of the translation.

1. THE PETERSBURG GAME

In the theory of probability, the "Petersburg Game" designates the following game[1] between two persons, A and B. Initially, A requires that B place a certain bet with A. Then they flip a coin. If it shows heads, then B receives 1 dollar from A, and the game is over. If it shows tails, it is flipped again. If the second throw turns up heads, then B receives 2 dollars from A, and the game ends. The coin is flipped until heads appears for the first time. If this happens at the nth throw (that is, if the first $n - 1$ throws all come up tails, but the nth throw is heads), then B receives 2^{n-1} dollars from A and the game ends. Depending on whether $n = 1, 2, 3, 4, \ldots$, B wins $2^0 = 1$, $2^1 = 2$, $2^2 = 4$, $2^3 = 8, \ldots$ dollars. What bet should B be willing to make?

What is B's mathematical expectation?[2] The probability that the first throw produces heads is 1/2, and since B in this case will win 1 dollar, this possibility contributes $(1/2)\cdot 1$ dollar to his expectation. The probability that heads first appears on the second throw so that B receives 2 dollars is 1/4, which means that this possibility contributes $(1/4)\cdot 2 = 1/2$ dollar. For any natural number n, the probability that heads first appears on the nth throw is $1/2^n$ and then B collects 2^{n-1} dollars. This contributes the amount of $(1/2^n)2^{n-1} = 1/2$ dollar to B's mathematical expectation, which therefore totals $1/2 + 1/2 + 1/2 \; \ldots \; + 1/2 + \ldots$ dollars, that is, is infinitely large.

Ever since the Petersburg Game was invented, it has been regarded as a paradox, which numerous investigators from various fields have attempted to resolve.

2. THE APPARENT PARADOX AND THE ACTUAL PROBLEM

Even the formulation of the problem is a matter of some dispute. What is usually considered paradoxical or at least to require explanation is the fact that the mathematical expectation of B is infinite. In the present paper, however, the position will be taken that this infinitude in itself is not paradoxical. If B were willing to pay any amount for the privilege of playing the game, then his behavior would conform, insofar as possible, to his infinitely large mathematical expectation. Despite the infinite mathematical expectation, nothing would seem to need further explanation.

Actually, however, the situation is quite different. B will not, if he is sane, risk all or even a considerable portion of his wealth in a Petersburg Game. Now the fact that B's actual behavior differs from that indicated by his mathematical expectation is not paradoxical in any *logical* sense. The discrepancy is a matter of *observation*; and a discrepancy between an observation and a devised formula never constitutes a logical paradox, just as conformity of an observation to a previously stated formula never is a matter of logical necessity. Still, the discrepancy under discussion is peculiar, and it is understandable that many investigators have wanted to find an explanation. The view taken in this paper is that the explanation of a fact consists in its classification within a larger group of facts. Since the question here deals with the economic behavior of normal people, one should expect from the start that the context of its explanation would be primarily psychological-economic, despite the mathematical interest which the problem may claim on historical grounds.

3. THE ALLEGED INCONSISTENCY OF THE GAME

Disregarding mathematically unsound commentaries on the game,[3] we first consider an attempted solution that blames the so-called paradox on the formulation of the problem.[4] According to the rules of the game, A may have to pay extremely high amounts; but he has only a finite fortune. B's mathematical expectation is thus illusory. In actuality B cannot expect to win more than A's entire wealth. To this may be added the fact that because each throw requires some time, the game must be stopped after a finite (though very high) number of throws even if heads has never appeared.

These and similar arguments, however, change the conditions of the game by introducing factors not mentioned in its definition. Furthermore, such explanations merely eliminate the infinitude of B's expectation without accounting for the discrepancy between expectation and behavior. But, as will be shown in Sections 7 and 8, the discrepancy also exists in other situations, where mathematical expectation is finite, and where one cannot possibly fall back on any inconsistency in the game.

4. A PROBABILITY-THEORETICAL SOLUTION

Another solution that has been proposed for the problem is this:[5] only if a game of chance is repeated many times is it reasonable to risk in each game the expected value. Indeed, only a repeated replaying of the game results in the approximate equality (mentioned in Note [2], p. 275 in the case of throwing dice) between winnings and bets, either one being roughly equal to the mathematical expectation. Applied to a single game, however, the concept of mathematical expectation loses its practicality. This is especially true for the Petersburg Game. The real meaning of the statement that B's mathematical expectation is infinite, according to this view, is that however high the bet, B will probably have an advantage over A (that is, B will win more than he has bet) provided A and B play the game sufficiently often and that B risks the same bet in every single game.

This unassailable argument is itself proof that the problem has no paradoxical elements. But it does not even touch the psychological-economic aspects which are responsible for the discrepancy between the mathematical expectation and the normal behavior in the case of a single game. The fact that a normal person playing a single game of chance may be willing to risk more or less than his mathematical expectation remains uninterpreted.

5. THE SOLUTION WITH UNBOUNDED VALUE FUNCTIONS

Other attempts at a solution have been based on the assumption that the gambler B evaluates his potential winnings not according to their absolute amount, but according to the utility which they have for him. This idea[6] has been used to analyze the Petersburg problem in two ways.

The better known version,[7] which will be discussed first, begins with the observation that B possesses a certain fortune when the game opens so that his winnings only constitute an addition to this wealth. The way a man evaluates additions to his fortune, however, depends on his total wealth. The greater B's wealth, the less highly does he evaluate a given sum. This supposedly explains that, while his mathematical expectation is infinite, his actual expectation is finite.

This idea has also been defined quantitatively.[8] To each addition D to wealth W is assigned a "subjective" value directly proportional to D and inversely proportional to W. It follows from this assumption that such an addition is attributed the subjective value of $c \cdot \log(1 + D/W)$, where c is a number which depends only on the evaluator (but is independent of D) and log indicates the natural logarithm (See the figure on p. 274).According to this formula, the addition kD to kW and the addition D to W are of equal subjective value for a person.

If the logarithmic formula is introduced into the Petersburg problem, then for each number $n = 1, 2, 3, \ldots$, the amount 2^{n-1} which B adds with the probability $1/2^n$ to his initial wealth W has for B the subjective value $c \cdot \log(1 + 2^{n-1}/W)$. His total subjective expectation is therefore

$$(c/2)\log(1 + 1/W) + (c/4)\log(1 + 2/W) + \ldots + \\ + (c/2^n)\log(1 + 2^{n-1}/W) + \ldots$$

It is easily seen that this series has a finite sum if $W > 0$, that is, when B starts the game with any wealth at all.[9] Hence B's subjective expectation is finite.

The basic idea of this attempted solution – that the value of an addition to wealth depends not only on its size, but on the wealth of the evaluator – is, of course, excellent and has played a great role in theoretical economics since its formulation by the founders of the subjective theory of value. But as far as the discrepancy between mathematical expectation and the behavior of normal people towards games of chance is concerned, that basic idea fails, as will now be shown, to offer a satisfactory explanation.

We begin by demonstrating that contrary to the views expressed for more than one hundred years in the textbooks on probability theory the solution of the Petersburg paradox according to the logarithmic formula for subjective value is unsatisfactory even on merely formal grounds. (We shall not discuss in this context whether the logarithmic evaluation describes the actual behavior of normal individuals, though this also might be questioned.) The solution is of a purely *ad hoc* character. There exist games similar to the Petersburg Game for which not only the mathematical but also the subjective expectation based on the logarithmic value function is infinite. Consider, for example, the following modified Petersburg Game. A coin is flipped until heads appears for the first time. If this occurs at the nth throw, then B wins from A the amount $We^{2n} - W$, where W is B's initial fortune. For all large n, perhaps even for all n, this amount exceeds 2^{n-1}, which B would receive in the Petersburg Game, so that his mathematical expectation in the modified game is of course infinite. But so is even B's subjective expectation based on the logarithmic value function. Indeed, for each natural number n, B has the probability $1/2^n$ of adding $We^{2n} - W$ to his wealth W – an addition of subjective value $c \cdot \log(1 + (We^{2n} - W)/W) = c \cdot \log e^{2n} = 2^n \cdot c$. This contributes, for each natural number n, the amount $(1/2^n)c2^n = c$ to his subjective expectation which, therefore, in total is $c + c + c + \ldots$ and hence infinite. Yet it is obvious that even in the modified Petersburg Game no normal person would risk his whole fortune or even a substantial amount. Thus, the solution of the Petersburg paradox through the use of the logarithmic formula for subjective evaluation is only apparent.

Furthermore, the unsatisfactory nature of this solution is not due to its specific casting in terms of the logarithmic function. Let more generally, $f(D)$ be the value that B attributes to the addition D to his fortune, where f is any function (dependent on B's wealth and perhaps other parameters) that is unbounded; that is to say, a function which, for sufficiently large D, assumes arbitrarily large values. Examples of such functions, besides that of Bernoulli, are Cramer's $\sqrt{D}$ and $c \cdot \text{loglog}(1 + D/W)$ and even the linear function $cW + c_1$, where $c > 0$. We now prove the

GENERAL THEOREM. *For the evaluation of additions to a fortune by any unbounded function there exists a game related to the Petersburg Game in which the subjective expectation of the risk-taker on the basis of this value function is infinite.*

It follows from the assumed unboundedness of the function f, that there is an addition D_1 with subjective value $f(D_1)$ of at least $2^0 = 1$ dollar, an addition D_2 with subjective value $f(D_2)$ of at least $2^1 = 2$ dollars, and for each natural number n, an addition D_n with subjective value at least 2^{n-1}. Now consider the following modified Petersburg Game. A coin is flipped until heads appears for the first time. If this happens on the nth toss, then B wins D_n dollars from A and the game ends. B's subjective expectation, given the valuation function f, clearly is

$$(1/2)f(D_1) + (1/4)f(D_2) + (1/8)f(D_3) + \dots,$$

which is at least equal to $(1/2) \cdot 1 + (1/4) \cdot 2 + (1/8) \cdot 4 + \dots$, thus infinite.

While the Petersburg problem calls for more than merely eliminating infinite expectations (Cf. Section 2), the above theorem shows that the introduction of an unbounded valuation function does not achieve even that.

6. THE SOLUTION WITH BOUNDED VALUE FUNCTIONS

The idea that the value of an amount of money does not depend only on its size has also been used in a second form, namely with the assumption that the function describing the subjective value is bounded,[10] that is, that there is a number W^* such that for every W, the value $f(W)$ is less than W^*. Under this assumption, B's expectation in the Petersburg Game, as well as in all Petersburg-type games considered in the previous section, is finite, since B's subjective expectation is smaller than $W^*/2 + W^*/4 + W^*/8 + \dots + W^*/2^n + \dots$, and this sum is equal to W^*.

Yet this solution cannot be regarded as satisfactory. (Note that the critical arguments advanced in what follows can also be raised against the unbounded valuation function.) First, introspection shows that our unwillingness to bet a substantial amount on the Petersburg Game is not only due to the fact that the possible large winnings are of lesser subjective value but also, and perhaps more importantly, to the consideration that it is highly unlikely that we shall win a very large amount. Second, there are games where under normal circumstances no-one would even be willing to bet that amount which corresponds to his subjective expectation on the basis of a bounded value function. Examples will be given in Section 7.

Thus even the introduction of bounded value functions does not eliminate the discrepancy between calculated and actual behavior. The idea that the value of goods does not depend solely on their quantities, is correct and important but insufficient to explain gambling behavior.

7. THE SOLUTION WHICH DISREGARDS SMALL PROBABILITIES

Another attempt at a solution begins with the observation that events with very small probability of occurrence are, for practical purposes, completely disregarded.[11] Consider, for example, a game where a single winning ticket of 1 billion dollars is drawn from a total of 100 million tickets. The expected value to the owner of each ticket is 10 dollars. Yet how many among 100 million people would risk 10 dollars in such a lottery?

Incidentally, this example shows that even in some games with completely finite conditions, normal behavior differs from that indicated by mathematical expectation. Moreover, this line of thought contributes to the criticism, started in Section 6, of the attempt to solve the Petersburg problem on the basis of a bounded value function. Consider, for example W^* as the bound of the valuation function f, where W^* is taken to be a rather large sum. (At the beginning of the 18th century, a sum equivalent to 10 million dollars was mentioned[12]). Most people would refuse to bet one dollar on a chance of 1 in 10,000,000 of winning an amount which has even a subjective value of 10 million dollars.

As an illustration, consider for every natural number k, a kth approximation to the Petersburg Game defined as follows: a coin is flipped until heads appears, but with the initial stipulation that at most k tosses will be made. B receives 2^{n-1} dollars from A when heads shows for the first time at the nth throw, but if heads does not show in k tosses, then B receives 2^{k-1} dollars, whether heads or tails shows at the kth toss. B's mathematical expectation is thus

$$1/2 + (1/4)\cdot 2 + \ldots + (1/2^n)\cdot 2^{n-1} + \ldots + (1/2^{k-1})\cdot 2^{k-2} + \\ + (1/2^k)\cdot 2^{k-1} + (1/2^k)\cdot 2^{k-1} = (k+1)/2$$

dollars. The greater k, the greater B's mathematical expectation, while B's willingness to risk does not under normal circumstances increase in the same proportion. In fact, the actual amount B is willing to risk for the kth game stays, however large k may be, below a not very high bound, while $(k+1)/2$ increases with k beyond all bounds. The larger k, the larger, therefore, the discrepancy between mathematical expectation and actual behavior. For the Petersburg Game itself, which is something like a limit of the approximating games, the discrepancy becomes infinite.

It is clear that by setting small probabilities equal to zero, all but a finite number of the infinitely many terms that make up B's infinite expectation in the Petersburg Game are reduced to zero. The resulting expectation is finite, and the allegedly paradoxical element is removed. A complete description of the behavior of normal people with regard to games of chance, however, cannot be achieved in this way, as may be seen by comparing this attempt at a solution with the one presented in the next section.

8. THE FUNDAMENTALS OF BEHAVIOR IN GAMBLING GAMES

Our explanation of the Petersburg paradox will consist in a general description of normal behavior in games of chance, in particular, of the behavior of a person promised to gain the amount D with the probability p or, as we shall say[13] the *chance* (p, D).

It must be noted at the outset that how a person evaluates a chance (p, D) is not an unequivocal question. One may ask either: "What is the largest amount that a man will pay in order to buy (p, D)?" or "What sure addition to his wealth does he equate in value to the chance (p, D)?" He must ask himself the first question when he has to purchase the chance, but he must answer the second question when he can choose between two gifts: of a sum with certainty, or of a chance. Of these two amounts, the first is usually smaller than the second. In this discussion we shall concern ourselves only with the first question (despite the fact that it involves some complications which do not arise in the second) since in nearly all cases the first is the basic question one has to ask oneself in daily life before entering into any gambling situation. We now list a number of important features which can be regularly observed in the valuation (according to the first definition) of chances.

First of all, the purchase of a chance presupposes a payment that is valued according to the laws of subjective value theory which deal with the estimation of certainty. They imply that in the case of very high probabilities (of winning) a man is unwilling to risk anything.[14] This statement seems peculiar only at first glance; a closer look shows that it is confirmed by experience. For example, a man would not bet a substantial part W' of his wealth in order to obtain a probability 99/100 of winning $100W'/99$ (or about 1.01 W') even though his mathematical expectation would be W'. He is unlikely to bet W' even for a 99/100 probability of winning 1.1 W' or 1.2 W'. For although the probability of losing the risked amount is small, the loss would be considerable. On the other hand, while the chances of winning are high, the net gain would be small, since even if the winnings are 1.1 W' or 1.2 W', the gain after subtracting the bet would be only $(1/10)W'$ or $(1/5)W'$.

The theorem on the undervaluation of chances with high probability should not be interpreted to mean that a businessman prefers relatively uncertain to almost certain transactions. While we have been dealing with the features of purchases of chances in gambling where one either wins or loses, the circumstances surrounding a businessman's decisions are usually such that, even if he does not win, he will get back some part of his original investment. The larger the part of the original investment that he can be sure to recover, the more will he prefer deals with a high probability of gain. In the sense of the theory of games of chance, one can regard as risked only that part of the investment which cannot be expected to be regained with certainty. If however, the businessman faces transactions where in the event of not winning he will lose a large part of his investment (and where, therefore, a large risk in the sense of that theory exists), then in conformity to the rules about behavior in gambling the businessman will considerably undervalue a very likely gain as compared to its expected value (which, in the case of high probabilities, is almost equal to the value of the gain itself). We shall see presently how these statements must be modified because of other considerations related to the wealth of the evaluator.

Second, it is clear that under normal circumstances a man will risk only a limited part of his total wealth to buy chances in games. If w is that part of his wealth which a man M is ready to spend on such chances, then this number, which varies from person to person, is usually much closer to 0 than to 1. There exist, however, cases where $w = 1$, that is, where M is ready to risk his whole fortune on a single card. For example, M may need an amount six times his whole estate for an expensive medical treatment or to settle a debt of honor, while life under his present circumstances may not appear worth living. In this case, he may well risk all his money in a game of dice, since he feels that if he wins he will have the opportunity to go on living, while if he loses he will not be any worse off than he was before the game when he had already given up the idea of living. This example leads us to consider, apart from M's wealth, the minimum amount U which he desires if he is to go on living. If his wealth W is considerably smaller than U, then $w = 1$. Normally, however, W is larger than U. Indeed, $U = 0$ since continuation of life is not usually tied to ownership of a certain amount of money, while W is greater than 0 since M has at least some money. Let U' be the amount which M needs in order to continue his present standard of living, and Z be the amount which he desires. If W is between U' and Z, then M will only risk amounts short of

$W - U'$. Furthermore, the smaller M's chances to obtain the desired difference $Z - W$ through work and saving, the closer w approaches its maximum value, $(W - U')/W$. At the same time, M will prefer to assume risks in which the possible gain is close to the difference $Z - W$.

Finally, the behavior of individuals in buying a chance (p, D) depends on the probability p. The first basic rule here is that people, in general, tend to undervalue very small probabilities as has been pointed out in Section 7. We can associate with each individual M a smallest probability $p_M(R)$ such that M will risk an amount R. When the probability p exceeds the value of $p_M(R)$ only slightly, then M will undervalue p inasmuch as he will risk an amount R on a chance (p, D) only if $p \cdot D$ is larger than R, or in other words, if the potential gain D is larger than R/p. At the same time, the difference $D_M(p) - R/p$ between the gain $D_M(p)$ which M demands in order to risk an amount R with the probability p, and the expected value R/p becomes greater as $p - p_M(R)$ decreases. The ratio $p_M(R)/R$ increases with R, that is, the larger the risk, the higher the minimal probability. Hence, the smaller p, the larger the difference between actual behavior and behavior conforming to the expected value of the chance (p, D).

We thus see that chances are undervalued both where the probabilities are very small (that is, close to 0) and where the probabilities are very high (that is, close to 1). Only chances with medium probabilities are valued in a way which begins to correspond to mathematical expectation. Chances with medium probabilities, however, will very often be overvalued. The confirmation of this statement can be found in the fact that in lotteries, roulette and similar games, risk-takers will knowingly and with pleasure bet more than their mathematical expectation. The difference makes up the expenses and profits of the organizer of the lottery or the owner of the casino.

The probabilities at which the maximum overvaluation takes place depend on the potential gain, on the wealth of the risk-taker, on the relationship of his fortune to the quantities U' and Z discussed above, and on other personal circumstances.

As far as the Petersburg Game is concerned, the difference between behavior according to mathematical expectation and actual behavior has now been completely explained in the sense of being included in a general description of behavior in game situations.

9. THE DESCRIPTION OF PERSONAL TRAITS

It has been repeatedly emphasized in what precedes that the numbers and functions involved in the description of behavior vary from person to person. These numbers and functions are characteristic of certain circumstances of the evaluator, in particular, of some personal traits of his.

Consider a man M for whom the minimum probability above which he is willing to take risks is smaller than for most other people, while the ratio of his evaluation of chances to their mathematical expectation increases with the probability faster than for other people in similar circumstances. M may be called a *gambler*. In a similar way, rather general statements about some numbers and some functions, as considered in the preceding section, characterize forms of *imprudence*, *cautiousness*, *pettiness*, and the like. Thus these vague terms can be endowed with a certain degree of precision.

One can for example, observe whether a man is willing to risk a larger part of his fortune for a certain chance than are others in similar circumstances. Two different aspects must be clearly distinguished: a) his evaluation of the uncertainty; b) the risk he is willing to take on the basis of that evaluation. If these two aspects can be separated at all, the first can be described by statements of the form: I bet a to b that the event E will occur; the second, by statements: My betting quotient for obtaining a gain G being a/b, I am willing to risk the amount R. If A estimates the likelihood of a gain almost always higher than does B in similar situations, then A may be called more *optimistic* than B. If A is willing in most cases to risk more than B, who lives in similar circumstances and has the same betting quotients, then A will be called more *daring* or *venturesome* than B.

10. METHODOLOGICAL REMARKS

This paper has primarily dealt with the Petersburg Game. But what has been done here for this special problem – the way it has been formulated, what has been considered as its explanation, and how it has been treated – obviously may be useful in dealing with many as yet unsolved problems or confusedly treated questions of social science.

Of particular importance is the fact that all the results established in Section 8 are purely empirical. For it is in this point that our theory differs, with regard to its claims, from many other theories – even from some that take subjective elements into consideration or that have been developed by mathematicians. For those theories stipulate certain formulae, such as the logarithmic evaluation, as *norms* for behavior, or they *recommend* only behavior according to those formulae. An often overlooked consequence of those views is the fact that no games exist that can be recommended to both partners. For even if the conditions of a game are formulated in such a way that the rules make it advisable for one player to accept it, the rules would stipulate that the other player should not offer the game. In contrast, our simple description of observed behavior with regard to games of chance, without arrogating any normative validity, takes into account the undeniable fact that under certain circumstances both partners gladly enter a game: that people *like* to take part in keno, lotteries and similar games, while sponsors gladly offer the games since they almost certainly will make a profit. The fact that, within certain probability brackets and in a certain relation to their fortune, people evaluate risks above their mathematical expectation, makes it possible that games of chance, lotteries, and the exchange of bets for chances materialize and lead to that satisfaction of both parties which results from every voluntary exchange of commodities.

It should further be stressed that in formulating general regularities in behavior with regard to games of chance we have confined ourselves to rather qualitative statements and have refrained from defining specific functions possessing those qualitative properties. This has three reasons: 1) The observed qualitative properties are shared by many, and even by many simple, functions so that there may be no reason to prefer a particular one to the others. 2) An exact description of actual evaluations is not supplied by any of these functions. 3) As mentioned in Section 8, the numbers and functions involved in the description vary from person to person. A comprehensive function, therefore, would include numerous parameters. In fact, the number of parameters would be so large, and the dependence on these parameters so complicated, that the general function would lack any transparency. And what would be gained by setting up such formulae? Could more than general remarks corresponding to the qualitative observations (ascertained directly from the empirical material) be predicated about the functions?

The mere assertion that a functional connection of some kind exists does not transcend the general qualitative statements that can be made without the introduction of a function symbol. Consider, for example, the following statement. The quantity of a commodity G_2 that a man is willing to spend for the acquisition of an addition D to his wealth W in a commodity G_1, is a function of D and W, say $f(D, W)$; if W is kept fixed, $f(D, W)$ increases with increasing D while, if D is kept fixed, $f(D, W)$ decreases with increasing W. Hence, if the function is assumed to have partial derivatives, then

$$\partial f(D, W)/\partial D > 0 \quad \text{and} \quad \partial f(D, W)/\partial W < 0.$$

These formulae do not in any way transcend the simple verbal statement that a larger addition to a fixed wealth is preferred to a smaller one, while the same addition is more appreciated by the poorer than by the richer man. Those who believe that by writing $f(D, W)$ they introduce a mathematical formulation superior to verbal expressions, are under a misapprehension, just as are, on the other hand, those who when seeing the mere symbol $f(D, W)$ suspect the use of mathematics which, because of ignorance or on the ground of general prejudices, they reject in the social sciences.[14a] Simple special functions, however, as the example of Bernoulli's logarithmic evaluation demonstrates, are not in the theory of value borne out by experience as they are in physics. In Section 8 we therefore have refrained from setting up such formulae for $f(D, W)$ and have followed rather the method of the Austrian school of economists – hopefully without being suspected of underestimating mathematics.

It may be appropriate to consider more closely the parallelism between statements about the evaluation of chances, as developed in Section 8, and propositions concerning the evaluation of commodities about which there is no uncertainty. Both express empirical regularities that can be ascertained in one of two ways: by observing the behavior of individuals or by collecting statements of theirs as to how they would act under certain circumstances. In order to obtain data concerning the evaluation of sure commodities one asks[15] a man what quantity, y, of a commodity G_2 he would be willing to spend in exchange for a quantity, x, of a commodity G_1. Comparing the information received from various people, despite many differences in detail, one discovers similarities in the general structure of the data, such as monotonic increase and a decreasing rate of increase for most individuals. Interpolating the finite set of data obtained from one person by a curve that is as smooth as possible, one arrives at a twice differentiable function $y = f(x)$ for which $f'(x) > 0$ and $f''(x) < 0$ for each x (without adding by this interpolation procedure to the immediate observations[16]). If one wishes to assemble corresponding material concerning risks one has to ask a man repeatedly how much he would be willing to spend for the acquisition of a chance, of course varying, in so doing, p. D and W. One can interpolate a so-called function of two variables or, graphically, embed the data in a surface. Here again one observes similarities between the data obtained from various people as to their general course. These similarities can be expressed as general qualitative regularities, some of which have been mentioned in Section 8. Exceptions from the regularities concerning the evaluation of the uncertain as well as of the certain are reflected in wrinkles of the surfaces and curves. These peculiarities of an individual in evaluating risks as well as sure commodities lend themselves to the description of personal traits.

Quantitative precision, even in a restricted way, for describing the dependence of the evaluation of a chance upon its probability and the potential gain can be achieved only in the case of games of chance. In other domains of economic actions, uncertainty also plays a very important role indeed but can rarely be made numerically precise. This is most obvious in the case of general economic and political uncertainty, however important their influence on economic actions may be. If some piece of real estate lends itself only to a special use, say, the development of a luxury hotel or an armament factory, then its evaluation will largely depend upon the evaluator's views on the economic development of the country or the prospect of war – thus on his views about uncertain circumstances. But even if he can make precise his personal judgment of likelihoods (for example by saying: I bet 2 to 1 that within a year the price index in country A will rise by 10 per cent, or: I bet 1 to 2 that within two years the countries B and C will be at war) one cannot speak of a probability in a stricter sense. Yet, even in some domains in which numerical probabilities cannot be formulated, one can compare the behavior patterns of various people with each other and with the average pattern.

In conclusion, it may be appropriate to point out the connection of the preceding remarks on value theory with studies of the will, as contained in treatises on ethics, psychology, and general philosophy. Most philosophical discussions of the motivation of human actions (especially studies of freedom of will) neglect or at least underemphasize the problem of uncertainty. For example, the statement that actions always follow the course of the strongest motive certainly is false unless in speaking of the strongest motive one takes the probability of its realization into account. Suppose someone lets me choose between two tickets: one offering a winning of 1,000 dollars with a probability of 1/1,000 and one offering a winning of 2,000 dollars with a probability of 1/4,000. The stronger motivation is certainly provided by the larger amount. Yet, apart from exceptional cases discussed in Section 8, I shall choose the first ticket. Of importance to our actions, besides the strength of the motive, is the probability of achieving the aim and thus the element of uncertainty.

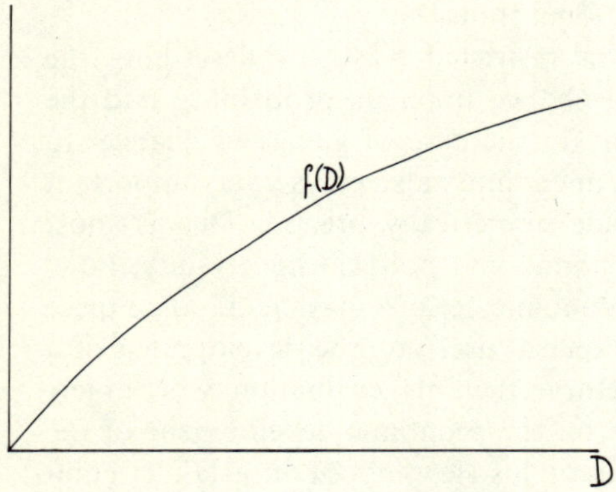

$f(D) = c \, . \log(1 + D/W)$
for $c = W = 1$
D. Bernoulli

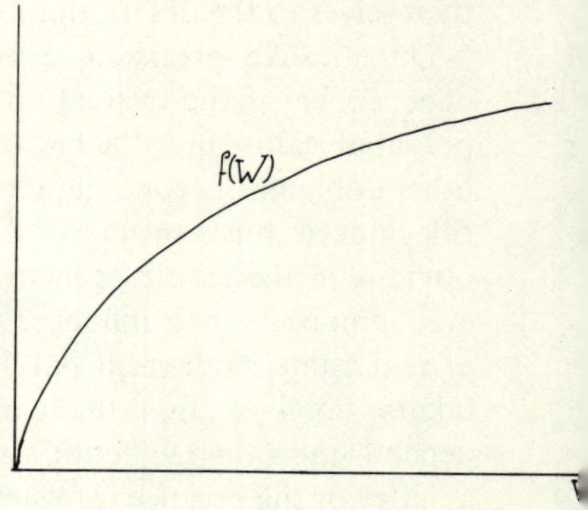

$f(W) = \sqrt{W}$
Cramer II

$$f(W) = \begin{cases} W \text{ for } W \leq W^* \\ W^* \text{ for } W > W^* \end{cases}$$

Cramer I

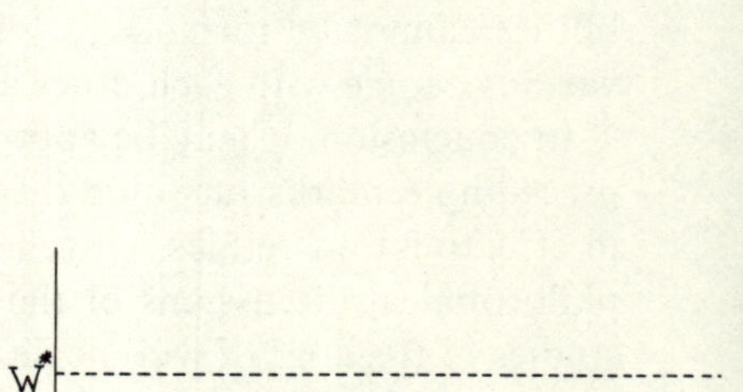

$$f(W) = \frac{W^*W}{W + W^*}$$

Timerding

NOTES

[1] This game was related by Nikolaus Bernoulli to the mathematician Montmort, who published it in his *Essai d'analyse sur les jeux de hazard*, Paris, 1713. It became better known through the treatise of Daniel Bernoulli, "*Specimen theoriae novae de mensura sortis*" in the *Commentarii Academia Petropol.*, V, 1738, p. 175. The game received its name because of the connection of the Bernoulli family with the City of St. Petersburg.

[2] When the probability is p that S units are won the mathematical expectation is defined as $p \cdot S$ units. In a game where there exists a probability p_1 of winning S_1, and also an independent probability p_2 of winning S_2, the mathematical expectation of the risk-taker is defined as $p_1 \cdot S_1 + p_2 \cdot S_2$. Consider for example the following game. A die is thrown once; the gambler receives 6 dollars if 1 shows, 1 dollar if either 2 or 3 or 4 shows, and nothing if 5 or 6 shows. Then his mathematical expectation is $(1/6) \cdot 6 + (1/2) \cdot 1 = 3/2$ dollars. If the dice are "honest" (that is, if thrown say 1200 times each of the point numbers, 1, 2, 3, 4, 5, and 6 appears in about 1/6 of the throws, or about 200 times), then winnings and losses will be about equal provided that the game is played a large number of times and that in each game the gambler bets his mathematical expectation of 3/2 dollars.

[3] In this category belong the remarks of d'Alembert, *Doutes et questions sur le calcul des probabilités*, *Mélange de Littérature, d'Histoire, et de Philosophie*, V, 1773; and of Béguelin, *Histoire de l'Acad. de Sciences et Belles-Lettres*, Berlin, 1767. These were already contradicted by Lichtenberg, *Physik. und Mathem. Schriften*, Vol. 4, 1806. For the whole discussion, compare Czuber, *Archiv. d. Math. u. Phys.*, **67**, 1882, p. 3 sq.

[4] The concepts described in this section were originated by Poisson, *Recherches sur la probabilité des Jugements en Matière Criminelle et en Matière Civile*, 1837, and are supported partially by Czuber, *op. cit.*, p. 19 ff. (despite the critical remarks on p. 13), and especially by Pringsheim in his notes to the German edition of the essay by D. Bernoulli (mentioned in note [1]) *Versuch einer neuen Theorie der Wertbestimmung von Gluecksfaellen*, Leipzig, 1896.

[5] This solution was given by Fries, *Versuch einer Kritik der Prinzipien der Wahrscheinlichkeitsrechnung*, Braunschweig, 1842, p. 116, and also by Czuber, *loc. cit.*, p. 15 ff.

[6] It seems to have appeared for the first time in the writings of the mathematician Cramer (whose contributions to the theory of linear equations are widely known). In 1728 he wrote a letter addressed to N. Bernoulli and published by D. Bernoulli about the question of the origin of the difference between mathematical calculation and normal estimation: "I think it stems from the fact that (in the theory) mathematicians evaluate money simply by its quantity, while reasonable people (in the real world) evaluate it according to its utility." D. Bernoulli formulated this idea independently (*op. cit.*) and also expressed it publicly for the first time. *Valor non est aestimandus ex pretio rei, sed ex emolumento quod unusquisque inde capessit. Pretium ex re ipsa aestimatur* (these words show incidentally that the basic idea of modern economic theory, which explains prices through the interplay of valuations by various individuals in barter, did not enter Bernoulli's mind), *omnibusque idem est, emolumentum ex conditione personae. Ita procul dubio pauperis magis refert lucrum facere mille ducatorum, quam divitis, etsi pretium utrique idem sit*. Buffon also supported this idea in his *Essai d'Arithmétique Morale* (supplement to Volume IV of the *Histoire Naturelle*, 1777, p. 72 ff.). *L'argent ne doit*

pas être estimé par sa quantité numérique. . . . Il s'en faut bien que les advantages qu'on tire de l'argent soient en juste proportion avec sa quantité: un homme riche a 100.000 *écus de rente, n'est pas dix fois plus heureux que l'homme qui n'a que* 10.000 *ecus*, and (*loc. cit.*, p. 89): *L'avare est comme le Mathématicien: tous deux éstimant l'argent par sa quantité numérique.* This idea was then emphasized by Laplace, *Théorie Analytique des Probabilités*, Paris 1812 (Oeuvres VII, Paris 1886, p. XIX sq.). *On doit distinguer dans le bien espéré sa valeur relative de sa valeur absolue: celle-ci est indépendente des motifs qui le font désirer, au lieu que la première croît avec les motifs*, and again by F. A. Lange (for example in *Die Arbeiterfrage*, 1865). In theoretical economics, the discovery of this idea in the 1870's introduced a new era. Good source material about the predecessors of the marginal utility theory can be found in Weinberger, *Die Grenznutzentheorie*, 1926, where, however, the theory itself is not completely presented and, in my opinion, unfairly criticized.

7 D. Bernoulli, *op. cit.*

8 At first probably by Cramer, who considered the valuation formula $C\sqrt{W}$, according to which a gain of $4W$ is evaluated only twice as high as W, and a gain of $9W$, three times as high (see the figure on p. 274). The famous definition which leads to the logarithmic valuation function was introduced under the title, *Mensura sortis*, by Bernoulli, *op. cit. Ita vero valde probabile est, lucrum quod vis semper emolumentum aferre summae bonorum reciprocae proportionale.* It was used by Laplace. *On ne peut donner de règle générale pour apprécier cette valeur relative: cependant il est naturel de supposer la valeur absolue en raison inverse au bien total de la personne interessée.* Following the latter, many mathematicians designate $c \cdot \log(1 + D/W)$ as the *moral value* of the increase D to wealth W, and the product of the moral value of a gain and its probability as its *moral expectation.* Oettinger (*Crelles Journ. f. d. reine u. angew. Math.* **36**, 1848, p. 26 and p. 300 ff.), proposes the terms *subjective value* and *subjective expectation*, which are used in the present article. Later the Weber-Fechner Law of Psychology proceeded from the general assumption that the sensation of an increase d to a stimulus s is directly proportional to the increase d and inversely proportional to the stimulus s and is, therefore, proportional to $\log(1 + d/s)$. Fechner, *Elemente der Psychophysik*, 1860, **I**, p. 236, **II**. p. 10, states that this formulation encompasses the Bernoulli valuation function, if one treats the size of wealth as the stimulus, and its value as the sensation. *Sensations* can never (and the expressions of sensations only rarely) be added and consequently, measured and compared on a numerical basis; at best they can be arranged according to their intensity. In other words, sensations cannot be mapped congruently, but at best topologically, on real numbers. The *expressions* of valuation sensations, however, (that is, the behavior in exchanging commodities) can be determined quantitatively. For example, one can determine, for various numbers x, the amount of money that a man will spend in order to acquire x units of a commodity. Marginal utility theory, if carefully formulated, can, therefore, well stand up to logical critique.

9 In order to make sure that this assumption be always satisfied, D. Bernoulli included in his definition of wealth the ability to earn something through work or by begging. Hence $W = 0$ only for a person deprived of even these opportunities.

10 Before considering the square root valuation, Cramer suggested, *loc. cit.*, that the value of money beyond a certain very high amount (he arbitrarily assumed 10 million dollars), no longer increases (see the figure on p. 274). He based his calculations on the following assumption. The subjective value $f(W)$, of an amount of money W, is equal to

W if W is smaller than 10 million dollars, and is equal to 10 million dollars whenever W exceeds that amount. (Clearly, neither this sudden change in the function at 10 million dollars nor the linear increase up to the 10 million dollar limit finds support from actual experience.) Buffon also states, *op. cit.* p. 72: . . . *c'est que l'argent, dès qu'on passe de certaines bornes, n'a presque plus de valeur réelle: et ne peut augmenter le bien de celui qui le possède: un homme qui decouvriroit une montagne d'or, ne seroit pas plus riche que celui qui n'en trouveroit qu'une toise cube.* An example of a steadily increasing bounded function which approaches asymptotically a certain amount, thereby combining the advantages of the logarithmic and the bounded functions, was given by Timerding in his essay on Bernoulli's theory of value (*Zeitschrift f. Math. u. Phys.* **47**, 1902, p. 337). It is the function which assumes for W the value $f(W) = A + b/(W + c)$, where a, b, c are numbers. If this function is 1) to assume the value 0 for $W = 0$, 2) to approach with increasing W asymptotically the value W^*, and 3) to have the derivative 1 for $W = 0$, then one arrives at the hyperbolic function

$$f(W) = \frac{W^* W}{W^* + W}.$$

The graphs on p. 274 represent the logarithmic function of Bernoulli, Cramer's parabolic square root function, the function first proposed by Cramer, and the hyperbolic function of Timerding.

[11] Buffon was probably the first (*op. cit.* p. 48 ff.) to use this fact in the analysis of the Petersburg Game, separately from the other fact that the subjective valuation of sums of money does not depend only on their size. None of the later commentators seems to have sufficiently emphasized this circumstance. Buffon distinguishes between mathematical certainty and moral certainty, mathematical impossibility and moral impossibility. Moral certainty is already determined in the case of a very large probability. *On sent bien que c'est un certain nombre de probabilités qui fait la certitude morale, mais quel est ce nombre?* He discusses in this connection the most important of fears, that of death. The probability that a fifty year old man may die within the next twenty-four hours is 1 in 10189. This probability is, according to Buffon, for practical purposes equal to 0. A probability below this limit fails to evoke any emotion and is not worth a thought. Without discussing Buffon's specific limit – which, incidentally, was criticized by Bernoulli – we shall consider the general assumption that probabilities below a certain level are set equal to zero.

[12] Compare footnote[10].

[13] [*Addition in this translation*] The (2-parametric) symbol (p, D) herein used for a chance is preferable to the symbol pD (used in the original paper) which may be confused with the symbol $p \cdot D$ designating the mathematical expectation.

[14] The essence of this remark is already contained in the work of D. Bernoulli.

[14a] [*Addition in this translation*] For a detailed comparison of the marginalism of the Austrian school and mathematical economics, Cf. [1973, 1].

[15] Compare D. Bernoulli, *op. cit.*, and Oettinger, *op. cit.*

[16] Naturally one only asks for the numbers y corresponding to a few numbers x thereby obtaining a finite number of points in the plane, which then can be connected by a curve that is as smooth as possible, or simply by a polygon. Even if the interviewed person reports a functional relation $y = f(x)$ or if he constructs a graphical relationship which

associates with every x a corresponding y, this is the result of a finite number of prior decisions by him to which he has applied interpolation. When asking him for the y corresponding to a finite number of x one should choose the x efficiently. That is, one should not necessarily distribute them evenly but, especially in the case of non-monotonic functions, should ask more questions in those ranges where the y show greater change.

[16] The empirical character of all these laws is evident from the fact that a world is imaginable in which most needs would increase with increasing satisfaction. In some cases, this relation may be observed even in our actual world, for example, with regard to narcotics.

[17] In another field, however, will-theoretical considerations of philosophers may be useful for the analysis of problems in value theory. Schlick correctly points out in his *Questions Concerning Ethics*, Vienna, 1930, that the statement "the actual human behavior occurs always in the direction of the strongest motivation" is not a tautology but empirical and that this point is illustrated by the fact that one may imagine a world where behavior is determined by those motives which are least satisfying. Several presentations of the theory of value treat the statement "we prefer the more valuable to the less valuable" as a tautology since the fact that something is more valuable to us can only be explicated by our preferring it. This, in some respects, extreme behavioristic conception (which also underlies, in a more or less vague form, the rejection of value theory as "metaphysics of price theory" as well as the criticism of a supposedly redundant doubling of concepts), I regard as incorrect. Through self-observation we arrive at the theorems about subjective value and establish a rather regular connection between these introspective experiences and the actual behavior that leads to the formation of prices.

CHAPTER 23

REMARKS ON THE LAW OF DIMINISHING RETURNS A STUDY IN META-ECONOMICS

This paper goes back to a conversation about the methodology of economics in which L. Mises claimed that certain propositions of economics can be *proved*. As an example, he mentioned the law of diminishing returns and referred me to the literature for the proofs. After I presented my findings to the Viennese Economic Society they were published in [1936, 6] and [1936, 6a]. In 1936, Mises wrote me that he learned a great deal from the paper. An English translation appeared in [1954, 3]. In what follows, that translation has been thoroughly revised and shortened and the mathematical parts have been relegated to paragraphs in small print.

1. INTRODUCTON

There have been several attempts to find the purely logical relations between laws about returns and other propositions of economics, especially between the law of diminishing returns on land and the theory of value.[1] Some outstanding theorists have claimed to prove it from generally admitted propositions about value by deduction.[2]

Much of this reasoning, however, fails to meet the requirements which logic places on a sequence of inferences intended to constitute a proof. The proposition to be proved is often stated without sufficient precision. Propositions are assumed, explicitly and implicitly, to have the same meaning though in fact they are by no means equivalent (see Section 2). Frequently there is a lack of precision in regard to the logical quantifiers; that is, as to whether a proposition is meant to apply to *all* or only *certain* levels of outlay. Yet it is only after such details have been settled that an assertion becomes precise enough to be subject to deductive treatment (see Sections 3 and 4). The proofs are not preceded by a full statement of the assumptions, though without a complete list it is impossible to determine whether a proof is correct, or even whether a proposition is capable of proof. Nor are the conclusions always precisely stated (see Section 4). Finally, some proofs contain crude mistakes, due to the incorrect formulation of the negations of somewhat more complicated statements (see Sections 5 and 6).

These criticisms may seem pedantic. But since similar remarks apply to many instances of reasoning in the social sciences, it may be useful to discuss in what follows one example[3] in detail and without vague arguments or a lighthearted use of such words as 'therefore', 'it follows' or 'since'.

The laws of return suit this purpose very well. On the other hand, empirical questions will not herein be considered. Thus, to prevent misunderstanding, the following should be made clear from the outset:

1. By saying that a proposition does not follow from certain assertions, we do not mean that it is not valid. For example, the proposition of diminishing product increments (the heart of the law of diminishing returns on land) will be shown not to be derivable from the laws of value theory from which it is sometimes inferred; yet we believe this proposition with proper qualifications to be valid, that is, capable of empirical confirmation.

2. Conversely, by saying that a proposition follows from certain others, we do not mean that it is valid. Even if a scientific proposition can be correctly derived from generally accepted propositions only experience can show whether or not the derived proposition is valid. If experience should demonstrate that it is not (as well may be the case despite all deductions), then the necessary conclusion is that some of the generally accepted propositions cannot be upheld without modification. Such situations occur frequently in physics: Conclusions are derived from general propositions; yet no physicist would on this ground neglect to test these conclusions empirically. If they are not empirically confirmed, then the general propositions must be altered. Much of the progress of physics is due to this procedure.

3. Finally, we emphasize that we attach little importance to the rather subtle logical relationships among the various laws about returns. We believe the crucial issue for economics to be whether or not these laws are empirically confirmable while we regard it as a secondary issue whether or not they follow from certain other propositions. It is, rather, some outstanding economists who have raised this issue by claiming to prove the law of diminishing returns on land logically, and thereby to make empirical tests superfluous.[4] All that we here say is that logical relationships, equivalences, deductions, proofs, etc. must be handled correctly.

Following a suggestion of Hilbert, modern logicians refer to the study of the logical relations between the statements of a theory as the corresponding *meta-theory*. In this terminology, the contents of the present paper can be described as a chapter in *meta-economics*.

2. THE LAW OF DIMINISHING RETURN ON LAND

The law of diminishing return on land, as formulated by the outstanding theorist Böhm-Bawerk,[5] reads as follows:

Additional applications of capital and labor on a piece of land increase the total product, but after a certain point this output increases relatively less than further costs. In other words: like increases of cost produce a decreasing increase of the product.

The meaning of Böhm-Bawerk's *second* formulation is clear. The assertion will be referred to as the

LAW OF DIMINISHING PRODUCT INCREMENTS,[6]
more precisely, of
DIMINISHING PRODUCT INCREMENTS FOR LARGE OUTLAYS. *An increase of cost outlay yields a smaller increase of the product when added to a larger outlay than when added to a smaller outlay, provided that both exceed a certain outlay level (which may depend on the amount of land used.)*

In all that follows, the product yielded by an outlay of y units of cost applied to x units of land (disregarding qualitative differences of land) is denoted by $E(x, y)$:

The increase of product due to adding h units of cost to the outlay of y units on the x units of land is then $E(x, y + h) - E(x, y)$. What is asserted is, in symbols, that for every amount of land x, there is a level y^* of outlay such that

$$E(x,y_2 + h) - E(x,y_2) < E(x,y_1 + h) - E(x,y_1),$$

provided $y^* < y_1 < y_2$.

Böhm-Bawerk's *first* formulation can, by means of the production function, E, be restated as follows: if the cost outlay rises from y_1 to y_2, then the ratio of $E(x, y_2)$ to $E(x, y_1)$ is less than the ratio of y_2 to y_1, provided both y_1 and y_2 exceed some level y^{**} (which may depend on x).

In symbols:

$$\frac{E(x, y_2)}{E(x, y_1)} < \frac{y_2}{y_1}, \text{ or equivalently, } \frac{E(x, y_2)}{y_2} < \frac{E(x, y_1)}{y_1},$$

provided $y^{**} < y_1 < y_2$ and $E(x, y_1) \neq 0$.

The number $E(x, y)/y$ is called the *average product* for a cost outlay y on x units of land. It is equal to the product per unit of cost for x units of land and y units of cost. Böhm-Bawerk's first formulation amounts to the

LAW OF DIMINISHING AVERAGE PRODUCT FOR LARGE OUTLAYS. *As the cost outlay rises, the average product of every piece of land falls after a certain point has been reached.*

Are the laws of diminishing product increment and of diminishing average product really formulations of the same law "in other words", as is asserted in the passage previously quoted? Not at all. There are possible product functions of which it can be shown by elementary algebra that one satisfies the law of diminishing product increments, but not that of diminishing average product, whereas the other satisfies the law of diminishing average product but not that of diminishing product increments. Hence far from being equivalent, neither of the two laws implies the other.[7]

Examples of such functions include

$$E'(x, y) = y\,\frac{2x^2 + y^2}{3x^2 + y^2} \quad \text{and} \quad E''(x, y) = y\,\frac{4x^2 + y^2}{3x^2 + y^2}\,.$$

Clearly, for any numbers y_1 and y_2,

$$\frac{E'(x, y_1)}{y_1} = \frac{2x^2 + y_1^2}{3x^2 + y_1^2} \quad \text{and} \quad \frac{E'(x, y_2)}{y_2} = \frac{2x^2 + y_2^2}{3x^2 + y_2^2}\,.$$

If $y_1 < y_2$, then, as one readily verifies, the second quotient is greater than the first, whence E′ does not satisfy the law of diminishing average product. One finds, analogously, that E″ does satisfy that law.

A similar, slightly more laborious algebraic calculation shows that E' satisfies the law of diminishing product increments while E'' does not. These assertions are also immediate consequences of the following expressions of the second partial derivatives of E' and E'':

$$\frac{\partial^2 E'}{\partial y^2} = 2x^2 y\,\frac{9x^2 - y^2}{(3x^2 + y^2)^3} = -\frac{\partial^2 E''}{\partial y^2}\,.$$

Indeed, for each x and each y such that $9x^2 - y^2 < 0$, that is, $y > 3x$ it follows that

$$\frac{\partial^2 E'}{\partial y^2} < 0 < \frac{\partial^2 E''}{\partial y^2}\,.$$

The laws of diminishing product increments and of diminishing average product can also be given simple geometric interpretations. In x, y, z-space, the product function E can be represented by a surface, $z = E(x,y)$, above the positive quadrant of the x, y-plane. For each point (x, y) in the plane, the point above it at the altitude $E(x,y)$ belongs to this surface. If x_0 is any specific number, then $z = E(x_0,y)$ is a curve on this surface and lying in a plane parallel to the y, z-plane. This curve represents the returns from x_0 units of land with various cost outlays. The *law of diminishing product increments for large outlays* amounts in geometric terms to the statement that for any x_0 beyond some cost outlay y^*, this curve is convex; that is, for any two of its points the curve lies above the chord connecting the two points. The *law of diminishing average product for large outlays* is equivalent to the statement that, beyond some point y^{**}, the curve $z = E(x_0,y)$ lies below the line connecting the point $(x_0, y, E(x_0,y))$ with the point $(x_0,0,0)$. For the slope of this line is equal to the average cost, $E(x_0,y)/y$, and the said property of the curve amounts to the statement that beyond some point the slopes of these connecting lines decrease. Figure 1 shows the curves obtained from the surfaces $z = E'(x,y)$ and $z = E''(x,y)$ for $x_0 = 1$ (representing the products for various outlays on 1 unit of land), that is,

$$z = y\frac{2 + y^2}{3 + y^2} \quad \text{and} \quad z = y\frac{4 + y^2}{3 + y^2}.$$

The second curve is convex from $y^* = 3$ on, although the slopes of the lines connecting points on this curve with the origin increase as the curve approaches the straight line $y = x$ asymptotically from below. The first curve ceases to be convex at $y^* = 3$ although the slopes of the connecting lines decrease as the curve approaches the line $y = x$ asymptotically from above.

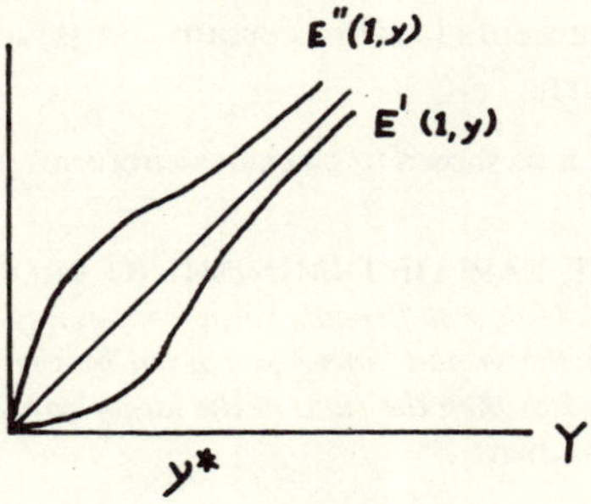

Fig. 1

The methods used in the foregoing examples, that is, the exhibition of simple functions with their algebraic and geometric properties, can often be used to prove the incorrectness of assertions about logical relationships between quantitative conditions, such as "conditions B_1 and B_2 are equivalent" or "condition B_2 follows from B_1". The proof can be given by exhibiting a function satisfying condition B_1 but not B_2, just as the functions E' and E'' have served to show that neither one of the laws of diminishing product increments and of diminishing average product follows from the other.

The question arises, which of these two propositions do economists mean when they speak of the law of diminishing return. Since the two propositions are frequently treated as equivalent, this is hard to determine. Böhm-Bawerk gives (*loc. cit.*[4]) the following example as an illustration.

If an outlay of 100 Fr. yields 100 units of products on a piece of land, then a doubling of this outlay to 200 Fr. increases the product, but not to 200 units. The increase is to a lesser figure, perhaps to 180 units. Tripling costs to 300 Fr. does not triple the product but increases it by still less, perhaps by another 70 units to a total of 250, and so on. In other words, if the first cost outlay of 100 Fr. yields 100 units of product, then a second outlay of the same amount does not add another 100 units to output but only 80 units; the third outlay may add 70 units, and so on in a diminishing sequence.

This example happens to illustrate both of the assertions in question. Nevertheless it is clear that Böhm-Bawerk here had in mind the law of diminishing product increments, and not the law of diminishing average product[8]. Otherwise he would not have stressed that the second outlay of 100 Fr. causes a smaller increment of product than the first, the third less than the second, "and so in a diminishing sequence". To illustrate the proposition of diminishing average product, it would have been sufficient if like additions of outlay yielded the same increase of product. Even increasing product increments (within certain limits) are compatible with a decreasing average return.

The following statements can be shown to be equivalent forms of the law of diminishing product increments.

SECOND FORM OF THE LAW OF DIMINISHING PRODUCT INCREMENTS FOR LARGE OUTLAYS. *If to a sufficiently large cost outlay two different increments are added, then the ratio of the product increase caused by the larger cost increment to that caused by the smaller is less than the ratio of the larger cost increment to the smaller.* In symbols: If $0 < h_1 < h_2$, then

$$\frac{E(x, y + h_2) - E(x, y)}{E(x, y + h_1) - E(x, y)} < \frac{h_2}{h_1}$$

for every x and every y exceeding some level y^*, which may depend on x (or be a function of x).

Each of the following three forms of the law has the preamble, *For each* x *there exists a number* y^* (*which may depend on* x) *with the following property* . . . (In order to indicate that the numbers y^* in the three cases need not be equal they will be denoted by y_3^*, y_4^*, y_5^*.)

THIRD FORM. *If* $y_3^* < y' < y'' < y'''$, *then*

$$\frac{E(x,y''') - E(x,y'')}{y''' - y''} < \frac{E(x,y''') - E(x,y')}{y''' - y'} < \frac{E(x,y'') - E(x,y')}{y'' - y'} .$$

FOURTH FORM. *If* x, y, h, *and* k *are any four positive numbers and* $y - h > y_4^*$, *then*

$$h\,E(x, y + k) + k\,E(x, y - h) < (h + k)\,E(x, y).$$

In the special case where $h = k$, the result is a particularly simple proposition which is equivalent to the entire assertion.

FIFTH FORM. *If* $y - h > y^*$, *then*

$$E(x, y + h) + E(x, y - h) < 2\,E(x, y).$$

The equivalence of all these assertions to the law of diminishing product increments can be easily proved (even without the use of differential calculus).

Distinctions similar to that between the law of diminishing product increment and the law of diminishing average product are necessary whenever economics deals with functions, as in the theories of price and value.

In a first reading, readers who are not mathematically oriented should proceed from here to Section 4 or even 5 and return to Section 3 or 4 only where references are made to its content.

3. PROPERTIES OF THE PRODUCTION FUNCTION

This section contains, for later reference, a list of various properties of functions – some belonging to all production functions, others only to certain ones. Some of these properties imply others; some of them are incompatible. All properties are stated with regard to products resulting from the combination of n goods $G_1, G_2, \ldots, G_n$; and $E(x_1, x_2, \ldots, x_n)$ is the number of units of the product obtained when x_1 units of G_1, x_2 of $G_2, \ldots, x_n$ of G_n are combined. The argument $(x_1, x_2, \ldots, x_n)$ is called the *outlay*. E is defined for n-tuples of nonnegative numbers.

The simplest general property of the production function is its

NONNEGATIVITY. $E(x_1, x_2, \ldots, x_n) \geq 0$ *for any possible outlay* $(x_1, x_2, \ldots, x_n)$.

Of particular importance is the

SUPERADDITIVITY OF THE PRODUCTION FUNCTION. *The sum of the products from two outlays is never larger than the product from the combined outlays.*

This property can also be stated as follows: The product from an outlay is never increased by dividing the outlay into two separate parts and then adding the products from the two parts. Otherwise this division would be carried out, and the combined product resulting from the two separate outlays would be considered to be the product of the initial outlay.

In symbols, Superadditivity reads

$$E(x'_1, x'_2, \ldots, x'_n) + E(x''_1, x''_2, \ldots, x''_n) \leq$$
$$\leq E(x'_1 + x''_1, x'_2 + x''_2, \ldots, x'_n + x''_n)$$

for any two possible outlays $(x'_1, x'_2, \ldots, x'_n)$ and $(x''_1, x''_2, \ldots, x''_n)$.

A simple consequence is obtained by setting

$$x_1' = x_1'' = x_1, x_2' = x_2'' = x_2, \ldots, x_n' = x_n'' = x_n.$$

It then follows that

$$2\,E(x_1, x_2, \ldots, x_n) \leq E(2x_1, 2x_2, \ldots, 2x_n).$$

It can be further be concluded that

$$3\,E(x_1, x_2, \ldots x_n) \leq E(3x_1, 3x_2, \ldots 3x_n')$$

and so on; that is,

$$k\,E(x_1, x_2, \ldots, x_n) \leq E(kx_1, kx_2, \ldots kx_n)$$

for each positive integer k.
We call this the

INTEGRAL SUPERHOMOGENEITY OF THE PRODUCTION FUNCTION. *For each positive integer* k, *the* k*-fold outlay yields at least the* k*-fold product.*

If here the integer k may be replaced by any real number $\lambda > 1$, then we say that E is *superhomogeneous.*

SUPERHOMOGENEITY. *If the amounts of all the input factors of production are increased in the same proportion, $\lambda > 1$, then the amount of output will be increased at least in the proportion λ*; equivalently, *if the amounts of all the input factors are diminished in the same ratio $\kappa < 1$, then the amount of product will be diminished at most in the same ratio, κ.*

In symbols: For every $x_1, x_2, \ldots, x_n$,

$$\lambda E(x_1, x_2, \ldots, x_n) \leq E(\lambda x_1, \lambda x_2, \ldots, \lambda x_n) \text{ for every number } \lambda > 1,$$

and

$$\kappa E(x_1, x_2, \ldots x_n) \geq E(\kappa x_1, \kappa x_2, \ldots \kappa x_n) \text{ for every } \kappa \text{ between 0 and 1.}$$

That the first formulation follows from the second can be readily seen by setting

$$\kappa x_1' = x_1, \ldots, \kappa x_n = x_n', 1/\kappa = \lambda.$$

Similarly, the second formulation follows from the first.

Figure 2 shows the graph of the function

$$f(x) = x^4 - 3.51x^3 + 4x^2.$$

This function can easily be proven to be superadditive (and hence integral superhomogeneous); but it is not superhomogeneous since, for instance,

$$\frac{3}{2}f(1) = 2.235 \quad \text{and} \quad f\left(\frac{3}{2}\right) = 2.221625, \text{ whence } \frac{3}{2}f(1) > f\left(\frac{3}{2}\right).$$

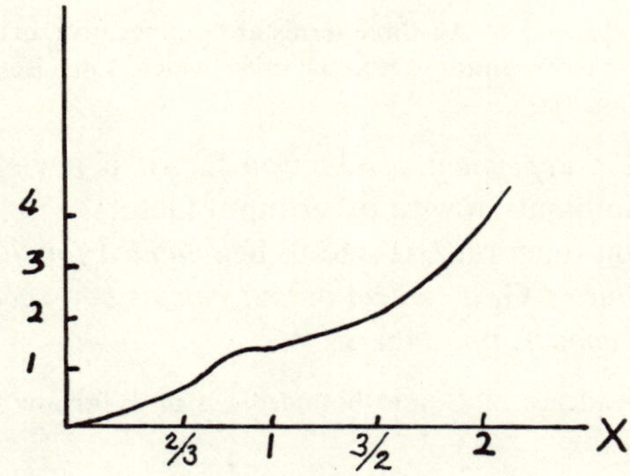

Fig. 2

Clearly, if f is the function just defined, then also

$$E(x_1, x_2, \ldots, x_n) = f(\sqrt{x_1^2 + x_2^2 + \ldots + x_n^2})$$

is integral superhomogeneous without being superhomogeneous.

While a superhomogeneous function clearly is integral superhomogeneous, the converse is not generally valid. The counterpart of superhomogeneity is

SUBHOMOGENEITY. *Increasing all input factors by the same multiple increases the product at most by that multiple.*

A function that is both super- and subhomogeneous is called *homogeneous. If the production function is homogeneous and if all the input factors of production are increased in the same ratio, the product increases in this same ratio.*

Dr. Karl Schlesinger[9] proved that, in conjunction with subhomogeneity, integral superhomogeneity implies superhomogeneity and hence homogeneity.

STRONG (AND WEAK) INCREASE. *The function E is said to be strongly (weakly) increasing with respect to G_1 if every increased outlay of G_1, ceteris paribus, yields an increased product (never leads to a decrease of product).*

(Some mathematicians use the term 'nondecreasing' instead of 'weakly increasing', and 'increasing' instead of 'strongly increasing'.)

If E is nonnegative and superadditive, then E is nondecreasing with respect to each of $G_1, G_2, \ldots, G_n$.

For example, if $x_1' < x_1''$, then because of superadditivity

$$E(x_1', x_2^0, \ldots, x_n^0) + E(x_1'' - x_1', 0, \ldots, 0) \leq E(x_1'', x_2^0, \ldots, x_n^0),$$

for any n numbers $x_2^0, \ldots, x_n^0$. All three terms are nonnegative, in particular, the middle term. Hence the first term cannot exceed the third, which is precisely what weak increase with respect to G_i asserts.

G_i is said to be a *dependent* production factor if it yields products only when used in combination with other input factors.

The production function E is said to be *bounded for the quantity of* G_i *if for any fixed input of* G_i *a certain output cannot be exceeded regardless of the quantities of cooperating inputs.*

In symbols, dependence of G_i and boundedness of E for any quantity of, say, G_i mean, respectively,

$$E(0, \ldots, 0, x_i, 0, \ldots, 0) = 0 \text{ for every number } x_i,$$

and the existence, for each number x_1^0 of some number $E^*(x_1^0)$ such that

$$E(x_1^0, x_2, \ldots, x_n) \leq E^*(x_1^0) \text{ for every possible outlay } x_2, \ldots, x_n.$$

We now formulate laws concerning product increments with regard to the first of n input factors. (Analogous statements may of course be formulated with regard to each of the n − 1 other factors.)

LAW OF DIMINISHING (NON-INCREASING) PRODUCT INCREMENTS FOR LARGE OUTLAYS. *The same increments* $(h_2, \ldots, h_n)$ *of outlays yield a smaller (never a larger) increment of product when added to a larger outlay than when added to a smaller outlay provided that the two outlays in question exceed a certain level.*

What meaning economists associate with occasional references to *increasing* return is in many cases unclear. The underlying idea seems to be capable of the following precise formulation:

LAW OF INCREASING (NON-DECREASING) PRODUCT INCREMENTS FOR SMALL OUTLAYS. *A not too large outlay on cooperating inputs yields a larger (never a smaller) increase of product when added to a larger outlay than when added to a smaller outlay provided the outlays do not exceed a certain level.*

In symbols, the former law (or, more precisely, pair of laws) may be expressed as follows: for every number x_1^0 there are numbers $x_2^*, \ldots, x_n^*$ (representing output levels which may depend on x_1^0) such that

$$\text{if } x_2^* < x_2' \leqq x_2'', \ldots, x_n^* < x_n' \leqq x_n'', \text{ then}$$

$$E(x_1^0, x_2'' + h_2, \ldots, x_n'' + h_n) - E(x_1^0, x_2'', \ldots, x_n'') <$$

$$(\leq)\ E(x_1^0, x_2' + h_2, \ldots, x_n' + h_n) - E(x_1^0, x_2', \ldots, x_n')$$

for any system of non-negative numbers $h_2, h_3, \ldots, h_n$, not all of which are equal to 0. The latter pair of laws assert that for every number x_1^0 there exist numbers (output levels) $x_2^{**}, \ldots, x_n^{**}$ such that if

$$x_2' \leqq x_2'' \leqq x_2^{**}, \ldots, x_n' \leqq x_n'' \leqq x_n^{**},$$

then inequalities concerning E opposite to the preceding, that is, with $>$ $(\geqq)$ instead of $<$ $(\leqq)$, hold for any system of non-negative numbers $h_2, \ldots, h_n$, not all of which are equal to 0. Of course, for the strong inequalities ($<$ and $>$) one must also exclude $x_2' = x_2'', \ldots, x_n' = x_n''$.

A few comments may be added. They are so trivial that the only excuse for bringing up these simple points is the fact that many authors have ignored them – a fact that has resulted in some of the controversies and misunderstandings with which the literature on return laws abounds. First, the proposition of increasing product increments for small outlays is not the negation of the proposition of diminishing (or non-increasing) product increments for large outlays. Both laws deal with product increments yielded by additional outlay – but to sufficiently *small* outlays in one case and to sufficiently *large* outlays in the other.

Second, neither of the two propositions is a consequence or a "logical complement" of the other, nor does either one "contain" the other. For any assertion concerning the effects of additions to sufficiently *large* outlays is compatible with any assertion about effects of additions to sufficiently *small* outlays. In particular, the proposition of diminishing product increments for large outlays is compatible with increasing product increments for additions to small outlays.

Third, the negation of the law of increasing product increments for small outlays is not the law of non-increasing product increments for small outlays. The latter proposition claims a certain inequality (namely, that like increments of outlay yield a not greater product increment when added to a larger outlay) *for all sufficiently small outlays*. The negation of the former proposition negates the opposite inequality for every sufficiently small outlay, whence there exist *some* small outlays for which the first inequality holds. Similarly, the negation of the law of diminishing product increments for large outlays claims only the existence of *some* large outlays for which a certain inequality holds; and this is much less than the assertion of the law of non-diminishing product increments for large outlays, which claims something about *all* large outlays.

Fourth, contrary to an opinion occasionally held in the literature, the proposition of increasing product increments for small outlays has nothing to do with the questions of whether or not the input factors are "infinitely divisible," whether they change continuously or in jumps, and the like. Our formulation of the proposition does not presuppose the assumptions needed for applications of the differential calculus and is quite independent of the question as to whether the system of all possible outlays includes all n-tuples of non-negative numbers $x_1, x_2, \ldots, x_n$ or merely some of them – even only a finite number of n-tuples.

4. WHAT FOLLOWS FROM THE BOUNDEDNESS OF THE PRODUCTION FUNCTION?

Several attempts have been made to derive the law of diminishing returns from the observation that the product of a piece of land is limited even with the largest outlays on costs. We now formulate two claims about the boundedness of the production function:

a) It implies neither of the main forms of the law of diminishing returns; but

b) It does imply parts of both laws, namely the following propositions:

I. *On every piece of land and for every outlay* y_1, *a fixed additional outlay,* h, *yields a larger product increment when added to* y_1 *than when added to a sufficiently large outlay.*

II. *On every piece of land, the average product of any given outlay is greater than the average product of any sufficiently large outlay.*

The difference between the demonstrable propositions and the laws themselves is not very great and seems to have been heretofore altogether overlooked – probably because of the lack of precision in the statements of the propositions with regard to quantifiers.

Comparing Proposition I and the law of diminishing product increments we see that the latter is stronger (where both outlays in question exceed a certain magnitude), insofar as it compares the product increment due to an addition to the smaller outlay with the increment due to the same addition to *any* larger outlay and not only to any larger outlay which is sufficiently large. Proposition I is stronger insofar as it compares product increments even in cases where one of the two outlays which are to be increased is not large. An analogous relation exists between Proposition II and the law of diminishing average product.

Claim a) that boundedness (even for a strongly increasing production function) does not imply the main forms of the law of diminishing returns is demonstrated by the following example, which is due to Dr. F. L. Alt. Only the products of various outlays in combination with 1 unit of land will be considered. For the sake of simplicity, we set $E(1, y) = f(y)$.

$$f(0) = 0,$$

$$f(1) = 1/2, \quad f(1 + 1/8) = 1/2 + 1/8,$$

$$f(2) = 3/4, \quad f(2 + 1/16) = 3/4 + 1/16,$$

$$f(3) = 7/8, \quad f(3 + 1/32) = 7/8 + 1/32,$$

and, for every positive integer n,

$$f(n) = 1 - 1/2^n, \quad f(n + 1/2^{n+2}) = 1 - 1/2^n + 1/2^{n+2}$$

Between the stated values, f is assumed to increase linearly so that $z = f(y)$ is represented by the broken line in Figure 3.

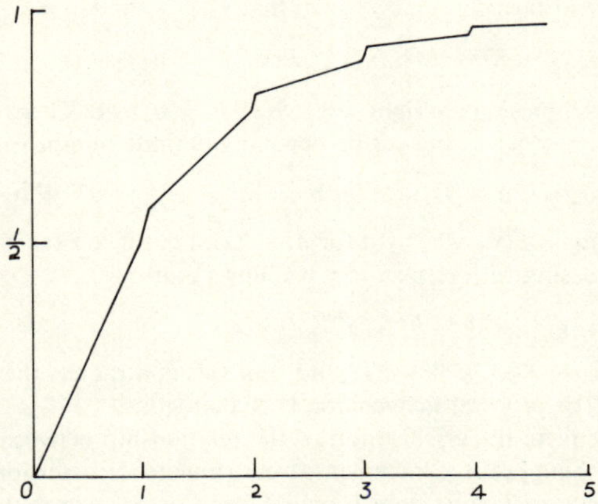

Fig. 3

Obviously, $f(y)$ is *bounded*, since with increasing y it asymptotically approaches 1; and it is *strongly increasing* (See Section 3).

Yet $f(y)$ fails to satisfy the law of diminishing product increments. As one readily verifies, the third form of the law does not hold for

$$y' = n + 1/2^{n+2}, y'' = n + 1, y''' = n + 1 + 1/2^{n+3}.$$

Nor does it satisfy the law of diminishing average product, since for each positive integer n,

$$f(n)/n < f(n + 1/2^{n+2})/(n + 1/2^{n+2}).$$

This completes the proof of claim a).

The propositions overlapping with the laws about returns can be formulated in symbols as follows:

I. For every x, for every y_1, and for every h for which $E(x, y_1 + h) \neq E(x, y_1)$, there exists a $y^*(> y_1)$ which may depend on y_1, x, and h such that for every $y_2 > y^*$

$$E(x, y_2 + h) - E(x, y_2) < E(x, y_1 + h) - E(x, y_1).$$

II. For every x and every y_1 for which $E(x, y_1) \neq 0$, there exists a $y^{**}(> y)$ which may depend on y_1 and x such that if $y_2 > y^{**}$, then

$$E(x, y_2)/E(x, y_1) < y_2/y_1.$$

The proof of Proposition I (of a greater logical complexity than the reasoning concerning the laws of diminishing return to be found in the literature) will not be used in the subsequent sections and thus may be skipped in a first reading of this paper. The proof is indirect. We assume that Proposition I be false and show that this assumption leads to a contradiction.

If the assertion I is false, there exist numbers x_0, y_1 and h such that $E(x_0, y_1 + h) \neq E(x, y_1)$ and there also exists some y_2 such that $y_2 > y_1$ and

$$E(x_0, y_2 + h) - E(x_0, y_2) \geqq E(x_0, y_1 + h) - E(x_0, y_1).$$

We denote the difference on the right side, which is $\neq 0$, by d. There exists a sequence of numbers $y', y'', \ldots, y^{(n)}, \ldots$ increasing beyond any finite number and such that

$$y' < y' + h < y'' < y'' + h < \ldots < y^{(n)} < y^{(n)} + h < \ldots$$

and $E(x_0, y^{(n)} + h) - E(x_0, y^{(n)}) \geq d$ for n + 1, 2, ... Since $E(x_0, y)$ has been assumed to be weakly increasing with respect to y it follows that

$$E(x_0, y^{(n)} + h) \leq E(x_0, y^{(n+1)})$$

for each n and hence $E(x_0, y^{(n)} + h) \geq nd$. But this contradicts the assumed boundedness of E for x_0. The proof of Proposition II is analogous.

In order to complete the clarification of the relationship between Proposition I and the law of diminishing product increments we formulate the assertion common to them.

For every x there exists a y^* (which may depend on x) such that, for every $y_1 > y^*$ and for every h for which $E(x, y_1 + h) \neq E(x, y)$, there exists some y' (which may depend on x, y_1 and h) such that for every $y_2 > y'$

$$E(x, y_2 + h) - E(x, y_2) < E(x, y_1 + h) - E(x, y_1).$$

In the original paper [1954, 3], Section 4 ends with the proof of a conjecture of K. Schlesinger's: that if $E(x, y)$ is bounded, nondecreasing in y, and satisfies the law of diminishing product increments it also satisfies the law of diminishing average product.

5. THE CLASSICAL PROOFS OF THE LAW OF DIMINISHING RETURNS ON LAND

We now turn to chains of propositions known in economic literature under the name of proofs of the law of diminishing returns on land. The present section is devoted to two arguments of classical economists; the three following sections deal with the reasoning of Wicksell (1909), Böhm-Bawerk (1912) and L. v. Mises (1933) – though, for the sake of exposition, in an order different from the historical.

The two classical proofs are indirect. They start from the assumption that the law is not valid and derive conclusions that must be rejected. Both proofs begin by inferring from that assumption the following intermediate conclusion: *The product of every piece of land combined with sufficiently large cost outlay could be increased to any level.*

The *first* proof further concludes that all the agricultural products needed by the entire population of a country could be produced on a small area, perhaps on a single acre. As everyone knows, this is not the case. Hence, the law of diminishing returns on land cannot be false; that is, it must hold.

The *second* proof infers from the intermediate conclusion that the demand for agricultural products could be met through outlays on the most fertile piece of land only, leaving less fertile land idle. But this contradicts experience, hence the law of diminishing returns on land cannot be false.

To begin with, the common intermediate conclusion cannot be accepted since it has been shown in Section 4 that boundedness of the production function implies neither diminishing product increments nor diminishing average product. This invalidates the first proof. This proof is also based on the conclusion that, if the product of a piece of land is unbounded, so is the product of every part of this piece of land. The validity of this conclusion rests upon problematic implicit assumptions which, however, will not here be discussed. The second proof actually uses only a part of the intermediate conclusion, namely that the product of the entire fertile land combined with sufficiently large cost outlays would meet the total demand for agricultural products. But even this cannot be correctly deduced from the negation of the law of return, as the following example shows. Suppose that 1) the total demand for agricultural products is 2 units; 2) the available fertile land is 1 unit; 3) E is Alt's function discussed in Section 4, which satisfies neither of the main laws of diminishing returns while $E(1, y) < 1$ for every y. In this case, the total demand of 2 units cannot be produced on the available fertile land.

Another error in the second proof is the conclusion that, if the entire demand for agricultural products can be met by using only the most fertile land, then only these pieces of land are actually used, and the rest is left idle. To see that this conclusion is not valid, assume, for the sake of simplicity, that there are only two grades of land, fertile and less fertile land, x units of which with a cost outlay of y units produce $E_1(x, y)$ and $E_2(x, y)\,(<E_1(x, y))$ units of product, respectively. If the total demand b for agricultural products can be met on fertile land, then there exist numbers x_1, y_1 such that $E_1(x_1, y_1) = b$. Perhaps one can also produce b on x_1 units of second grade land, but in this case a cost outlay y_2 exceeding y_1 might be needed to achieve $E_2(x_1, y_2) = b$; and this process would be less advantageous than the production on fertile land. But if enough second grade land is available, then it is possible that $E_2(x_2, y_2) = b$ for some large x_2 and a number y_2 *less* than y_1, in which case this process would be more advantageous than the production on fertile land. This case is not likely to occur but it certainly is compatible with all other assumptions thereby demonstrating the conclusion of the second proof to be invalid.

6. MISES' PROOF OF THE LAW OF DIMINISHING RETURNS[10]

This proof, too, is indirect and begins with the inference of the same intermediate conclusion which marks the first step in the classical proofs. It then continues: Let x^* be the number of available units of land; since, for every positive integer n, $E(x^*/n, y)$ is unbounded with respect to y, the entire demand can be met by using only 1/nth of the available land. The remaining $(1 - 1/n)$th part of the land is useless. Therefore, land is a non-economic (i.e., free) good. But since land actually is an economic good, the law of returns cannot be incorrect.

This continuation of the intermediate conclusion cannot be accepted. It is quite possible that a good is economic even though every quantity of the good yields an arbitrarily large product when used with sufficiently large quantities of other inputs, provided the latter are economic goods. If, for every n, $E(x^*/n, y)$ is unbounded with respect to y, then for every integer n there exists some number y_n such that $E(x^*/n, y_n) = b$, where b is the total demand for agricultural products. But, in general, $y_1, y_2, \ldots, y_n$ are not equal; they increase, $y_1 < y_2 < \ldots < y_n$. Therefore, it cannot be concluded that it is a matter of indifference whether x^* units of land or only x^*/n are at one's disposal. It is true that under the above assumptions the demand b can be met either on x^* or on x^*/n units of land. But insofar as a smaller cost outlay of y_1 is preferable to the larger y_n, that is, insofar as the cost inputs are economic goods, land may very well be an economic good.

While the second classical proof ignores the possibility that larger cost outlays can sometimes be avoided by using larger areas of land, Mises overlooks the possibility that under some circumstances the use of smaller areas of land must be compensated by larger outlays of other goods.

What can be proved from the arguments criticized in this section is the following rather obvious consequence of the boundedness of product: *even with the addition of any quantities of free* (non-economic) *cooperating inputs* (like air, sunshine, sea water near the coast), *the product of a part of the available quantity of an economic good cannot be increased without limit.* For, suppose that by means of sufficiently large applications of free cooperating inputs, the product of any part of the available land could be raised without limit. Since the greater or lesser use of free goods is of no economic significance, the availability of the remaining part of the land would be of no economic significance, and land, too, would be a free good.

7. BÖHM-BAWERK'S PROOF[11]

"Let us assume that 100 units of product can be produced on one acre by expending 100 units of cost (capital and labor). The law of diminishing returns on land now states that doubling the costs applied to this acre of land will increase the product, but will not quite double it. Or in mathematical form, if 1 H + 100 C yields a product of 100 P, then 1 H + 200 C yields *less* than 200 P. Or in another, completely equivalent mathematical form, 2(½H + 100 C) yields less than 2(100 P); and ½H + 100 C yields less than 100 P. Now I ask, is it something which requires a proof, or is it a foregone conclusion which approaches pure truism to assert that the smaller amount ½ H + 100 C of the input cannot serve to produce as much product as the larger amount H + 100 C ? . . . If someone wishes to deny it, he must claim that the availability of more or less land is a matter of indifference in production."

In a general and precise form, it is assumed that $E(x, y)$ is 1) subhomogeneous (See p. 287), 2) increasing with respect to x; that is, $x' > x''$ implies $E(x', y) > E(x'', y)$. It is contended that for $\lambda > 1$

$$E(x, \lambda y)/\lambda y < E(x, y)/y \quad \text{or} \quad E(x, \lambda y) < \lambda E(x, y).$$

And the proof runs as follows. According to assumption 1), $\lambda E(x, y) \geqq E(\lambda x, \lambda y)$ holds for every $\lambda > 1$. According to assumption 2), $E(\lambda x, \lambda y) > E(x, \lambda y)$, since $\lambda > 1$ and $x < \lambda x$. Hence the asserted inequality.

Assumptions 1) and 2) imply the law of diminishing average product for every cost outlay. From the point of view of empirical evidence, this conclusion goes too far. For small outlays the law does not seem to be confirmed by experience. For some productive processes outside the field of agriculture, to which Böhm-Bawerk's formal arguments are, of course, equally applicable, economists assert a proposition of non-decreasing or even of increasing average product for small cost outlays.

What then can be concluded from the considerations of this section? Evidently, in the cases just mentioned, where the law of diminishing average product does not hold for all cost outlays, assumptions 1) and 2) cannot both be valid for small outlays. Either an increase of all inputs in the same proportion yields a more than proportional increase in product or use of more land does not, *ceteris paribus*, lead to an increase in product.

Two assumptions which are weaker than 1) and 2) and imply only the law of diminishing average product *for large outlays* are

1) *Subhomogeneity for large outlays:* for every x there exists an outlay level y^* (depending on x) such that $\lambda E(x, y) \geq E(\lambda x, \lambda y)$ for all $y \geq y^*$ and $\lambda > 1$;

2) *Increase with respect to x for large outlays:* for every x there exists an outlay level y^{**} (depending on x) such that $x < x'$ implies $E(x, y) < E(x', y)$ for every $y > y^{**}$.

8. WICKSELL'S PROOF

"The validity of the law of returns on land does not require experimental proof, but presents itself as a logical postulate or corollary. For if it could be shown that twice the labor and twice the capital applied to a piece of land would yield twice the product, an even better result would be achieved *per unit of capital and labor* if the labor and capital at hand were concentrated on half of the formerly used area, the remainder to be used for grazing, forestry or some other productive purpose which requires no labor."

In a general and precise form, it is assumed a) that E is subhomogeneous, b) that E is superadditive, c) that land is an *independent* input: it yields a product even without the use of any cooperating input; in symbols: $x \neq 0$ implies $E(x, 0) > 0$.

The proof runs as follows. If $\lambda > 1$, then application of assumption a) to $E(x/\lambda, y)$ yields

$$E(x, \lambda y) \leq \lambda E(x/\lambda, y).$$

By c), $E(x - x/\lambda, 0) > 0$. Hence

$$E(x, \lambda y) < \lambda(E(x/\lambda, y) + E(x - x/\lambda, 0));$$

and from b) it follows that $E(x, \lambda y) < \lambda E(x, y)$.

That assumptions a), b), c) imply the law of decreasing average return can also be seen from the fact that a) is identical with assumption 1) of Section 7, while b) and c) imply assumption 2). Indeed, if $x' < x''$, then $E(x', y) + E(x'' - x', 0) \leq E(x'', y)$ and since $E(x'' - x', 0) > 0$ it follows that $E(x', y) < E(x'', y)$.

9. THE IMPOSSIBILITY OF PROVING DIMINISHING PRODUCT INCREMENTS

The law of diminishing product increments is not a consequence of the properties of the product function herein considered. This is demonstrated by a production function E^* which does not satisfy that law even though E^* has the following properties. E^* is *homogeneous* (i.e., both sub- and superhomogeneous); *increasing* for fixed x with increasing y; *nondecreasing* for fixed y with increasing x; *bounded* with respect to x for fixed y, and with respect to y for fixed x; and *superadditive*. Moreover, both input factors are *dependent*, that is, $E^*(x, 0) = E^*(0, y) = 0$.

The function is defined in the first quadrant of the (x, y)-plane by specifying, for each integer $k = 0, 1, 2, \ldots$, the value of E^* at each point of the sector between the rays $y = kx$ and $y = (k + 1)x$, namely as follows:

$$E^*(x, y) = kx/(k + 2) + y/(k + 1)(k + 2) \quad \text{for} \quad kx \leq y \leq (k + 1)x.$$

Thus

$$E^*(x, y) = y/2 \text{ between } y = 0 \text{ and } y = x;$$

$$E^*(x, y) = x/3 + y/6 \text{ between } y = x \text{ and } y = 2x;$$

$$E^*(x, y) = 2x/4 + y/12 \text{ between } y = 2x \text{ and } y = 3x;$$

and so on. On the ray $y = kx$, clearly $E^*(x, kx) = kx/(k + 1)$; in particular, $E^*(x, 0) = 0$. For each y, if x approaches 0, then so does $E^*(x, y)$. Hence E^* is continuous for $x = 0$ if its definition is supplemented by $E^*(0, y) = 0$ for each y.

It is easy to verify that E^* has all the asserted properties. That E^* does not satisfy the law of diminishing product increments is illustrated by the example

$$E^*(1, \tfrac{1}{2} + \tfrac{1}{4}) - E^*(1, \tfrac{1}{2}) = E^*(1, \tfrac{1}{3} + \tfrac{1}{4}) - E^*(1, \tfrac{1}{3}) \; (= \tfrac{1}{8}).$$

10. ON NON-INCREASING PRODUCT INCREMENTS

LAW OF NON-INCREASING PRODUCT INCREMENTS. *An increase of cost outlay added to a larger outlay never yields a larger increase of product than when added to a smaller cost outlay.*

This law can be stated in four other forms corresponding to the forms of the law of diminishing product increments (See Section 2). It can also be restricted to cost outlays above a certain level.

Using Alt's function described in Section 4 one can easily show that increasing product increments are compatible with most of the properties of production functions herein discussed. But the conjunction of homogeneity and superadditivity implies nonincreasing product increments, even though (according to the example in Section 9) it does not imply diminishing product increments.

Indeed, if E is superadditive, then for each x, y, h, k such that $y - h \geq 0$

$$E(hx, h(y + k)) + E(kx, k(y - h)) \leq E((h + k)x, (h + k)y).$$

Because of superhomogeneity, the term on the right side does not exceed $(h + k)E(x, y)$. Because of subhomogeneity the sum on the left side is not less than $hE(x, y + k) + kE(x, y - h)$. Hence

$$hE(x, y + k) + kE(x, y - h) \leq (h + k)E(x, y).$$

But this is the fourth form of the law of nonincreasing product increments.

According to the remark of Schlesinger (See Section 3), subhomogeneity in conjunction with superadditivity is sufficient for the derivation of this law since they imply superhomogeneity. Indeed, by setting $h = k$, $hx = x'$, $ky = y'$, $h^2 = k^2 = hk = h'$ in the inequality implied by superadditivity one obtains

$$E(x', y' + h') - E(x', y' - h') \leq E(2x', 2y')$$

and using subhomogeneity

$$E(x', y' + h') - E(x', y' - h') \leq 2E(x', y'),$$

which is the fifth form of the law of nonincreasing product increments.

11. ON THE RELATIONSHIP BETWEEN THE LAWS OF DIMINISHING AVERAGE PRODUCT AND OF DIMINISHING PRODUCT INCREMENTS

That diminishing product increments for any outlay are a consequence of diminishing average product for any outlay is incorrect, as can be seen from the example in Section 9. But the converse is valid.

The law of diminishing product increments for any outlay implies the law of diminishing average product for any outlay.

This theorem is of interest since in Section 2 it was shown that, if the two laws of diminishing returns are limited to large outlays, the theorem is not valid. The theorem can be proved by elementary means and without assuming differentiability properties of the production function. A simple proof using the second mean value theorem of integral calculus can be found in [1954, 3], p. 475 ff.

12. ECONOMIC GOODS AND THE LAWS OF RETURN

Almost all so-called proofs of the laws of return that can be found in the literature use the assumption that the input factors of production are economic goods. Actually, however, *the economic character of the inputs bears no relation to the question of whether the product increments are nonincreasing*. Without superadditivity and subhomogeneity of the production function, the law of nonincreasing product increments need not hold even if the inputs are economic goods. On the other hand, if the product function is superadditive and subhomogeneous, the law does hold even for noneconomic inputs.

We examine briefly the special case in which one of the means of production $G_1, G_2, \ldots, G_n$ say G_1, is a noneconomic good, that is, is available in larger quantity than can be used. We can assume that the product is an economic good, since otherwise the production in question would not materialize. Let $E(x_1, x_2, \ldots, x_n)$ be the maximum number of units of the economic good that can be produced by using x_1 units of G_1, x_2 units of $G_2, \ldots, x_n$ units of G_n. At least one of $G_2, \ldots, G_n$ must be an economic good.

It is quite possible that G_1, even though it is a noneconomic good, leads to product increments; for example, even a noneconomic good G_1 may be essential for production. It is possible that $E(0, x_2, \ldots, x_n) = 0$ for each combination $x_2, \ldots, x_n$ and yet $E(x_1, x_2, \ldots, x_n) > 0$ for sufficiently large x_1. Water in a region in which it is plentiful is in this sense an essential noneconomic input in many productions.

Under the assumption of subhomogeneity, considerations similar to those in Section 5 demonstrate that $E(x, x_2, \ldots, x_n)$ is bounded for every fixed combination $x_2, \ldots, x_n$ – at least if the economic goods among $G_2, \ldots, G_n$ are of no use except in the production in question. The methods of this paper also point the way to a series of conclusions regarding the relationship of the upper bound of the production function to various other data of the problem. The arguments concerning non-increasing product increments (Section 10) apply to G_1 just as much as to any other goods, including the economic goods among $G_2, \ldots, G_n$.

Nor is the economic character of goods of direct importance for the law of diminishing average product. All that is relevant is a purely technical property of a good: that its more extensive use in the production causes an increase of product. This assumption must be adjoined to that of subhomogeneity for large outlays (alternatively, for any outlays) in order to derive the law of diminishing average products for large outlays (for any outlays, respectively). Without assuming the technical property one can prove only nonincreasing average product.

13. CONCLUDING REMARKS

From the point of view of methodology, the present paper can perhaps lay claim to being the first instance in economics of a clear separation between the question of logical interrelations among various propositions and the question of empirical validity. Which of the various forms of the laws of return and of the other propositions that we have discussed are empirically valid – these questions can be decided only empirically, by observation. How the various propositions are related, which ones are consequences of others – these and similar questions are purely logical and have nothing to do with experience.

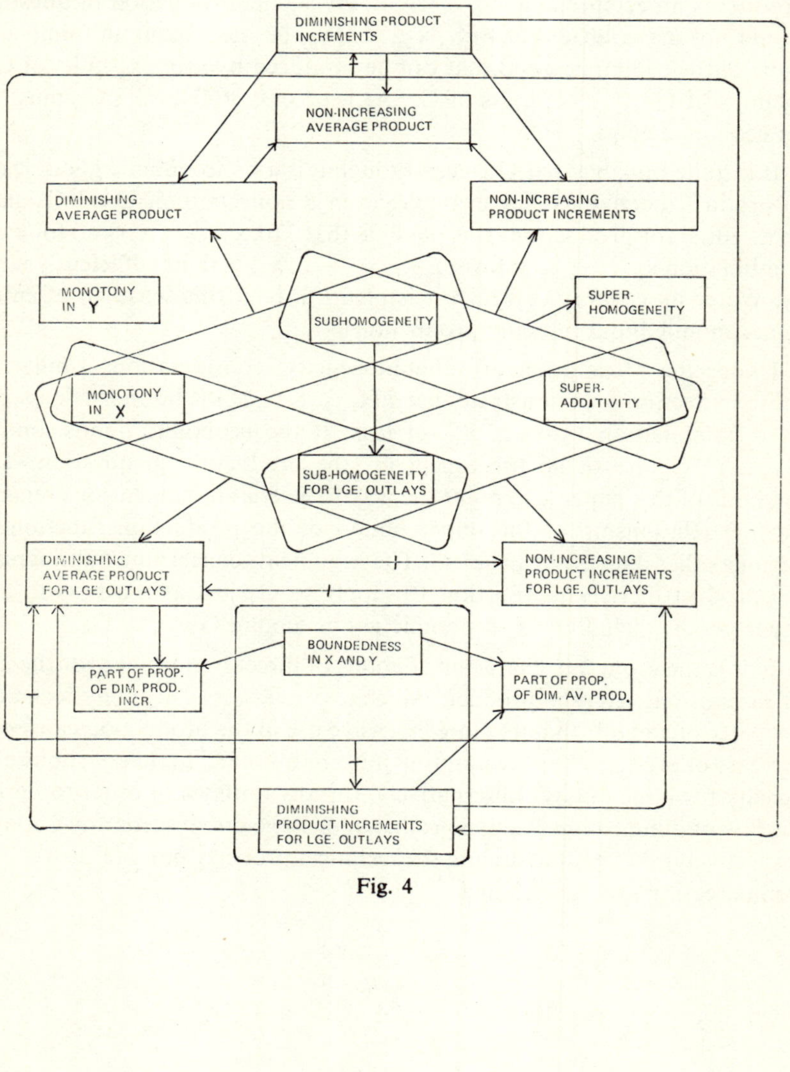

Fig. 4

In view of the numerous subpropositions into which the law of diminishing returns had to be divided, we give a graphic summary of some of the most important relationships between these propositions. In the figure, an arrow means that the proposition or group of propositions within a box at the beginning of the arrow implies the proposition or group of propositions to which the arrow points; an arrow with a line across it means that logical implication does not hold. For the sake of greater transparency of the picture, arrows which would be consequences of those shown in the figure are omitted. If p, q, r are three propositions, then $p \rightarrow r$ is omitted if $p \rightarrow q$ and $q \rightarrow r$ are shown. Similarly, if $p \rightarrow r$ is shown, then $(p\&q) \rightarrow r$ is omitted. Crossed arrows are shown sparsely – mainly where in the literature or in discussions implications were claimed erroneously. The words 'for any outlay' have been omitted. Only the qualification 'for large outlays' appears. The theorem proved by Schlesinger has been included.

NOTES

[1] Numerous references to the literature preceding 1931 can be found in the interesting article by O. Morgenstern in Vol. 2 of *Zeitschrift für Nationalökonomie*.

[2] We shall not discuss efforts to prove the law on the basis of the natural sciences. Besides the work of J. Essler, the reader is referred to the more recent article by P. Klaeb in *Weltwirtschaftliches Archiv*. Vol. 42, p. 248, which contains numerous references to the literature.

[3] For discussions of the general relationship between logic and problems of the social sciences, cf. [1936, 7], [1934, 9] and [1974, 1].

[4] E. v. Böhm-Bawerk, *Gesammelte Schriften*, Vienna 1924, vol. 1, p. 195 sq.

[5] *ibid.*, p. 193 sq.

[6] We avoid the misleading name "proposition of diminishing returns on land" since the returns do not diminish absolutely as the outlay is increased. On the contrary, returns increase as outlay increases. The designation "law of diminishing increase of product" was given by Th. Brinkmann, *Grundriß der Sozialökonomik* **VII** (1922), p. 32.

[7] As I discovered after completing this article, Edgeworth had clearly recognized that two different propositions were being confused under the title "law of diminishing returns", "an ambiguity, the existence of which in so familiar a subject might have been deemed impossible" (*Papers related to Pol. Econ.*, Vol. I, p. 61). But since Edgeworth does not relate the separation of the two propositions to the question of proving them, I felt that I should present the issue in a form adapted to the discussions in the later sections of this paper.

[8] It appears that the same can be said of the formulation by F. X. Weiss in the article *Abnehmender Ertrag* in *Handwörterbuch der Staatswissenschaft*, 4th ed., Vol. I, p. 12.

[9] *Ein Beitrag zu Menger's Bemerkungen zu den Ertragsgesetzen. Zeitschrift f. Nationalökonomie* **7** (1936), 398–400.

[10] *Grundprobleme der Nationalökonomie*, Jena 1933, p. 145–59.

[11] Cf. loc. cit.[4] p. 198; also Strigl, *Zeitschrift f. Nationalökonomie*, **8**, p. 216 sq.

PART VIII

GULLIVER'S INTEREST IN MATHEMATICS

Chapter 24 is a reprint of [1959, 2], which appeared in the British *Mathematical Gazette*. The paper represented part of the work made possible by a grant from the Carnegie Foundation of New York for the development of my approach to mathematics. Chapter 25 contains [1960, 1], published when R. D. James was the editor of the *American Mathematical Monthly*. The paper is here abbreviated in order to avoid repetitions of Chapter 24. After reading the Gulliver story in the *Gazette*, the mathematics students of Cambridge invited me to speak to them when they heard that I would be visiting England, and then published the talk in their yearly, *Eureka*, [1960, 4]. It is reprinted with a few minor changes in Chapter 26. This story is independent of the other two.

CHAPTER 24

GULLIVER IN THE LAND WITHOUT ONE, TWO, THREE

About the year 1700, in the course of heretofore unrecorded travels, Gulliver met islanders who used a rather odd arithmetical vocabulary. They counted:

stix, stixpair, stixtrip, four stix, five stix, six stix, and so on; and they wrote

(1) $\quad |, \quad |^p, \quad |^t, \quad 4|, \quad 5|, \quad 6|$, and so on.

The symbol | and the word *stix* also denoted kauri sticks – the insular currency and principal object of applied mathematics.

The symbols $|$, $|^p$, $|^t$ and the corresponding words for the first three numbers had, as Gulliver learned, always been in use. But beyond stix, stixpair, and stixtrip, the islanders originally counted: four, five, six, and so on (and they wrote: 4, 5, 6, and so on). The combination of these two types of original symbols, however, created difficulties. For instance, while one could write

(2) $\quad | + |^p = |^t \quad$ and $\quad 4 + 5 = 9$,

symbols such as $| + 4$ and $|^p + 5$ as well as the words stix plus four and stixpair plus five appeared to be incongruous. As a remedy, mathematicians proposed a uniform symbolism; and they achieved uniformity by assimilating the higher numerals to the lowest three. After this reform, (1) became the official sequence of insular numerals.

Ever since, sums were indicated by

(3) $\quad |^p + |^t, \quad | + 4|, \quad |^p + 5|, \quad 4| + 5|, \ldots$

Stix plus four stix (in contrast to stix plus four) sounded unobjectionable and so did stixpair plus five stix. For instance, addition tables included the formulae

(4) $\quad | + |^p = |^t, \quad | + 4| = 5|, \quad |^p + 5| = 7|, \quad 4| + 5| = 9|.$

But the islanders never wrote

(5) $\quad 4| + 5| = (4 + 5)|.$

Aiming at uniformity, they avoided (5) because they could not write

(6) $|^{p} + 5| = (|^{p} + 5)|$.

As a result, even though the meaning of (5) differs from that of the last formula (4), the contents of formulae such as (5) were rarely expressed at all.

Difficulties arose for the islanders in their higher mathematics, that is, in multiplication. In advanced treatises, mathematicians wrote, for example,

(7) $4 \cdot 5| = 20|$

(read: four times five stix equal twenty stix); that is to say, they denoted the first factor of the product by the original numeral 4, and not by the official 4|. But products such as

(8) $5 \cdot |^{t}$, $6 \cdot 4|$, $4 \cdot 5|, \ldots$

(whose first factors exceeded stixtrip) were the only ones that could be designated in a direct way. Eschewing any incongruity, mathematicians refrained from using the numerals |, $|^{p}$, and $|^{t}$ as the first terms in designations of products. They never wrote, e.g.,

(9) $|^{p} \cdot 4|$ or $|^{t} \cdot |$ or $| \cdot |$,

nor did they ever say: stixpair times four stix or stixtrip times stix or stix times stix.

Gulliver, who attributed these difficulties to lacunae in the insular system of numerals, wrote in his diary: "*Landed on an island without One, Two, Three.*"

Of course mathematicians also discussed products with small first factors – but in an indirect way, namely, by introducing for that very purpose (where their ordinary numerals for small numbers failed) other symbols – usually the letters h and k. But even the ad hoc use of such letters did not make it possible to express facts about multiplication in simple formulae such as (5). Mathematicians were forced to resort to implications. For instance, they wrote:

(10) If $h| = |$, then $h \cdot h| = h|$ and thus $h \cdot h| = |$.

(read: if h stix equal stix, then h times h stix equal h stix and thus stix) and, more generally,

(11) If $k| = |$, then $k \cdot n| = n| = n \cdot k|$ for any number $n|$ (or, as some wrote, for any number n).

Ad-hoc-symbols also played a great role in the formulation of problems. Pages and pages of textbooks were devoted to exercises such as

(12) Find k^2 if $k = |^p + |^t$; $k = | + 4|$; $k = 4| + 5|$; $k = 5 \cdot 6|$.

Naturally under the circumstances, multiplication was a topic taught only in universities and, even there, teachers as well as students grumbled a good deal about the fact that few beginners other than those majoring in mathematics acquired more than a merely mechanical facility.

Not everything went smoothly in applied mathematics, especially in counting objects other than sticks. The islanders wrote

(13) 4 arrows, 5 arrows, . . . and not 4| arrows, 5| arrows, . . .

in this connection, too, reverting to the original numerals 4, 5, . . . But, of course, they refrained from writing

(14) | arrow, $|^p$ arrows, and $|^t$ arrows

and from saying: stix arrow, stixpair arrows, and stixtrip arrows. What they did say and write was: arrow, arrowpair, and arrowtrip. Hence, when Gulliver bought four arrows for six kauri sticks, he received the following bill

4 arrows . . . 6|.

But when he bought two arrows for three sticks, the bill read:

arrowpair . . . $|^t$.

How simple would everything be, Gulliver thought, if the islanders had achieved uniformity *by assimilating their lower to their higher numerals* (and not the other way round); in other words, if they had introduced symbols, say,

(15) 1, 2, 3

for the first three numbers – for the numbers *themselves* and not for groups of sticks – and kept the original numerals 4, 5, 6, . . . for the others. This minor reform would even have permitted the islanders to retain the ancient symbols |, $|^p$, and $|^t$, to which some of them were strongly attached. Of course, these symbols (in accordance with their spoken counterparts: stix, stixpair, and stixtrip) would describe one stick, two sticks, and three sticks; in other words, they would serve as synonyms of 1|, 2|, and 3|, respectively – facts that might be expressed in the formulae:

(16) $1| = |$, $2| = |^p$, $3| = |^t$.

In the same spirit, the formula $| + |^p = |^t$ might be retained as an expression of the fact that

(17) $1| + 2| = 3|$.

The statement (17) about sticks would reflect a relation between three pure numbers, namely,

(18) $1 + 2 = 3$.

Gulliver developed these ideas in a paper "*Arithmetic without Stix.*" He showed that, along the lines indicated, one would arrive at the following results:

(a) Unified symbols for sums. Formulae, acceptable to semanticists, would include (18) and, instead of (4),

$$1 + 2 = 3,\ 1 + 4 = 5,\ 2 + 5 = 7,\ 4 + 5 = 9.$$

(b) Unified symbols for products, such as $1 \cdot 3$, $2 \cdot 4$, $4 \cdot 5$.
The simple formula $1 \cdot 1 = 1$ would replace the cumbersome implication (10).

(c) The awkward implication (11) would be replaced with the transparent law

$$1 \cdot n = n = n \cdot 1 \quad \text{for any number } n.$$

(d) The problems (12) could be formulated without symbols introduced ad hoc, namely, by saying simply:

$$\text{find } (2 + 3)^2,\ (1 + 4)^2,\ (4 + 5)^2,\ (5 \cdot 6)^2.$$

(e) The separation of numerals and symbols for sticks would lead to a unification of applied mathematics, satisfactory to philosophers of science; e.g., the afore-mentioned bills would read:

4 arrows . . . $6|$; 2 arrows . . . $3|$.

There would be no reason for shunning formula (5) concerning sticks, since it would have the analogue

$$2| + 5| = (2 + 5)|,$$

instead of (6).

The reception of these ideas on the island was mixed. Several outstanding mathematicians, including the discoverer of the celebrated formula $| + |^p = |^p + |$, said Gulliver's paper was trivial. Noted educators, on the other hand, called the criticism of their numerals extremely far-fetched and Gulliver's introduction of the symbols 1, 2, 3 (or, as they said, his new notation) a *tour de force*. But the two groups concurred in one point: that the idea of *teaching* terms other than stix, stixpair, and stixtrip was simply ridiculous. Again, some philosophers of science hailed Gulliver's ideas as a great simplification of the existing system. Two professors of higher mathematics ventured the prediction that, in terms of 1, 2, 3, multiplication would one day be taught in secondary schools. But these voices were silenced by the powerful organization of the Island's Major Mathematicians of Real Talent and Learning – briefly, the Immortals – whose judgment naturally carried greater weight as being more detached, since no Immortal had seen a freshman in decades. The Immortals were reported to feel that (a) even though no one had ever taken the trouble to write down the ideas propounded by Gulliver, all of them had been clear to all mathematicians all the time; (b) the so-called difficulties, pointed out by him, were merely didactic; they had never in the least bothered any Immortal nor, as far as Immortals could place themselves into the position of minor intellects, even mortal research mathematicians.

Gulliver's schedule called for his departure from the island before the dispute was settled. But the discussion aroused in him a broader interest in mathematics and, upon his return to England in the early 1700's, he familiarized himself with the latest progress – the theory of functions and fluents. He studied

(1′) $x, x^2, x^3, \sqrt{x}, \log x, \tan x, \ldots$

(read: the functions x, x square, x cube, square root of x, . . .)

Among the functions, as Gulliver noticed with surprise, only the more complicated ones had self-contained symbols, such as $\sqrt{\ }$, log and tan. The small raised 3 and 2 in the symbols for cube and square could not occur by themselves. These two functions as well as the first power (assuming for any number x the values x^3, x^2, and x, respectively) were referred to by those very values.

The combination of the symbols $\sqrt{\ }$, log, and tan with the symbols x, x^2, and x^3 would have led to difficulties, e.g., in designating products of functions. Mathematicians might have written

(2′) $x \cdot x^2 = x^3$ as well as $\sqrt[6]{\ } \cdot \sqrt[3]{\ } = \sqrt{\ }$,

but they could not very well write $x^2 \cdot \sqrt[3]{\ }$ nor, for that matter, $x^2 + \sqrt[3]{\ }$ or $x + \log$.

These difficulties were overcome by a uniform notation, and uniformity was achieved by assimilating the symbols for the more complicated functions to those for the simpler ones; that is to say, mathematicians also designated the functions $\sqrt{\ }$, log, and tan by their values for x; and (1′) represented the official symbols for functions.

Sums of functions were denoted by

$$x + x^2,\ x^2 + \sqrt{x},\ \log x + \cos x;$$

products, by

$$(3') \qquad x^2 \cdot x^3, \quad x \cdot \frac{1}{\sqrt{x}}, \quad \sqrt[6]{x} \cdot \sqrt[3]{x}, \quad \log x \cdot \tan x$$

(the last three dots were sometimes omitted); for instance,

$$(4') \qquad x \cdot x^2 = x^3, \quad x \cdot \frac{1}{\sqrt{x}} = \sqrt{x}, \quad \sqrt[6]{x} \cdot \sqrt[3]{x} = \sqrt{x}.$$

But mathematical treatises lacked formulae such as $\log x + \cos x = (\log + \cos)x$; $\log x \cdot \cos x = (\log \cdot \cos)x$ or

$$(5') \qquad \sqrt[6]{x} \cdot \sqrt[3]{x} = (\sqrt[6]{\ } \cdot \sqrt[3]{\ })x.$$

Aiming at uniformity, mathematicians avoided (5′) because they could not write

$$(6') \qquad (x^2 \cdot \sqrt{\ })x = x^2 \cdot \sqrt{x} \text{ or } (x + \log)x = x + \log x.$$

As a result, even though the meaning of (5′) differs from that of the last formula (4′), the contents of formulae such as (5′) were rarely expressed at all.

Further difficulties arose in substitution of functions into others, also called composition of functions. For instance, mathematicians wrote

$$(7') \qquad \sqrt[3]{\sqrt[6]{x}} = \sqrt[18]{x};$$

that is to say, they denoted the function into which they substituted another function by $\sqrt[3]{\ }$ and not by its official symbol $\sqrt[3]{x}$. But results of substitution such as

$$(8') \qquad \sqrt[3]{\log x}, \quad \log \tan x, \quad \tan x^2$$

were the only ones that could be designated in a direct way. Mathematicians could not designate the results of substituting the functions log x,

$\tan x$ and x^2 into simple functions such as x^2, $2x + 1$, and x, respectively, by writing

(9′) $\quad x^2 \log x, \quad (2x + 1) \tan x, \quad xx^2$ or

$x^2(\log x), \quad (2x + 1)(\tan x), \; x(x^2)$.

Nor could they write xx or $x(x)$ for the result of substituting the function x into itself.

Of course mathematicians also discussed the results of substitutions into simple functions – but in an indirect way, namely by introducing for that very purpose (where their ordinary symbols for power functions failed) other symbols, e.g., the letters h and k. But even the ad hoc use of such letters did not make it possible to express facts about substitution in simple formulae such as (5′). Mathematicians were forced to resort to implications; for instance,

(10′) If $h(x) = x$, then $h(h(x)) = h(x)$ and thus $h(h(x)) = x$

and, more generally,

(11′) If $k(x) = x$, then $k(f(x)) = f(x) = f(k(x))$ for any function $f(x)$ (or, as some wrote, for any function f).

Pages and pages of textbooks were filled with problems formulated in terms of ad-hoc-symbols, the letter f being a favorite; e.g.

(12′) Find the differential quotient of $f(x)$ if

$f(x) = x + x^3$; $f(x) = x^2 \log x$; $f(x) = \log x \cos x$;
$f(x) = \log \tan x$.

Even applied mathematics was somewhat affected by the difficulties of the functional notation. The results of substituting s, v, and t (the distance travelled, the volume, and the time) into the square root, the logarithm, and the cosine (in other words, the square root of the distance, the logarithm of the volume, and the cosine of the time) were denoted by

(13′) $\sqrt{s}, \quad \log v, \quad \cos t$,

respectively. But the results of substituting the time into the function x^2, or the volume into the function $2x + 1$ could not be denoted by

(14′) $x^2 t$, $(2x + 1)v$ or $x^2(t)$, $(2x + 1)(v)$.

Yet it seemed to Gulliver that everything would have been simple if uniformity had been achieved *by assimilating the symbols for the simpler functions to those for the more complicated ones* instead of the other way round; in other words, if mathematicians had introduced symbols for the former, say,

(15′) j, j^2, and j^n (for any number n)

as designations of the identity function (or first power), the second power, and the nth power, respectively – the functions *themselves*, not their values – and had used the symbols $\sqrt{}$, log, cos, etc. for the others. They might have retained x, x^2, and x^n as symbols for the values that the functions (15′) assume for x; that is to say, as synonyms of $j(x)$, $j^2(x)$, and $j^n(x)$, respectively – a fact that might be expressed in the formulae:

(16′) $j(x) = x, j^2(x) = x^2$, and $j^n(x) = x^n$ for any n.

In the same spirit, the formula $x \cdot x^2 = x^3$ might have been retained as an expression of the fact that

(17′) $j(x) \cdot j^2(x) = j^3(x)$ for any number x.

The statement (17′) about numbers would reflect a relation between three functions, namely,

(18′) $j \cdot j^2 = j^3$.

Rescuing the identity function and the power functions from anonymity and dealing with functions themselves instead of dealing with their values would, it seemed to Gulliver, lead to the following results:

(a′) Unified symbols for sums, products, and quotients of functions, such as

$$j^3 + \cos, \quad j \cdot j^2, \quad \frac{\sin}{j}, \quad \log \cdot \tan, \quad \sqrt{} \cdot \tan, \quad \sqrt[6]{} \cdot \sqrt[3]{}$$

(the dot for multiplication would never be omitted). Examples of acceptable formulae would include (18′) and, instead of (4′)

$$j \cdot j^{-\frac{1}{2}} = j^{\frac{1}{2}} \text{ and } \sqrt[6]{} \cdot \sqrt[3]{} = \sqrt{} \text{ or } j^{1/6} \cdot j^{1/3} = j^{1/2}.$$

There would be no reason for shunning the formula (5′) about numbers, since it would have the following analogs

$$(j^2 \cdot \sqrt{})x = j^2 x \cdot \sqrt{x} \text{ and } (j + \log)x = jx + \log x,$$

instead of (6′).

(b′) Unified symbols for the results of substitution could be set up by mere juxtaposition of function symbols; for instance,

$$\log \tan, \quad j^2 \cos, \quad j^2 \log, \quad \cos j^2, \quad (2j+1)j^2.$$

In particular, the simple formula

$$jj = j,$$

(the analog of $0+0=0$ and $1 \cdot 1 = 1$) would replace the cumbersome implication (10′).

(c′) The awkward implication (11′) would be replaced with the transparent law

$$jf = f = fj \quad \text{for any function } f,$$

the analogue of the laws

$$0+n=n=n+0 \text{ and } 1 \cdot n = n = n \cdot 1 \quad \text{for any number } n.$$

(d′) The problems (12′) could be expressed without symbols introduced ad hoc, namely, by writing simply (if **D** stood for the derivative):

$$\text{find } \mathbf{D}(j+j^3), \quad \mathbf{D}(j^2 \cdot \log), \quad \mathbf{D}(\log \cdot \sin), \quad \mathbf{D}(\log \tan).$$

(e′) Thanks to the clean separation of symbols for functions connecting fluents and the symbols for the fluents themselves, applications of the function concept could be unified. Of course such a separation is possible where the connecting functions have symbols, as the sine function in the classical description of a certain harmonic oscillator:

(19′) $\quad s = \sin t.$

This formula indicates that the position (in a certain unit) is connected with the time (in a certain unit) by the sine function. On the other hand, clean separation is impossible where the connecting function is anonymous as in the law for falling objects

$$s = 16t^2.$$

This formula looks simpler than

(20′) $\quad s = 16j^2(t)$

to him who is used to taking the identity and power functions for granted. But only formulae such as (20′) and (19′), in which functions are cleanly separated from fluents, make it possible to apply pure mathematics to observable material automatically and by uniform procedures following articulate schemes.

Consider, e.g., the basic scheme for applying the derivative of a function to the rate of change of one fluent with regard to another. This scheme reads:

$w = f(u)$ implies $\mathbf{d}w/\mathbf{d}u = (\mathbf{D}f)u$ for any two fluents u and w and any differentiable function f, where $(\mathbf{D}f)(u)$ is the result of substituting the fluent u into the function $\mathbf{D}f$.

According to differential calculus,

$$\mathbf{D}\sin = \cos \quad \text{and} \quad \mathbf{D}(16j^2) = 32j.$$

Hence the scheme yields, by purely substitutive procedures (which could be easily carried out by machines),

$$s = \sin t \text{ implies } \frac{\mathbf{d}s}{\mathbf{d}t} = \cos t$$

and

$$s = 16j^2(t) \text{ implies } \frac{\mathbf{d}s}{\mathbf{d}t} = 32j(t).$$

Since, according to (19′) and (20′), the antecedents in these implications are valid, so are the consequents. Identifying $\mathbf{d}s/\mathbf{d}t$ with the velocity v, one thus has proved

$$v = \cos t \quad \text{and} \quad v = 32j(t), \text{ respectively.}$$

Gulliver intended to describe his experiences in the Land without One, Two, Three in letters to Newton, to the successors of Descartes, to Leibniz, and to the Bernoullis. One of these great minds, rushing from one discovery to the next, might have paused for a minute's reflection upon the way their own epochal ideas were expressed. It is a pity that, because of Gulliver's preparations for another voyage, those letters were never written.

CHAPTER 25

GULLIVER'S RETURN TO THE LAND WITHOUT ONE, TWO, THREE

Gulliver's interest in advanced mathematics can be traced to adventures not mentioned by Swift and only recently recorded.[1] They took place on an island which, because of the current sophisticated symbols for numbers used there, the traveller called the Land without One, Two, Three. Upon his return to Europe, Gulliver studied, both in England and on the Continent, what then (that is, in the early 1700's) was the most modern topic of investigations: the functions

$$x, x^2, x^3, \sqrt{x}, \cos x, \log x, \ldots .$$

The more the traveller penetrated into analysis, the more he admired the ingenious theory. Yet he was puzzled by several features of those sophisticated symbols for functions, which somehow reminded him of the numerals in the Land without One, Two, Three – features of analysis that, incidentally, are still apparent today. He had proposed to the islanders a simplification of their system of numerals as a solution of difficulties with their symbols. So he was curious about whatever had become of the arithmetic in the Land without One, Two, Three; and the next time he was in the vicinity, several years after his first stay, Gulliver revisited the island.[2]

The ideas that Gulliver had propounded in a pamphlet "*Arithmetic without Stix*," circulated on the island widely, though for various reasons, only in handwritten copies. The Immortals were unimpressed and declared Gulliver's ideas to be trivial; the publishers' referees, on the contrary, complained about their abstruseness – some called stix-free numerals outright incomprehensible; and everyone considered it as completely out of the question that, in terms of 1, 2, 3, numbers could ever be *taught*. Nor, obviously, could publishers regard a work whose title included the word *arithmetic* (in contrast to, say, the *Abstract Stix Theory and its Generalizations*) as a prestige book.

Toward the end of Gulliver's first stay, however, two little-known professors – mortal but alive – had reported about experimental courses on stix-free multiplication. Their experiences had been satisfactory – in fact, so much so that they considered using the method in college courses. Approaching the pedagogical problem in a strictly scientific spirit, they designed large-scale experiments which, they felt, should be arranged by the Insular Arithmetic Foundation – an organization sponsored by the island's ruler for various purposes including the improvement of arithmetic education. So they wrote to the director of the I.A.F. pointing out that Gulliver's ideas would make multiplication available to wider strata of the learning population; but when Gulliver left they had not yet received an answer.

Even though on Gulliver's return to the Land without One, Two, Three most islanders still counted stix, stix-pair and so on, he found arithmetic in a state of fermentation. The island was swamped with beautifully printed new textbooks most of which discussed, in one form or another, the insular numerals and tried to improve them. (Even his formerly only hand-written pamphlet "*Arithmetic without Stix*" had been planographed.) Besides the symbol $n|$, there appeared various other notations for the amount of n monetary units, *e.g.*,

$$(n, |), \qquad (|, n), \qquad |n, \qquad \cdot|n.$$

"Whatever other difficulties may beset analysis back home," Gulliver thought, "the value that the function f assumes for x is denoted by $f(x)$; and, thank goodness, analysts have not to cope with other symbols such as

$$(f, x), \qquad (x, f), \qquad xf, \qquad \cdot xf."$$

(Of course, Gulliver was thinking of his contemporaries only, without anticipating all future developments.)

More important than those notational efforts was the fact that many new textbooks stressed the conceptual distinction between amounts of stix and numbers, *e.g.*, the *amounts* $|^p$ and $5|$, and the *numbers* $|^p$ and $5|$. Only some of the island's Major Mathematicians kept emphasizing that the traditional equivocation, being perfectly obvious, had never in the least confused any Immortal, whence they concluded in an impressive, if unusual, display of humility that it could not confuse anyone. Then there were a few who declared that using a symbol in discrepant ways required courage which, they proclaimed, mathematicians ought to display.

Most mortal arithmeticians who had recognized the equivocal usage as such, however, tried to avoid it; and they attempted to express the distinction between numbers and numbers of stix in visible form. In fact, they tried very hard. Some reformers proposed to write $(|)^p$, $5(|)$, and $n(|)$ for the amounts and $(\cdot)^p$, $5(\cdot)$, and $n(\cdot)$ for the numbers. Others rendered the distinction by writing $|^p$, $5|$, $n|$ for the amounts, and $[|]^p$, $5[|]$, $n[|]$ for the numbers.

"Who back home would want to write $(x)^2$, $\log(x)$, and $f(x)$ for the values, and $(\cdot)^2$, $\log(\cdot)$, and $f(\cdot)$ for the functions?" Gulliver wrote in his diary. "Or x^2, $\log x$, $f(x)$ for the values, and $[x]^2$, $\log[x]$, $f[x]$ for the functions? If they wanted to improve things, they would rescue the identity function from its anonymity by introducing a symbol for it, say j. Thereafter they would write $j^2(x)$ or x^2, $\log x$, and $f(x)$ for the function values, and j^2, log and f for the functions. In fact, if they decided to discriminate between functions and numbers *typographically*, say, by using roman type for references to numbers, and either italics or boldface letters for functions (as in j^2, *log*, f or **j**,^{2} **log**, **f**), then they might even save the parentheses in many symbols for function values and write j^2x, *log* x, and fx or **j**^{2}x, **log** x, and **f**x."

Still other insular arithmeticians claimed that the words *five stix* (or $5|$) were abbreviations of the words

the number of the amount of five stix

but that this phrase was so cumbersome that people got tired of repeating it and rather referred to the number, elliptically and with mental reservations, as *five stix*. Gulliver, on the other hand, was an outspoken man with an idiosyncrasy against mental reservations, particularly in mathematics. Besides, from his point of view, the number in question could be satisfactorily denoted by the single word *five*, whence the words *five stix* were redundant or hyperbolic rather than abbreviated and elliptic.

"Back home," the traveller wrote, "the symbol $\log x$ might be considered as an abbreviation or elliptical rendering of the phrase

the function whose value for x is $\log x$.

But at the same time the symbol is also redundant and hyperbolic as a designation of the logarithmic function or the function log. Certain lacunae in a symbolism create *hyperbolic ellipses*."

A last group of reformers treated pure numbers as *ratios*. They replaced the traditional numerals $5|$ and $|^p$ by the symbols

$5| \to |$ and $|^p \to |$ (pronounced: five stix to stix and stix-pair to stix) –

a notation that they regarded as very suggestive, if in need of a supplementary explanation, namely, that

$$(|^p \to |)| = |^p \quad \text{and} \quad (|^p \to |) \text{ canoes} = \text{canoe}^p.$$

The followers of this line of thought trusted that the elegant symbols $| \to |$, $|^p \to |$, and $|^t \to |$ (for the numbers that Gulliver proposed to denote by the uncouth numerals 1, 2, 3) would undoubtedly be adopted as standard.

"Some of the proposals in the Land without One, Two, Three aiming at an improvement of the traditional numerals are sophisticated indeed," Gulliver wrote in his diary. "Their analogue in the realm of functions would be the proposal that, instead of designating the logarithmic and square and identity functions by simple symbols such as *log* and j^2 and j (and their values for x, by $\log x$ and $j^2(x)$ and $j(x)$), one should rather denote the functions by

$$x \to \log x \quad \text{and} \quad x \to x^2 \quad \text{and} \quad x \to x$$

and explain that

$$(x \to \log x)(x) = \log x \quad \text{and} \quad (x \to x^2)(t) = t^2$$

$$\text{and } (x \to x)(c) = c$$

– an inconceivable proposal."

(In this case again Gulliver spoke only about his contemporaries and failed to foresee all that would be written 250 years later.)

The two lively mortals still had not heard from the I.A.F., but had, nonetheless, gone ahead since they considered their educational experiments to be in the public interest. They wrote a second report which, in contrast to the first, was deemed worthy by the *Arithmetical Reviews* of a mention of its title. In this report the authors described how they made some bright youngsters in a secondary school (yes, a secondary school!) learn something until then unheard of on the island – a multiplication table:

$$1 \cdot 1 = 1, \qquad 1 \cdot 2 = 2, \qquad 1 \cdot 3 = 3, \ldots$$

$$2 \cdot 1 = 2, \qquad 2 \cdot 2 = 4, \qquad 2 \cdot 3 = 6, \ldots$$

$$3 \cdot 1 = 3, \qquad 3 \cdot 2 = 6, \qquad 3 \cdot 3 = 9, \ldots$$

. . . .

For students trained in the traditional way, on the other hand, it was only in their senior year in college that they caught a few glimpses of the advanced arithmetic of stix, such as:

$$\text{If } h| = |, \quad \text{then } h\cdot h| = |, \quad h\cdot|^{p} = {}^{p}|, \quad h\cdot|^{t} = |^{t}, \ldots$$

$$\text{If } h| = | \text{ and } k| = |^{p}, \quad \text{then } k\cdot h| = |^{p}, \quad k\cdot|^{p} = 4|, \quad k\cdot|^{t} = 6|, \ldots$$

$$\text{If, moreover, } m| = |^{t}, \quad \text{then } m\cdot h| = |^{t}, \quad m\cdot|^{p} = 6|, \quad m\cdot|^{t} = 9|, \ldots$$

Such profund results naturally could not be degraded to the level of memorizing material; nor were students expected to remember these sophisticated formulations. The two independent mortals showed Gulliver not only the new books that completely ignored the numerals 1, 2, 3, but also some texts freely using them as though 1, 2, 3 had always been current. Besides, the discoverers of the celebrated formula $| + |^{p} = |^{p} + |$, who had called *Arithmetic without Stix* trivial, had begun a reform of the insular symbolism – of course a nontrivial reform. The *Arithmetical Reviews*' report about their work was condensed in the *Insular Daily Press* under the headlines: OUR NUMERALS REVOLUTIONIZED, IMMORTALS REPLACE $|^{p}$ AND $|^{t}$ BY II AND III.

The situation thus was not without promising signs. So, after a week's stay, the traveller wrote in his diary:

"Leaving the Land with embryonic One, Two, Three. Returning to the World without Identity and Power Functions."

NOTES

[1] See the preceding chapter.

[2] The original paper contains here a review of the islanders' symbolism and Gulliver's suggestions, described in the preceding chapter.

CHAPTER 26

GULLIVER IN APPLYLAND

Before "he married Mary Burton, second daughter . . . of a hosier in Newgate Street," young Gulliver, according to Swift,[1] "learned navigation and other parts of mathematics useful to those who intend to travel." Mathematics for its own sake began to interest Gulliver, a few years later, because of his adventures (not mentioned by Swift and only recently recorded in the *Mathematical Gazette*[2]) on an island that he called Land without One, Two, Three. Upon his return to England he began to study what then – in the early 1700's – was the most advanced branch of mathematics – analysis: the theory of functions and fluents. On a later trip to the Land without One, Two, Three[3], Gulliver discovered a neighbouring isle called Applyland – a name that had originated in the preoccupation of the insular arithmeticians with scientific applications.

The Applylanders said that an object was *multiplicatively connected* with another object if the former was a multiple of the latter. If they wished to indicate that a pound was multiplicatively connected with an ounce but did not intend to specify the number establishing the said connection (i.e., the number expressing *which* multiple of an ounce was equal to a pound), then they wrote

$$(1) \qquad lb. = lb. \times oz.$$

On the other hand, they would not write

$$(2) \qquad yd. = yd. \times oz.,$$

since a yard is not multiplicatively connected with an ounce at all. In order to indicate that a yard is some multiple of a foot as well as some multiple of an inch, the Applylanders wrote

$$(3) \qquad yd. = yd. \times ft. = yd. \times in.$$

"Suppose," Gulliver wrote in his diary, "someone back home wished to express the fact that the weight w of an object is functionally connected with its mass m, but did not intend to specify the function that established the said connection. Following the example of the Applylanders he would have to write

(1′) $w = w(m)$.

Denoting the velocity of the falling object by v he would not, on the other hand, write

(2′) $v = v(m)$,

since v is not a function of m at all. In order to indicate that the velocity in ft./sec. is some function of the distance in feet travelled as well as some function of the time in seconds elapsed, such a man would write

(3′) $v = v(s) = v(t)$.

Just as the symbol *yd.* in (3), the letter v in (3′) would have three discrepant meanings: a specific fluent (the velocity in ft./sec.); a specific function (since $v = 8\sqrt{s}$, the function $8\sqrt{\ }$); and another specific function (since $v = 32\,t$, 32 times the identity function)."

A few days later, Gulliver added a postscript:

"Rereading the preceding entry I am reminded of an incident involving the older sister of my wife, Mary Burton. She is married to a man whose first name happens to be Lemuel like mine. Once, at a little gathering in Newgate Street, someone mistakenly introduced me as the husband of my sister-in-law, whereupon the latter exclaimed: 'Oh no. This Lemuel is the Lemuel of Mary Burton. And the Lemuel of this here (sic!) David,' she added pointing to my older brother David, who was sitting next to me. She thus used 'Lemuel' in three ways: as my name; for 'the husband of'; and for 'the brother of' – all this in one sentence. Her way of talking comes to my mind when I see formulae such as (3) or (3′).

Newton never would write anything of that kind."

In this last remark, Gulliver was of course right but he was too optimistic to foresee that, 200 years later, formulae such as (3′) would become rather common.

Even though Gulliver considered the matter unimportant in itself, he recalled it as a symptom of a confusion of objects with numbers connecting objects, of which he found further evidence in the Applylandish theories of the predecessors of numbers and differences between two lengths or between two amounts.

The *immediate predecessor* of a number was indicated by an asterisk after the numeral. The numerals, each of which included the symbol | (pronounced: stix), had been taken over from the Land without One, Two, Three (See Chapter 24). Hence the Applylanders wrote: $(6|)^* = 5|$, $(5|)^* = 4|$, though, on conceptual grounds, some preferred $6^*|$ to $(6|)^*$, and $5^*|$ to $(5|)^*$. There were no numerals $3|$, $2|$, $1|$; the islanders wrote $|^t$, $|^p$, $|$ (pronounced: stixtrip, stixpair, stix). So everyone wrote $(|^t)^* = |^p$ and $(|^p)^* = |$; and they said: the predecessor of stixpair is stix. The *remainder* of a length B after the subtraction of a smaller A was denoted by B – A. For example, they wrote *yd.* – *in.* and *m.* – *cm.*

To Gulliver's bewilderment, the Applylanders considered the asterisk and the minus sign, as applied to lengths, as synonymous. "Why, those symbols refer to altogether different worlds," he said. "The former belongs to the logical realm of numbers, the latter to the domain of observables. Besides, the predecessor is associated with a single number, and the difference, with a pair of observables." He put these ideas in a little paper in which he also articulately connected predecessors and differences:

If an amount or a length B is connected with an amount or a length A by a number, then B – A is connected with A by the predecessor of that number; in a formula, if $B = nA$, then $B - A = n^*A$. Thus, since

$$yd. = 36\ in. \quad \text{and} \quad m. = 100\ cm.$$

it follows that

$$yd. - in. = 36^*in. = 35\ in. \text{ and } m. - cm. = 100^*cm. = 99\ cm.$$

Even for numbers[4], since $35| = 7 \times 5|$, it follows that

$$35| - 5| = 7^* \times 5| = 30|; \quad \text{and} \quad n| - | = n^*| \text{ for any n.}$$

"It is distressing," wrote the Immortal recommending that Gulliver's paper be rejected, "to see a man who in his youth mastered useful parts of mathematics waste his time on minutiae of terminology and trivia of notation. Certainly the paper is not up to the standards of the *Applylandish Achievements*."

At the same time, Gulliver wrote in his diary:

"The Applylander's odd identification of predecessors of numbers with differences between two observables has no parallel in analysis back home. The closest analogue that I can invent is as follows: For each function, denote the derivative by a prime next to the function symbol[5] as in $(\sin x)' = \cos x$, although, on conceptual grounds, some might prefer $\sin' x$ to $(\sin x)'$. But unless a permanent symbol were introduced for the identity function[6] they would have to write $(x^5)' = 5x^4$.

The rate of change of s with respect to t (the distance travelled with respect to the time elapsed) is denoted by $\mathrm{d}s/\mathrm{d}t$. If s is connected with t by the sine function, then $\mathrm{d}s/\mathrm{d}t$ is connected with t by the derivative of the sine function; in symbols: $s = \sin t$ implies $\mathrm{d}s/\mathrm{d}t = \sin' t = \cos t$. If j were used for the identity function, one might even write: $j^{15} = j^5(j^3)$ implies $\mathrm{d}j^{15}/\mathrm{d}j^3 = j^{5\prime}(j^3) = 5j^4(j)^3$ and $\mathrm{d}f/\mathrm{d}j = f'$ for any differentiable f. But would people back home make the mistake of considering the prime, $'$, and d/d to be synonymous? Could they overlook that the derivative is defined for a single function while the rate of change is associated with a pair of fluents or functions? Could they ever confuse fluents introduced by Newton with functions introduced by Leibniz?"

And in his optimism, Gulliver concluded this entry in his diary by writing with a flourish:

"NEVER."

NOTES

[1] Swift, *A Voyage to Lilliput*, Chapter I.

[2] See Chapter 24.

[3] See Chapter 25.

[4] It will be remembered (loc. cit.[2]) that in writing 4 ft. and 4 × 5 | the islanders used only the stix-free part of the numeral 4|, just as in writing sin t and sin log x we use only the x-free part of the traditional symbol sin x for the sine function.

[5] Here, Gulliver anticipated Lagrange's famous concept of *fonction dérivée* and Lagrange's notation for the derivative.

[6] In possession of a symbol, say, j for the identity function, one would of course write simply $\sin' = \cos$ and $j^{5\prime} = 5j^4$.

KARL MENGER: PRINCIPAL DATES

1902	born in Vienna
1920–1924	studied at the University of Vienna; Ph.D. in Mathematics
1925–1927	docent at the University of Amsterdam
1927–1936	professor of geometry at the University of Vienna
1930–1931	visiting lecturer at Harvard University and The Rice Institute
1931–1937	founder of the *Ergebnisse eines Mathematischen Kolloquiums*
1937–1946	professor of mathematics at the University of Notre Dame, Indiana
1937–1946	founder of *Reports of a Mathematical Colloquium*, 2nd series, and *Notre Dame Mathematical Lectures*
1946–1971	professor of mathematics, Illinois Institute of Technology, Chicago
1951	visiting lecturer at the Sorbonne, Paris
1961	guest lecturer in several European universities
1964	visiting professor, University of Arizona and Ford Institute, Vienna
1968	visiting professor, Middle East Technical University, Ankara
1971	professor emeritus in Chicago

FIELDS OF RESEARCH

I. Theory of Curves and Dimension Theory
II. Geometry of General Metric Spaces (Convexity. Geodesics. Characterization of Euclidean Sets)
III. A General Theory of Length and the Calculus of Variations
IV. Coordinate-Free Treatment of Curvature
V. Probabilistic Metric
VI. Algebra of Geometry. Normed Lattices. Comprehensive Treatment of Diverse Mathematical Structures. Duality
VII. New Foundations for the Bolyai-Lobachevsky Geometry
VIII. Algebra of Functions. Interconnection of Algebraic Systems
IX. A New Approach to Calculus
X. Analysis of the Idea of Variables. A Theory of the Application of Analysis to Science
XI. Multiderivatives
XII. Analytic Functions
XIII. Foundations of Mathematics and Logical Tolerance
XIV. Economics: Introduction of Probability into the Theory of Value. The Logic of the Laws of Return
XV. Ethics. Formal Studies of Human Relations

KARL MENGER BIBLIOGRAPHY

This bibliography lists all writings of Karl Menger published before 1976. English translations (E. tr.) of works originally written in German, but not translations into other languages are included. Newspaper articles and book reviews are omitted.

Abbreviations of the titles of some journals:

AM	*Annals of Mathematics*
AMM	*American Mathematical Monthly*
AW	*Anzeiger, Wiener Akademie der Wissenschaften, Mathematisch-Naturwissenschaftliche Klasse*
BAMS	*Bulletin of the American Mathematical Society*
CRP	*Comptes Rendus, Académie des Sciences, Paris*
ErgMK	*Ergebnisse eines Mathematischen Kolloquiums*
FM	*Fundamenta Mathematicae*
JbDMV	*Jahresbericht der Deutschen Mathematiker-Vereinigung*
MA	*Mathematische Annalen*
Mh	*Monatshefte für Mathematik und Physik*
PA	*Proceedings, Royal Academy of Science, Amsterdam*
PUSA	*Proceedings, National Academy of Science, U.S.A.*
RepMC	*Reports of a Mathematical Colloquium*

1921–1923

1. 'Zur Dimensions- und Kurventheorie. Unveröffentlichte Aufsätze aus den Jahren 1921–1923 (Published in *Mh* **36** (1929), 411–432.).
   I. . 'Zur Theorie der Punktmengen' (Fall 1921);
   II. 'Über die Dimensionalität von Punktmengen' (January 1922);
   IIa. 'Ergänzung' ('Supplement') to 'Über die Dimensionalität von Punktmengen' (February 1922);
   III. 'Über die Dimensionalität von Punktmengen' (Fall 1922);
   IV. 'Zur Theorie der Punktmengen II' (Fall 1923).
2. 'Über die Dimensionalität von Punktmengen, I. Teil', *Mh* 33 (1923) 148–160.
3. 'Einleitung' ('Introduction') to Carl Menger, *Grundsätze der Volkswirtschaftslehre*, 2. Auflage, Wien.

1924

1 'Einige Überdeckungssätze der Punktmengenlehre', *Sitzungsber. Akad. d. Wissensch. Wien, Math.—Naturw. Kl.* 133, 421–444.

2. 'Über die Dimension von Punktmengen', *PA* 27, 639–643.
3. 'Über die Dimension von Punktmengen, II. Teil', *Mh* 34, 137–161.

1925

1. 'Grundzüge einer Theorie der Kurven', *PA* 28, 67–71.
2. 'Grundzüge einer Theorie der Kurven', *MA* 95, 277–306.

1926

1. Zur allgemeinen Kurventheorie', *FM* 10, 96–115.
2. 'Das Hauptproblem über die dimensionelle Struktur der Räume', *PA* 30, 138–144.
3. 'Allgemeine Räume und Cartesische Räume I, II', *PA* 29, 476–483 and 1125–1128.
4. 'Bericht über die Dimensionstheorie', *JbDMV* 35, 113–150.
5. 'Über geodätische Linien in allgemeinen metrischen Räumen', *PA* 29, 166–169.
6. 'Über reguläre Baumkurven', *MA* 96, 572–582.
7. 'Zur Entstehung meiner Arbeiten über Dimensions- und Kurventheorie', *PA* 19, 1122–1124.

1927

1. 'Bemerkungen zur zweiten Untersuchung über allgemeine Metrik', *PA* 30, 710–714 (See 1928, 2, II).
2. 'Die Haupttheoreme der Dimensionstheorie', *JbDMV* 36, 8–12.
3. 'Zusammenhangsstufen und Cantorsche Mannigfaltigkeiten', *PA* 30, 705–709.

1928

1. 'Bemerkungen zu Grundlagenfragen, I–IV', *JbDMV* 37:
   I. 'Über Verzweigungsmengen', 213–226;
   II. 'Die mengentheoretischen Paradoxien', 298–302;
   III. 'Über Potenzmengen', 303–308;
   IV. 'Axiomatik der endlichen Mengen und der elementargeometrischen Verknüpfungsbeziehungen', 309–325.
2. 'Untersuchungen über allgemeine Metrik, I–III', *MA* 100:
   I. 'Theorie der Konvexität', 75–113;
   II. 'Die Euklidische Metrik', 113–141;
   III. 'Eine n-dimensionale Metrik', 142–163.

2a. Abstracts of 1928, 2 in *AW* 65, 14–17 and in *ErgMK* 1, 20–23.

3. 'Konvexitätstheoretische Notizen: Über konvexe Hüllen; Über Vollkonvexität; Über Verallgemeinerungen des Zwischenbegriffes; Über den Begriff der Konkavität', *AW* 65, 154–156 and *ErgMK* 1, 23–26.
4. 'Die Metrik des Hilbertschen Raumes', *AW* 65, 159–160 and *ErgMK* 1, 26–27.
5. 'Die Halbstetigkeit der Bogenlänge', *AW* 65, 278–281.
6. 'Ein Theorem der Topologie', *AW* 65, 4 and *ErgMK* 1, 2–3.
7. 'Dimensionstheoretische Notizen: Ein Satz zur Ergänzung der Dimensions-definition; Über schwachdimensionale Mengen; Über n-dimensionale Gewächse', *AW* 65, 5–8 and 11–12, and *ErgMK* 1, 6–8.
8. 'Dimension und Zusammenhangsstufe' (with **W. Hurewicz**), *MA* 100, 618–633.
9. *Dimensionstheorie*, Leipzig & Berlin, iv + 319 pp.

1929

1. 'Über die Dimension von Punktmengen, III. Teil. Zur Begründung einer axiomatischen Theorie der Dimension', *Mh* 36, 193–218; abstr. in *AW* 66, 148–149.
2. 'Zur Frage nach der Herleitung des Dimensionsbegriffes aus Forderungen', *AW* 66, 87–90.
3. 'Allgemeine Räume und Cartesische Räume III. Beweis des Fundamentalsatzes', *PA* 32, 330–340.
4. 'Dimensionstheoretische Notizen: Ein Zerlegungssatz für rational-dimensionale Mengen; Über die nirgends dichten Teilmengen des R_n; Über die Summe regulärer Kurven', *AW* 66, 84–87 and *ErgMK* 1, 6–8.
5. 'Über eine neue Definition der Bogenlänge. Weitere Verallgemeinerung des Längenbegriffes', *AW* 66, 23–25.
6. 'Ein allgemeiner Trennungssatz', *AW* 65, 266–288 and *ErgMK* 1, 16–17.
7. 'Die allgemeine Kurventheorie', *Forschungen und Fortschritte*, 5, No. 20, 232–233.

1930

1. 'Axiomatische Einführung des Dimensionsbegriffes', in *Comptes Rendus du Premier Congrès des Mathématiciens des Pays Slaves, Warszawa* 1929, pp. 57–65.
2. 'Untersuchungen über allgemeine Metrik. IV. Zur Metrik der Kurven', *MA* 105, 466–501.
3. 'Entwurf einer neuen Theorie des Masses', *ErgMK* 2, 6–10 and *AW* 67, 12–16.
4. 'Remarques sur la Théorie Axiomatique de la Dimension' (with **C. Kuratowski**), *Mh* 37, 169–174.
5. 'Bemerkungen über dimensionelle Feinstruktur und Produktsatz', *Prac Mat.-Fizycz.* 37, 77–90.
6. 'Otto Schreier', *Mh* 37, 1–6.
7. 'Eine dimensionstheoretische Bemerkung von O. Schreier', *Mh* 37, 7–12; abstr. in *ErgMK* 2, 1.
8. 'Antwort auf eine Note von Brouwer', *Mh* 37, 175–182.
9. 'Über den n-Beinsatz in lokal-zusammenhängenden Kontinua' (with **G. Nöbeling**), *AW* 67, 86–88.
10. 'Eine neue Definition der Bogenlänge', *ErgMK* 2, 11–12.
11. 'Das Botenproblem', *ibid.*, p. 12.
12. 'Bemerkungen von Mazurkiewicz und Tarski über die Axiomatik des Dimensionsbegriffes', *ibid.*, p. 16.
13. 'Über Mengensysteme', *ibid.*, pp. 16–17.
14. 'Probleme der allgemeinen metrischen Geometrie', *ibid.*, pp. 20–22.
15. 'Bericht über die mengentheoretischen Überdeckungssätze', *ibid.*, pp. 23–27.
16. 'Über plättbare Dreiergraphen und Potenzen nichtplättbarer Graphen', *ibid.*, pp. 30–31 and *AW* 67, 30–31.
17. 'Einführung des Komplexen in die allgemeine Metrik und metrische Untersuchungen in abstrakten Gruppen', *ibid.*, pp. 34–38 and *AW* 67, 39–44.
18. 'Über die sogenannte Konstruktivität bei arithmetischen Definitionen', *AW* 67, 257–258.
19. 'Der Intuitionismus', *Blätter für Deutsche Philosophie* 4, 311–325.

1931

1. 'New Foundations of Euclidian Geometry', *Amer. Jl. of Math.* 53, 721–745.
2. 'Beiträge zur Gruppentheorie I. Über einen Abstand in Gruppen', *Math. Zeitschrift* 33, 396–418.
3. 'Bericht über metrische Geometrie', *JbDMV* 40, 201–219.
4. 'Some Applications of Point-Set Methods', *AM* 32, 739–760.
5. 'Remarks Concerning the Paper of W. L. Ayres on the Regular Points of a Continuum', *Transactions Am. Math. Soc.* 33, 663–667.
6. 'Ein Problem von Blaschke', *AW* 68, 75–76.
7. 'Eine neue Kennzeichnung der Geraden', *ibid.*, pp. 103–105.
8. 'Bericht über ein mathematisches Kolloquium 1929/30', *Mh* 38, 17–38.
9. 'Über den Konstruktivitätsbegriff (Zweite Mitteilung)', *AW* 68, 7–9.
10. 'Eine elementare Bemerkung über die Struktur logischer Formeln', *ErgMK* 3, 22–23.

1932

1. *Kurventheorie*, edited in collaboration with **G. Nöbeling**, Leipzig, iv + 376 pp.; repr. New York 1967.
2. 'Neuere Methoden und Probleme der Geometrie', in *Verhandlungen des internationalen Mathematiker-Kongresses, Zürich* 1932, I, pp. 310–323.
3. 'Über die Hinweise auf Brouwer in Urysohns Mémoire', Wien (published by the author).
4. 'Eine Zuschrift von Karl Menger an Hans Hahn betreffend Mengers Antwort auf eine Note von Brouwer in Band 37 der *Monatshefte*', *Mh* 40, 233.
5. 'Über die lokale Dimension von Mengensummen', *ErgMK* 4, 1–2.
6. 'Eine NZ-Kurve', *ibid.*, p.7.
7. 'Über eine Limesklasse', *ibid.*, p. 9.
8. 'Diskussion über die Verzweigungsordnung von Flächenpunkten', *ibid.*, pp. 11–12.
9. 'Zur mengentheoretischen Behandlung des Tangentenbegriffes und verwandter Begriffe', *ibid.*, p. 23.
10. 'Ein Überdeckungssatz für F_σ', *ibid.*, pp. 40–41.
11. 'Eine neue Kennzeichnung der Geraden', *ibid.*, pp. 41–43.

1933

1. 'Die neue Logik', in *Krise und Neuaufbau in den Exakten Wissenschaften. Fünf Wiener Vorträge*, Leipzig & Wien, pp. 94–122 (E. tr. 1937, 6).

1a. 'Vorwort' ('Preface', unsigned) to *Krise und Neuaufbau*, pp. i–ii.

2. 'Über den imaginären euklidischen Raum', *Tohoku Math. Jl.* 37, 475–478.
3. 'Zur Begründung einer Theorie der Bogenlänge in Gruppen', *ErgMK* 5, 1–6.
4. 'Eine Bemerkung über die Potenzen schwach eindimensionaler Mengen', *ibid.*, pp. 9–10.
5. 'Rein imaginäre Räume, indefinite Metriken und verwandte Probleme', *ibid.*, pp. 16–17.
6. 'Neuer Aufbau der Vektoralgebra', *ibid.*, pp. 27–29.
7. 'Diskussion über koordinatenlose Differentialgeometrie' (with **K. Gödel** and **A. Wald**), *ibid.*, pp. 25–26.

1934

1. 'Ist die Quadratur des Kreises lösbar?', in *Alte Probleme – Neue Lösungen in den exakten Wissenschaften. Fünf Wiener Vorträge, Zweiter Zyklus*, Leipzig & Wien, pp. 1–28.

1a. 'Nachwort' ('Postscript', signed K.M.) to *Alte Probleme – Neue Lösungen*, pp. 117–122.

2. 'Bericht über neueste Ergebnisse der metrischen Geometrie', in *Comptes Rendus du Deuxième Congrès des Mathématiciens des Pays Slaves, Praha* 1934, pp. 116–117.

3. 'Eine Bemerkung über Längenmengen', *ErgMK* 6, 1–2.

4. Remarks in the Discussion of K. Schlesinger's Modification of the Walras – Cassel Production Equations and A. Wald's Solution of the Modified Equations, *ibid.*, pp. 18–20.

5. 'Ein Satz über endliche Mengen mit Anwendungen auf die formale Ethik', *ibid.*, pp. 23–26.

6. 'Bernoullische Wertlehre und Petersburger Spiel', *ibid.*, pp. 26–27.

7. 'Hans Hahn', *ibid.*, pp. 40–44.

8. 'Das Unsicherheitsmoment in der Wertlehre. Betrachtungen in Anschluss an das sogenannte Petersburger Spiel', *Zeitschrift für Nationalökonomie* 5, 459–485 (E. tr. 1967, 1).

8a. 'Bemerkungen' ('Remarks') to 'Das Unsicherheitsmoment in der Wertlehre', *Zeitschrift für Nationalökonomie* 6, 283–285 (E. tr. 1967, 1).

9. *Moral, Wille und Weltgestaltung*, Wien, iv + 143 pp.

1935

1. 'Hans Hahn', *FM* 24, 317–320.

2. 'Metrische Geometrie und Variationsrechnung', *FM* 25, 441–458.

3. 'Sull' Indirizzo di Idee e sulle Tendenze Principali del Colloquio Matematico di Vienna', *Ann. Scuola Norm. Sup., Pisa* 4, 1–13.

4. 'Sur un Théorème Général du Calcul des Variations', *CRP* 201, 705–707.

5. 'Algebra der Geometrie. Zur Axiomatik der projektiven Verknüpfungsbeziehungen', *ErgMK* 7, 11–12.

6. 'Über die φ-Metrik und φ-Bogenlänge', *ibid.*, pp. 13–14.

7. 'Programmatisches zur Anwendung der metrischen Geometrie auf die Variationsrechnung', *ibid.*, pp. 40–51.

1936

1. 'New Foundations of Projective and Affine Geometry. Algebra of Geometry' (in collaboration with **F. Alt** and **O. Schreiber**), *AM* 37, 456–482.

2. 'Calcul des Variations dans les Espaces Distanciés Généraux', *CRP* 222, 1007–1009.

3. 'Courbes Minimisantes Non-Rectifiables et Champs Généraux des Courbes Admissibles dans le Calcul de Variations', *ibid.*, pp. 1648–1650.

4. 'Metric Methods in Calculus of Variations', in *Congrès International des Mathématiciens, Oslo* 1936, II, pp. 45–46.

5. 'New Ways in Differential Geometry', *ibid.*, pp. 171–173.

6. 'Bemerkungen zu den Ertragsgesetzen', *Zeitschrift für Nationalökonomie* 7, 25–26 (E. tr. 1954, 3).
6a. 'Weitere Bemerkungen zu den Ertragsgesetzen', *ibid.*, pp. 388–397 (E. tr. 1954, 3).
7. 'Einige neuere Fortschritte in der exakten Behandlung sozialwissenschaftlicher Probleme', in *Neuere Fortschritte in den exakten Wissenschaften. Fünf Wiener Vorträge. Dritter Zyklus*, Leipzig & Wien, pp. 103–132.
8. 'Vorwort' ('Preface') to Friedrich Waismann, *Einführung in das mathematische Denken. Die Begriffsbildung der modernen Mathematik*, Wien, pp. iii–iv. (E. tr. *Introduction to Mathematical thinking* by T. J. Benach, New York, 1951).

1937

1. 'Metric Methods in Calculus of Variations', *PUSA* 23, 224–250.
2. 'What Is the Calculus of Variations and What Are Its Applications?', *Scientific Monthly* 44, 250–253; repr. in *The World of Mathematics* II, edited by J. R. Newman, New York 1956, pp. 886–890.
3. 'A Note on a Previous Paper "New Foundations of Projective and Affine Geometry" ' with **F. Alt**. (See 1936, 1), *AM* 38, 450.
4. 'La Géométrie des Distances at Ses Relations avec les Autres Branches des Mathématiques', *L'Enseignement Mathématique* 35, 348–372.
5. 'An Exact Theory of Social Relations and Groups', in *Report of Third Annual Research Conference on Economics and Statistics, Cowles Commission forResearch in Economics, Colorado Springs*, pp. 71–73.
6. 'The New Logic', *Philosophy of Science* 4, 299–336 (E. tr., with revisions, of 1933, 1 by H. B. Gottlieb and J. K. Senior).

1938

1. 'Non-Euclidian Geometry of Joining and Intersecting', *BAMS* 44, 821–824.
2. 'Axiomatique Simplifiée de l'Algèbre de la Géométrie Projective', *CRP* 206, 306–308.
3. 'A Foundation of Projective Geometry', *Proceedings, Indiana Academy of Science* 47, 189–191.
4. 'A Symposium on the Algebra of Geometry and Related Subjects', *Science* 87, 324.
5. 'An Abstract Form of the Covering Theorems of Topology', *AM* 39, 794–803.
6. 'New Foundation of Non-Euclidian, Affine, Real Projective and Euclidian Geometry', *PUSA* 24, 486–490.
7. 'Nouvelle Base pour le Développement de la Géométrie de Bolyai et Lobatchefski', *CRP* 206, 458–460.
8. 'Une Forme Abstraite du Théorème de Borel-Lebesgue Généralisé', *ibid.*, pp. 563–565.
9. 'Introduction' to L. M. Blumenthal, *Distance Geometries*, University of Missouri Studies 13, Columbia, Missouri, pp. 7–10.
10. 'An Exact Theory of Social Groups and Relations', *American Journal of Socioiogy* 43, 790–798.

1939

1. 'A Logic of the Doubtful. An Optative and Imperative Logic', *RepMC* 1, 53–64.
2. 'A Theory of Length and Its Applications to the Calculus of Variations', *PUSA* 25, 474–478.
3. 'On Cauchy's Integral Theorem in the Real Plane', *ibid.*, pp. 621–625.
4. 'On Necessary and on Sufficient Conditions in Elementary Mathematics', *School, Science and Mathematics*, 631–642.
5. 'On Linear Sets in Metric Spaces' (with **A. N. Milgram**), *RepMC* 1, 16–17.

1940

1. 'Three Lectures on Mathematical Subjects', *The Rice Institute Pamphlet* 27:
   I. 'Analysis and Metric Geometry', 1–40
   II. 'On Algebra of Geometry and Recent Progress in Non-Euclidian Geometry', 41–79;
   III. 'Topology without Points', 79–107.
2. 'On Shortest Polygonal Approximations to a Curve', *RepMC* 2, 33–38.
3. 'On Green's Formula', *PUSA* 26, 660–664.
4. 'Redundancies in the Classical Treatment of the Cauchy-Riemann Conditions', *RepMC* 2, 45–48.

1942

1. 'Statistical Metrics', *PUSA* 28, 535–537.
2. 'Projective Generalization of Metric Geometry' (abstract), *BAMS* 48, 833.

1943

1. 'What Is Dimension?' *AMM* 50, 2–7.
2. 'Differential Equations', in *Practical Mathematics*, edited by R. S. Kimball, New York, Issue 9, vol. 8, pp. 513–540.

1944

1. *Algebra of Analysis*, Notre Dame Mathematical Lectures 3, Notre Dame, Indiana, 50 pp.
2. 'Tri-Operational Algebra', *RepMC* 5/6, 3–10.
3. 'Projective Generalizations of Metric Geometry', *ibid.*, pp. 60–76.
4. 'On the Teaching of Differential Equations', *AMM* 51, 392–395.
5. 'Why Study Mathematics?', *Al-Geo-Trig* (Central Catholic High School, Toledo, Ohio) 7, 5–6.
6. 'On the Relation between the Calculus of Probability and Statistics', in Notre Dame Mathematical Lectures 4, Notre Dame, Indiana, pp. 44–53.

1945

1. 'Définition Intrinsèque de la Notion de Chemin', *CRP* 221, 739–741.
2. 'Methods of Presenting e and π', *AMM* 52, 28–33.

1946

1. 'New Projective Definitions of the Concepts of Hyperbolic Geometry', *RepMC* 7, 20–28.
2. 'General Algebra of Analysis', *ibid.*, pp. 46–60.
3. 'Analysis without Variables', *Jl. Symbol. Log.* 11, 30–31.

1948

1. 'The Topology of the Triangle Inequality', *Revista de Ciencias* 50, 155–156.
2. 'Stieltjes Integrals Considered as Lengths', *Annales, Société Polonaise Math.* 21, 173–175.
3. 'Fundamental and Applied Research in Geometry', *Illinois Tech Engineer* 13, 13–14 and 38–46.
4. 'Independent Self-Dual Postulates in Projective Geometry', *RepMC* 8, 81–87.
5. 'Aristotelian Demonstration and Postulational Method' (with **Y. R. Simon**), *The Modern Schoolman*, 25, 190–192.

1949

1. 'What Paths Have Length?', *FM* 36, 109–118.
2. 'Modern Geometry and the Theory of Relativity', in *Albert Einstein: Philosopher–Scientist*, The Library of Living Philosophers VII, edited by P. A. Schilpp, Evanston, Illinois, pp. 459–474.
3. 'Generalized Vector Spaces I. The Structure of Finite-Dimensional Spaces', *Canadian Jl. of Math.* 1, 94–104.
4. 'Self-Dual Fragments of the Plane', *AMM* 56, 545–546.
5. 'La Géométrie Axiomatique de l'Espace Projectif', *CRP* 228, 1273–1274.

1950

1. 'The Projective Space', *Duke Math. Jl.* 17, 1–14.
2. 'The Mathematics of Elementary Thermodynamics', *American Jl. of Physics* 18, 89–103.

1951

1. 'Probabilistic Theories of Relations', *PUSA* 37, 178–180.
2. 'Ensembles Flous et Fonctions Aléatoires', *CRP* 232, 2001–2003.
3. 'Espaces Vectoriels Généraux, Topologies Triangulaires, Transformations Linéaires Généralisées, *CRP* 232, 2176–2178.
4. 'Probabilistic Geometry', *PUSA* 37, 226–229.

1952

1. 'The Formative Years of Abraham Wald and His Work in Geometry', *Annals of Math. Statistics* 23, 14–20.

2. 'A Topological Characterization of the Length of Paths', *PUSA* 38, 66–69.
3. 'Une Théorie Axiomatique Générale des Déterminants', *CRP* 234, 1941–1943.
4. *You Will Like Geometry. A Guide Book for the Ill. Inst. of Tech. Geometry Exhibition at the Museum of Science and Industry*, Chicago, 36pp.
5. *Calculus. A Modern Approach* (mimeographed), Chicago, xxv + 255 pp.

1953

1. 'On the Design of Grouping Problems and Related Intelligence Tests', *Jl. of Educational Psychology* 44, 275–287.
2. 'The Ideas of Variable and Function', *PUSA* 39, 956–961.
3. *Calculus. A Modern Approach* (mimeographed), 2nd enlarged edition, Chicago, xxiii + 304 pp.

1954

1. 'Variables de Diverses Natures', *Bulletin Sciences Math.* 78, 229–234.
2. 'Is Calculus a Perfect Tool?', *Jl. of Engineering Education* 45, 261–264.
3. 'The Logic of the Laws of Return. A Study in Meta-Economics', in *Economic Activity Analysis*, edited by O. Morgenstern, New York, pp. 419–481 (E. tr. of 1936, 6 and 6a by O. Eckstein).
4. 'Tossing a Coin', *AMM* 61, 634–636.
5. *Géométrie Générale*, Mémorial d. Sci. Math. 124, Paris, 81 pp.
6. 'On Variables in Mathematics and in Natural Science', *British Jl. Philosophy of Science* 5, 134–142.
7. 'A Simple Definition of Analytic Functions and General Multifunctions', *PUSA* 40, 819–821.
8. 'Remarks to L. Sommer's Translation of Bernoulli's "Exposition of a New Theory on the Measurement of Risk" ', *Econometrica* 22, pp. 28, 31–32 and 34.

1955

1. 'Random Variables from the Point of View of a General Theory of Variables', in *Proceedings, Third Berkeley Symposium on Math. Statistics and Probability* II, Berkeley & Los Angeles, pp. 215–229.
2. 'The Behavior of a Complex Function at Infinity', *PUSA* 41, 512–513.
3. 'Generalized Derivatives and Expansions' (with **S. S. Shü**), *ibid.*, pp. 591–595.
4. *Calculus. A Modern Approach*, Boston, xviii + 354 pp.

1956

1. 'On the Formulation of Certain Arithmetical Questions' (with **H. J. Curtis**), *The Mathematics Teacher* 49, 528–530.
2. 'Calculus 1950 – Geometry 1880', *Scripta Mathematica* 22, 89–96 and 203–206.
3. 'What Are Variables and Constants?', *Science* 123, 547–548.
4. 'Why Johnny Hates Math', *The Mathematics Teacher* 49, 578–584.
5. 'What Are *x* and *y*?', *Math. Gazette* 40, 246–255.

1957

1. *The Basic Concepts of Mathematics. A Companion to Current Textbooks on Algebra and Analytical Geometry. Part I. Algebra*, Chicago, vii + 93 pp.
2. 'Multiderivatives and Multi-Integrals', *AMM* 64, 58–70; *Supplement to L. R. Ford on His Seventieth Birthday* (Sixth H. E. Slaught Memorial Paper).

1958

1. 'Optimal Differences in Computing Probable Derivatives', *Journal de Mathématiques Pures et Appliquées* 38, 245–252.
2. 'Rates of Change and Derivatives', *FM* 46, 89–102.
3. 'Is w a Function of u?', *Colloquium Mathematicum* 6, 41–47.
4. 'New Approach to Teaching Intermediate Mathematics', *Science* 127, 1320–1323.

1959

1. 'An Axiomatic Theory of Functions and Fluents', in *Symposium on the Axiomatic Method*, edited by L. Henkin, P. Suppes and A. Tarski, Amsterdam, pp. 454–473.
2. 'Gulliver in the Land without One, Two, Three', *Math. Gazette* 43, 241–250.
3. 'On Probabilistic Metrics and Numerical Metrics with Probability 1' (with **B. Schweizer** and **A. Sklar**), *Czechoslovak Math. Jl.* 9, 459–466.
4. 'Mensuration and Other Mathematical Connections of Observable Material', in *Measurement, Definitions and Theories*, edited by C. W. Churchman and P. Ratoosh, New York, pp. 97–128.

1960

1. 'Gulliver's Return to the Land without One, Two, Three', *AMM* 67, 641–648.
2. 'Introduction' to Ernst Mach, *The Science of Mechanics*, 6th American edition, La Salle, Ill, pp. iii–xxi.
3. 'Gulliver in Applyland', *Eureka*, No. 23, 5–8.
4. 'A Counterpart of Occam's Razor in Pure and Applied Mathematics. Ontological Uses', *Synthese* 12, 415–428.

1961

1. 'A Counterpart of Occam's Razor in Pure and Applied Mathematics. Semantic Uses', *Synthese* 13, 331–349.
2. 'What Is Length?' (with **F. Henmueller**), *Philosophy of Science* 28, 172–177.
3. 'The Algebra of Functions: Past, Present, Future', *Rendiconti di Matematica* 20, 409–430.
4. 'Frammenti Piani Autoduali e Relative Sostituzioni', *Rendiconti Acc. Naz. di Lincei* 30, 713–717.
5. 'Variables, Constants, Fluents', in *Current Issues in the Philosophy of Science*, edited by H. Feigl and G. Maxwell, New York, pp. 304–313 and 316–318.
6. 'Eine Algebra der Funktionen', *Nachr. Österr. Math. Ges.* 15 No. 68/69, 99–100.

1962

1. 'A Group in the Substitutive Algebra of the Calculus of Propositions', *Archiv. d. Math.* 13, 471–478.
2. 'On Compositive Functions of Matrices', *Annali di Matematica Pura e Applicata* 58, 69–84.
3. 'Function Algebra and Propositional Calculus', in *Self-Organizing Systems*, edited by M. C. Yovits, G. T. Jacobi and G. D. Goldstein, Washington, D.C., pp. 525–532.

1963

1. 'A Self-Dual Theory of Real Determinants' (with **F. Kozin**), *Publ. Math. Debrecen* 10, 123–127.
2. 'Postulates for the Substitutive Algebra of the 2-Place Functors in the 2-Valued Calculus of Propositions' (with **M. Schultz**), *Notre Dame Jl. on Formal Logic* 4, 188–192.
3. 'Calculus of Variations', in *Harper Encylopedia of Science* I, New York & Evanston, pp. 195–196.

1964

1. 'On Substitutive Algebra and Its Syntax', *Zeitschr. f. math. Logik und Grundlagen d. Math.* 10, 81–104.
2. 'Superassociative Systems and Logical Functions', *MA* 157, 278–295.

1965

1. 'A Characterization of Weierstrass Analytic Functions', *PUSA* 54, 1025–1026.
2. 'Analytische Funktionen', *Wissensch. Abh. Forschungen d. Landes Nordrhein-Westfalen* 33, 609–612.
3. 'Une Caractérisation des Fonctions Analytiques', *CRP* 261, 4968–4969.

1966

1. 'Weierstrass Analytic Functions', *MA* 167, 177–194.
2. 'Two Theorems on the Generation of Systems of Functions' (with **H. I. Whitlock**), *FM* 50, 229–240.

1967

1. 'The Role of Uncertainty in Economics', in *Essays in Mathematical Economics in Honor of O. Morgenstern*, edited by M. Shubik, Princeton, pp. 211–231 (E. tr. of 1934, 8 and 8a by W. Schoellenkopf and W. G. Mellon).

1968

1. 'General Algebraic Equations and Functional Equations', *Equationes Mathematicae* 1, 281.

1970

1. 'Introduction', 'Projective and Related Structures' (Part 2) and 'Curve Theory' (Part 4), in *Studies in Geometry* (with **L. M. Blumenthal**), San Francisco, pp. iv–vii, 135–223 and 391–506.
2. 'Mathematical Implications of Mach's Ideas: Positivistic Geometry. The Clarification of Functional Connections', in *Ernst Mach. Physicist and Philosopher*, Boston Studies in the Philosophy of Science VI, edited by R. S. Cohen and R. J. Seeger, Dordrecht, pp. 107–125.

1971

1. 'The New Foundation of Hyperbolic Geometry', in *A Spectrum of Mathematics*, edited by J. C. Butcher, Auckland, pp. 86–97.
2. 'The Geometry Relevant to Modern Education', in *Educational Studies in Mathematics* IV, edited by H. Freudenthal, Dordrecht, pp. 1–17. Reprinted in *The Teaching of Geometry at the Pre – College Level*, edited by H. G. Steiner, Dordrecht, pp. 225–241.

1972

1. 'Bericht über metrische Geometrie', in *Geometrie*, Wege der Forschung 177, edited by K. Strubecker, Darmstadt, pp. 371–395.
2. 'Österreichischer Marginalismus und mathematische Ökonomie', *Zeitschrift f. Nationalökonomie* 32, 19–28.

1973

1. 'Austrian Marginalism and Mathematical Economics', in *Carl Menger and the Austrian School of Economics*, edited by J. R. Hicks and W. Weber, Oxford, pp. 38–60 (Expanded E. tr. of 1972, 2 by the author).

1974

1. *Morality, Decision and Social Organization. Toward a Logic of Ethics*, Vienna Circle Collection VI, pp. vi and 118 edited by Henk. L. Mulder, D. Reidel, Dordrecht Boston, 115 pp. (E. tr. with minor revisions of [1934, 9] by the author based on work by E. van der Schalie, and a 1974 postscript pp. 93–115).

INDEX OF NAMES

Abbott, Dr., 133
Ackermann, W., 45, 66
Adams, E. W., 144, 152
d'Alembert, J. L. R., 275
Alexandroff, P. S., 233f., 241f.
Alkhowarizmi, 108, 135
Alt, F. L., 17, 224, 291, 293, 297
Archimedes, 43, 167
Aristotle, 18ff.
Arrow, K., 260
Artin, E., 134

Baire, R., 49
Beethoven, L. v., 242
Béguelin, 275
Bellman, R., 232, 234
Bergmann, G., 4, 221, 223
Bergson, H., 36, 46, 52, 217
Bernays, P., 45
Bernoulli, D., 5f., 259, 264, 274ff.
Bernoulli, N., 275
Birkhoff, Garrett, 4, 220f.
Birkhoff, G. D., 199
Böhm-Bawerk, E. v., 280f., 284, 293, 295f., 301
Bohr, N., 234
Bolzano, B., 248, 251
Bolyai, J., 219, 225
Boole, G., 4, 20, 44
Borel, E., 39, 48f., 53, 56, 58
Braithwaite, R. B., 259
Bridgman, P. W., 17, 512
Brinkmann, 301
Brouwer, L. E. J., 2, 5, 7, 13, 36S., 39f., 50–56, 79, 83ff., 86f., 238–255
Burali-Forti, C., 27, 89
Buffon, G. L. L. de, 275, 277
Bush, R. R., 229

Cantor, G., 24f., 48, 88, 210, 213–216, 231, 237
Carnap, R., 1, 4, 11–16, 45, 61, 134, 260
Cauchy, A., 172
Chuquet, N., 131f.
Church, A., 45
Courant, R., 134
Cramer, G., 5, 264, 275f.
Curtis, H. J., 3, 185, 219
Czuber, E., 275

Danzig, D. van, 243
Dedekind, R. 000
Descartes, R., 110–118, 127f., 134, 154, 213, 314
Deuticke, F., 17
Diophantes, 167

Edgeworth, F. Y., 301
Einstein, A., 203, 243
Ekstein, H., 143
Essler, J., 301
Euclid, 4, 199, 220, 225f., 228
Euler, L., 33f., 43, 167

Fechner, G. T., 276
Feigl, H., 14
Fermat, P. de, 33ff., 213
Ford, L. R., 207
Fraenkel, A., 2, 45f., 246
Franklin, Ladd, 44
Fréchet, M., 214, 241
Frege, G., 22f., 44, 71, 144, 157
Fried, B. D., 161
Fries, J. F., 275
Frink, O., 184

Galanter, E., 229
Galileo, 110f., 114f., 196, 202
Gauss, C. F., 167, 225, 244, 249
Gentzen, G., 33, 45
Gerneth, D. C., 67
Gödel, K., 1, 13, 17, 32–35, 39, 45, 54

Hadamard, J., 148
Hahn, H., 12, 14f., 17, 89, 237, 240, 249, 251f.
Hardy, G. H., 252
Hausdorff, F., 79, 86f., 214, 242, 246, 251
Heisenberg, W., 226
Herbart, J. F., 217
Herbrand, J., 30, 45, 52
Heyting, A., 38, 45, 86, 246, 253
Hilbert, D., 13, 29f., 35, 45, 54f., 216, 249, 253, 280
Huntington, E. V., 215
Hurewicz, W., 216, 245f., 248, 250, 253
Husserl, E. 12, 46, 215

James, R. D., 303
Janiszewski, Z., 251
Jaśkowski, St., 66
Johnson, L. W., 7
Jordan, C., 237, 252

Kalaba, R., 232, 234
Kant, I., 19, 36, 225
Kaufmann, F., 12, 15, 215
Kepler, J., 131
Kerékjártó, B. v., 249, 252
Keynes, J. M., 220
Kittel, Ph., 7, 260
Klaeb, P., 301
Klein, Felix, 252
Klein, Fritz, 221
Kolmogoroff, A. N., 117
König, D., 253
Kowalski, R., 7, 46
Kraft, V., 11f.
Kronecker, L., 35ff., 47f., 50f., 54, 240
Kuhn, H. W., 259

Lagrange, J. L., 172, 323
Lange, F. A., 276
Langford, C. H., 45
Laplace, P., 276
La Vallée Poussin, C. de, 134, 175
Lebesgue, H., 39, 48f., 243
Leibniz, G. W. v., 19, 24, 167, 172, 191f., 314, 323
Lewis, C. I., 12, 31, 45
Lichtenberg, G. C., 275
Liouville, J., 34
Lobachevsky, N. I., 219, 225
Loor, B. de, 51
Luce, R. D., 229
Łukasiewicz, J., 1, 22, 30, 45, 61f., 66, 71, 77, 91
Lusin, N., 39, 49, 85

Mach, E., 2, 136, 202, 225f., 229, 240, 260
Mally, E., 96ff.
Mark, H., 17
Masereel, S., 243
Marshak, J., 260
Mayer, H., 259
McCulloch, W. S., 133
McKinsey, J. C. C., 134
Mehlberg, J., 135, 139, 143
Milgram, A. N., 228
Minkowski, H., 4, 217
Mises, L. v., 259, 279, 293, 295
Möbius, F., 200
Molière, 105, 107, 133
Molk, J., 58
Montmort, P. R. de, 275
Moore, R. L., 228
Morgenstern, O , 259f., 301
Mulder, H., 7
Musset, A. de, 216

Neumann, J. v., 45, 220f., 224, 259f.
Neurath, O., 12ff.
Newton, I., 106, 114, 134, 139, 149, 169, 314, 321, 323
Nicod, J. G., 215, 228
Nöbeling, G., 15, 17
Noether, E., 241

Occam, William of, 2f., 7, 105f., 133, 162
Oettinger, 276f.
Osgood, W. F., 137

Pasch, M., 216
Peano, G., 24, 33, 210, 237
Peirce, C. S., 20f., 23, 71

Pincherle, S., 133
Poincaré, H., 5, 13, 29, 31, 36, 39, 41f., 47f., 50f., 55, 167, 227–230, 239f., 243, 248, 251, 259
Polya, G., 184
Popper, K., 11, 14ff.
Post, E., 30, 91
Pringsheim, A., 58, 275
Pythagoras, 43, 142, 167, 217

Quine, W. V., 162

Ramsey, F. P., 27, 259f.
Reichenbach, H., 45
Reidemeister, K., 245
Riemann, B., 225, 234, 249
Russell, B., 2, 13, 24–28, 35, 44ff., 71, 90, 109, 116, 135, 177–180, 183, 217

Samuelson, P., 260
Sarten, G., 163
Schlesinger, K., 292, 298, 301
Schlick, M., 3, 12, 14f., 254, 260, 278
Schlick Kreis: See Vienna Circle.
Schreier, O., 240, 252
Schreiber, O., 224
Schröder, E., 20f., 44
Schrödinger, E., 232, 234
Schweizer, B., 7, 77, 229, 234
Senior, J. K., 18, 198
Šerstnov, A. N., 229, 234
Sheffer, H. M., 2, 71
Sierpiński, W., 49, 51, 58, 86
Skinner, B. F., 173
Sklar, A., 7, 77, 229, 234
Skolem, Th., 45, 56
Spearman, Ch., 198
Steenrod, N. E., 161
Stone, M. H., 95, 228
Strigl, R. v., 301
Study, E., 116, 134
Sugar, A. ,134
Suppes, P., 134
Suslin, M., 49, 86
Swift, J., 6, 315, 320

Tarski, A., 27, 45, 92, 138
Thirring, H., 17, 249
Timerding, 274, 277
Tintner, G., 260
Townsend-Johnson, P., 7
Tucker, A. W., 259

Urysohn, P., 214, 216, 240f., 245, 247f., 250, 252, 255

Van Danzig: See Danzig
Vienna Circle, 1f., 4, 7, 11f., 14f., 61, 161, 260
Vieta, F., 112f.

Waismann, F., 4, 12, 14f., 45
Wald, A., 17, 215f., 229, 234
Wallman, H., 228, 248, 253
Weber, E. H., 270
Weinberger, O., 276
Weierstrass, K., 108, 138, 153, 155f.
Weininger, O., 243
Weiss, F. X., 301
Weitzenböck, R., 246
Weyl, H., 13, 36, 50f., 53, 134, 246
Whitehead, A. N., 44f., 179, 215, 228
Whitlock, A. I., 78
Wicksell, K., 293, 296
Wigner, E., 143
Wittgenstein, L., 1, 3, 7, 12, 14f., 21, 61, 153–157
Wundheiler, A., 152, 161

Zadeh, L., 232, 234
Zermelo, E. 36, 48, 50, 53

VIENNA CIRCLE COLLECTION

1. OTTO NEURATH, *Empiricism and Sociology*. Edited by Marie Neurath and Robert S. Cohen. With a Section of Biographical and Autobiographical Sketches. Translations by Paul Foulkes and Marie Neurath. 1973, xvi + 473 pp., with illustrations. ISBN 90-277-0258-6 (cloth), ISBN 90-277-0259-4 (paper).

2. JOSEF SCHÄCHTER, *Prolegomena to a Critical Grammar*. With a Foreword by J. F. Staal and the Introduction to the original German edition by M. Schlick. Translated by Paul Foulkes. 1973, xxi + 161 pp. ISBN 90-277-0296-9 (cloth), ISBN 90-277-0301-9 (paper).

3. ERNST MACH, *Knowledge and Error. Sketches on the Psychology of Enquiry*. Translated by Paul Foulkes. 1976, xxxviii + 393 pp. ISBN 90-277-0281-0 (cloth), ISBN 90-277-0282-9 (paper).

4. MARIA REICHENBACH and ROBERT S. COHEN, *Hans Reichenbach: Selected Writings, 1909-1953 (Volume One)*. 1978, in press. ISBN 90-277-0291-8 (cloth), ISBN 90-277-0292-6 (paper). *Hans Reichenbach: Selected Writings, 1909-1953 (Volume Two)*. 1978, in press. ISBN 90-277-0909-2 (cloth), ISBN 90-277-0910-6 (paper). Sets: ISBN 90-277-0892-4 (cloth), ISBN 90-277-0893-2 (paper).

5. LUDWIG BOLTZMANN, *Theoretical Physics and Philosophical Problems. Selected Writings*. With a Foreword by S. R. de Groot. Edited by Brian McGuinness. Translated by Paul Foulkes. 1974, xvi + 280 pp. ISBN 90-277-0249-7 (cloth), ISBN 90-277-0250-0 (paper).

6. KARL MENGER, *Morality, Decision, and Social Organization. Toward a Logic of Ethics*. With a Postscript to the English Edition by the Author. Based on a translation by E. van der Schalie. 1974, xvi + 115 pp. ISBN 90-277-0318-3 (cloth), ISBN 90-277-0319-1 (paper).

7. BÉLA JUHOS, *Selected Papers on Epistemology and Physics*. Edited and with an Introduction by Gerhard Frey. Translated by Paul Foulkes. 1976, xxi + 350 pp. ISBN 90-277-0686-7 (cloth), ISBN 90-277-0687-5 (paper).

8. FRIEDRICH WAISMANN, *Philosophical Papers*. Edited by Brian McGuinness with an Introduction by Anthony Quinton. Translated by Hans Kaal (Chapters I, II, III, V, VI and VIII and by Arnold Burms and Philippe van Parys. 1977, xxii + 190 pp. ISBN 90-277-0712-X (cloth), ISBN 90-277-0713-8 (paper).